Salvadori's STRUCTURE IN ARCHITECTURE

The Building of Buildings

Fourth Edition

Robert A. Heller, PhD., PE

Deborah J. Oakley, AIA, PE

PEARSON

Boston   Columbus   Indianapolis   New York   San Francisco   Hoboken
Amsterdam   Cape Town   Dubai   London   Madrid   Milan   Munich   Paris   Montréal   Toronto
Delhi   Mexico City   São Paulo   Sydney   Hong Kong   Seoul   Singapore   Taipei   Tokyo

Editor-in-Chief: Andrew Gilfillan
Product Manager: Anthony Webster
Program Manager: Holly Shufeldt
Project Manager: Rex Davidson
Editorial Assistant: Nancy Kesterson
Team Lead Project Manager: Bryan Pirrmann
Team Lead Program Manager: Laura Weaver
Director of Marketing: David Gesell
Senior Product Marketing Manager: Darcy Betts
Field Marketing Manager: Thomas Hayward
Procurement Specialist: Deidra M. Skahill

Creative Director: Andrea Nix
Art Director: Diane Y. Ernsberger
Cover Designer: Deborah Oakley
Cover Image: Deborah Oakley
Full-Service Project Management: George Jacob, Integra Software Services
Composition: Integra Software Services
Printer/Binder: R. R. Donnelley
Cover Printer: Phoenix Color
Text Font: 10/12 Times LT Pro

Unless otherwise indicated herein, any third-party trademarks that may appear in this work are the property of their respective owners and any references to third-party trademarks, logos or other trade dress are for demonstrative or descriptive purposes only. Such references are not intended to imply any sponsorship, endorsement, authorization, or promotion of Pearson's products by the owners of such marks, or any relationship between the owner and Pearson Education, Inc. or its affiliates, authors, licensees or distributors.

Library of Congress Cataloging-in-Publication Data
Names: Salvadori, Mario, author. | Oakley, Deborah, author.
| Heller, Robert A., author.
Title: Salvadori's structure in architecture : the building of buildings /
Deborah Oakley, Robert Heller, Mario Salvadori.
Other titles: Structure in architecture | Structure in architecture
Description: Fourth edition. | Boston : Pearson, [2017] | Includes index.
Identifiers: LCCN 2015028454 | ISBN 9780132803205 | ISBN 0132803208
Subjects: LCSH: Structural analysis (Engineering) | Architecture.
Classification: LCC TA646 .S33 2017 | DDC 729—dc23 LC record available
at http://lccn.loc.gov/2015028454

23 2020

ISBN 10:   0-13-280320-8
ISBN 13: 978-0-13-280320-5

BRIEF CONTENTS

TABLE OF CONTENTS

FOREWORD

When *Structure in Architecture* first appeared in 1963, it awakened architects to a qualitative, conceptual understanding of structures that was lacking, as engineers had always described structures clouded with mathematics. Here was an important new path that showed how structures work rather than how they are computed. Not only architects but engineers themselves and the general public were able for the first time to learn from this innovative approach. A strength of this book is that it demonstrates that even the most complicated-looking structure can be deconstructed to reveal its elementary roots: beams, columns, frames, trusses, and shells, whose actions can be conceptually understood, clarifying the way in which the whole structure works.

In the 50 years since the first edition of this book was published, a vastly expanded catalog of available structural types has appeared; new materials have been developed, new shapes have been introduced, and, above all, advances in computing technology have allowed architects and engineers the freedom to conceive designs never before possible. A new edition was therefore inevitable.

Mario Salvadori was my teacher, my mentor, and then my partner in Weidlinger Associates. Together, we wrote four books on structural design, failures, and seismicity. All were descended from the approach Mario conceived in *Structure in Architecture* to explain technical concepts using simplified language, making them accessible to readers of any age. I am honored, and it gives me great pleasure to introduce both new readers and readers of previous editions of *Structure in Architecture* to this new edition. The classic work has now been greatly improved to bring the original into the twenty-first century with updated graphics and structural examples, as well as a revised text to reflect recent advances in structural typology. This new edition will undoubtedly stand for the next decades as the go-to reference to understanding how structures work.

Matthys Levy

PREFACE

It has been 30 years since Mario Salvadori updated the last edition of *Structure in Architecture*. On its initial publication in 1963, it was one of the first and only books of its kind to introduce the principles of structures to architectural students in a largely nonmathematical manner. The variety of textbooks of this genre has grown and changed dramatically since that time, and contemporary publishing practices have dramatically evolved as well. Now long out of print and superseded by many newer books presenting rich graphic content, *Structure in Architecture* has not only been surpassed in popularity by other texts, but the presentation may also seem dated or unappealing to contemporary students. Nevertheless, it remains an outstanding work of one of the most influential individuals in the area of architectural structures education. Rather than relegate it to the bin of history, a new edition to perpetuate its legacy was called for. This edition thus presents a substantial revision of the graphic presentation, while retaining the clarity of, and expanding on, the original text.

ON MARIO SALVADORI

To better understand the history, place, and authority of this text, it is helpful to understand briefly something about Mario Salvadori. Throughout his career, he wrote voluminously and taught extensively on the topic of architectural structures, as well as engaging in a number of conference discussions about the nature of the dialog between architects and engineers. Holding Italian doctorate degrees in mathematics *and* civil engineering, for nearly 50 years he taught in both schools of civil engineering and architecture at Columbia University in New York, rising to be one of the most distinguished faculty members of that era.

Fourteen years into his teaching career, while continuing his academic appointment, he joined the practice of the brilliant Hungarian engineer Paul Weidlinger. There, too, Dr. Salvadori distinguished himself by becoming a partner in 1963 and later chairman of Weidlinger Associates, thus impacting the design of many important structures, conducting numerous forensic structural investigations, as well as shaping the careers of generations of young engineering practitioners. For his lifetime of contributions, he was widely honored by engineering, architecture, and academic societies alike.

Of all the achievements of an illustrious career, however, Mario Salvadori was most proud of his work teaching science and math to inner-city children in the New York City region, using buildings and bridge structures as a springboard. The last three decades of his life were increasingly dedicated to this personal educational mission. The legacy of this work lives on today in the form of the Salvadori Center (salvadori.org), a nonprofit educational center that he established in 1987—an organization dedicated to the mission of educating children in what is now referred to as STEM, for Science, Technology, Engineering, and Math. Clearly decades ahead of his time, Mario was active with the Center until the very end of his life, passing away in 1997 at the age of 90.

ABOUT *STRUCTURE IN ARCHITECTURE*

Along with the works produced in his dual careers of academia and practice, Mario Salvadori wrote also for the lay audience. His most popular books such as *Why Buildings Stand Up* and *Why Buildings Fall Down* (coauthored with Matthys Levy) have been in print continuously since their first publication in 1980 and 1992, respectively. These can be seen as later-career books very much influenced by his work with children, written in a manner accessible to anyone with no formal training beyond basic schooling.

The first edition of *Structure in Architecture*, in contrast, came much earlier (1963), yet Dr. Salvadori had at this point been teaching at Columbia for nearly twenty-five years and this was already his fourth published book. Two subsequent editions in 1975 and 1986, plus ten foreign-language translations, attest to the interest and worldwide popularity of the book. Unlike the later popular texts, however, *Structure in Architecture* went deeply into principles that are important for architects to understand, though never with much mathematics.

The issue of just how technical an engineering education an architectural student requires has been a matter of debate for decades. A polarity exists even within the community of educators who teach and research in architectural structures: On the one hand, there are those who firmly believe that calculations are the basis for the study while, on the other hand, there are also those who feel quite the opposite. With his unique talents, Mario Salvadori was able to successfully bridge these two disparate worlds and recognize the commonality between the two. He was able to translate arcane principles of mathematics and science into simple language that—quite literally—even young children could understand. Dr. Salvadori believed that the conceptual approach was a vital starting point for (or at least concurrent study with) a more technical study. He was thus able to engage many architecture students who would otherwise have had no interest in the more technical aspects of architectural design.

THE INTENT OF THE FOURTH EDITION

Deborah Oakley was approached by Pearson Education to undertake the project as a new coauthor, joining with Robert Heller to revisit the manuscript for an update of this classic book. As noted previously, the objective was to appeal to a new audience, while retaining all of the strong points of the earlier editions. The organizational structure of the new edition has been largely retained from the previous. Rather than making any drastic changes to the text and examples of the third edition, we consider the book to be a mid-to-late career watermark of one of the most celebrated architectural technology educators, and thus important to conserve the spirit of the previous editions. Editing of the text was a shared effort between Deborah Oakley and Robert Heller, while the acquisition of photos and creation of the majority of illustrations and 3D models were by Deborah Oakley.

The approach was to strike a balance between what should be retained and/or expanded, and what should be updated or removed. With respect to the written text, it remains largely that of Mario Salvadori, with additions and alterations ranging from minimal to significant, depending on the chapter. Regarding example projects, where possible we retained those that are iconic and clearly illustrate fundamental principles, while replacing with contemporary projects some that had been superseded. Visually, the greatest difference will be seen in the graphics and computer renderings by the new author, and in the color photographs. Many illustrations have been provided with extensive supplemental captioning. Original 2D illustrations have been greatly expanded and are full color for best clarity. Some readers who know the earlier editions may miss the simple line drawings conceived by Robert Heller and executed by Felix Cooper. Unfortunately, the originals for these are lost to the sands of time. More than a few of them, however, live on in updated reproductions.

Among the more noticeable changes are that the text has been made gender neutral in language, following current practices and reality. Errors or omissions that we identified have been corrected, and contemporary topics have been added in various chapters. This edition therefore renews a classic volume with a new look and feel and more recent examples. In so doing, we intend for it to reclaim a place in the canon of modern architectural structures texts, and to reintroduce Mario Salvadori to a contemporary audience, a new generation of students, and even educators.

There are numerous books on architectural structures that feature extensive use of calculations, but far fewer that explain complex principles to new students using a largely nonmathematical, conceptual approach. With the updated graphic presentation, this book can be studied at the image and caption level first, and then more deeply in the text itself. This is a text that any intelligent individual with an understanding of elementary trigonometry and algebra should be able to pick up and learn from on his or her own. It remains an excellent preparatory or companion book for a numerically based study.

Looking to the future, we envision not only updates to examples but also branching into new media and learning resources of the digital age. But whatever may come to pass with future editions, one thing remains constant: The reason that a more than 50-year-old book such as this can still remain relevant in the twenty-first century is that the fundamental principles of structure have not changed. In fact (at the risk of oversimplification), it can be said that they have most elementally not changed since the time of Newton. Thus, Dr. Salvadori's voice remains vibrantly alive in this work, as it has in perpetuity with his several other works oriented toward the lay audience. We hope that the spirit of Mario Salvadori approves of the new edition, and that new generations are introduced to his work.

Deborah Oakley,
Las Vegas, Nevada

Robert Heller,
Burlington, North Carolina
June 2015

NEW FOURTH EDITION HIGHLIGHTS

- Entirely new graphics package:
  - Previous line illustrations updated with over 150 full color photographs, nearly 500 new full color rendered illustrations by Deborah Oakley, and extensive new image captioning
  - Many completely new illustrations added throughout the book to best demonstrate fundamental concepts
  - Designed to be accessible and attractive to the current generation of architectural students in a media-saturated world:
    - Big ideas can be grasped by studying the images and captions.
    - An in-depth understanding comes by studying the text with the images and experimenting with end-of-chapter exercises.
- Broken into three overall sections for better comprehension of organizational structure:
  - Part I: Fundamental Concepts (Chapters 1–5)
  - Part II: Structural Forms (Chapters 6–9)
  - Part III: Beyond the Basics (Chapters 10–15)
- New example structures illustrated throughout text
- Expanded content with enhanced text discussion and related graphics on critical topics such as beam behavior, moment of inertia, redundancy, and so on
- New at the end of each chapter:
  - Summary key ideas of chapter
  - Thought questions and simple exercises for further reflection
  - List of recommended key references of similar subject matter

FEATURES RETAINED FROM PREVIOUS EDITIONS

- Intuitive, nonmathematical approach
- Geared as an introductory text for beginning architectural students, students of technical schools, and interested laypeople
- Most of the historical examples, since they represent milestone accomplishments
- Most of the original text by Mario Salvadori

About metric units in the text It has been more than 40 years since the U.S. congress passed the Metric Conversion Act of 1975, and yet the country continues to use U.S. Customary (a version of British Imperial) units. When first written, this text was all in U.S. Customary units, and illustrations were created using whole numbers. This presents a quandary to the current edition. With an international audience, we cannot ignore the fact that as of 2015 all but two other countries (Liberia and Myanmar) have adopted SI units (*Système International d'Unités*, the international standard), and yet the text is directed primarily at a U.S. audience. SI units therefore accompany the U.S. units parenthetically and have been rounded to the nearest whole number equivalent (or no more than one decimal place). It is not an optimal solution, but it is also one that Mario Salvadori himself used in some of his other popular works. We continue to hope that a future edition may be wholly in SI units and thus dispense with this temporary workaround.

ACKNOWLEDGMENTS

Deborah Oakley would like to thank Pearson Education for the opportunity of undertaking this project and also the many individuals who have contributed photographs throughout the text (credit is provided with image captions). Thanks are extended to my colleague Vincent Hui at Ryerson University and his students for some of the initial 3D models in Chapter 6, as well as to my graduate assistants at the University of Nevada, Las Vegas School of Architecture; in particular Vincent del Greco, who also worked on some early Chapter 6 models, and Adam Bradshaw, for conducting photo research and assisting with the final image preparation. Special thanks are extended to Terri Meyer Boake of the University of Waterloo for the many photographs, as well as mentorship and friendship over the years. I would also like to thank my father, Donald Oakley—a writer by trade—for final proofreading. And the most important thanks of all go to the many students who have taught me how to teach structures to architecture students.

Robert Heller wishes to extend his appreciation to Deborah J. Oakley for her diligence, drive, and ingenuity in making this work, a tribute to the memory of his late friend Dr. Mario G. Salvadori, possible. Coauthoring the first edition of *Structure in Architecture* gave him the impetus to teach structural mechanics for 50 years.

AUTHOR BIOGRAPHIES

Deborah J. Oakley, AIA, PE
Deborah Oakley has been teaching structures to architecture students for nearly 20 years. She is an associate professor at the School of Architecture at the University of Nevada, Las Vegas, where she also teaches design studio classes. Uniquely qualified as both a Registered Architect and Professional Engineer, she came to academia with education and experience in fields of both civil (structural) engineering and architecture. She is a passionate crusader for the intergration of architecture and structure, including associated educational endeavors in the field. She is a founding member, past president, and board member of the Building Technology Educators' Society (BTESonline.org), the only North American academic organization of architectural educators focused on construction and structural technology education and research. Prior appointments have been as an assistant professor at the University of Maryland School of Architecture, Planning and Preservation and at Philadelphia University School of Architecture and Design. Her current work involves conducting Discipline-Based Education Research in the area of architectural structures pedagogy.

Robert A. Heller, PhD, PE
Robert Heller received his education at Columbia University. After earning a PhD in engineering mechanics, he joined the Faculty of Columbia's Department of Civil Engineering. There he was Mario Salvadori's colleague and eventually became his coauthor. After leaving Columbia, Heller was appointed Professor of Engineering Science and Mechanics at Virginia Tech. In that capacity, he developed new courses on probabilistic structural mechanics and reliability and service life of structures and courses for architects.

His series of educational videos entitled "Mechanics of Structures and Materials" has been widely used in Schools of Architecture and Engineering. Heller's research work on the Service Life of Solid Rocket Propellants and on Aircraft Fatigue for the Department of Defense has been published in numerous scientific journals.

FOREWORD TO PREVIOUS EDITIONS

In this thoughtfully written book, Professor Salvadori endeavors to eliminate one of the most serious gaps between theory and practice in the field of structures. His aim is to build a bridge between the more or less conscious intuition about structure, which is common to all mankind, and the scientific knowledge of structure, which gives a fair representation of physical reality on the basis of mathematical postulates.

No one doubts that the bridging of this gap is possible and that, if achieved, it would be extremely useful.

In order to invent a structure and to give it exact proportions, one must follow both the intuitive and the mathematical paths.

The great works of the distant past, built at a time when scientific theories were nonexistent, bear witness to the efficiency and power of intuition.

Modern theories are incessantly and progressively developed, and their refinement is illustrated by the construction of ever greater and more daring structures.

If structural invention is to allow the efficient solution of the new problems offered daily by the ever-growing activity in the field of construction, it must become a harmonious combination of our personal intuition and of an impersonal, objective, realistic and rigorous structural science.

In other words, theory must find in intuition a force capable of making formulas alive, more human and understanding, and of lessening their impersonal technical brittleness. On the other hand, formulas must give us the exact results necessary to obtain "the most with the least," since this is the ultimate goal of all human activities.

Through always clear and, at times, most elementary examples, Professor Salvadori's book tends to unify these two viewpoints (I was almost going to say, these two mentalities), which must be cast into a unique synthesis if they are to give birth to the essential unity of all great structures.

Future architects will find it particularly useful to study this book in depth and to meditate upon it, since even if they can entrust the final calculation of a structure to a specialist, they themselves must first be able to invent it and to give it correct proportions. Only then will a structure be born healthy, vital and, possibly, beautiful.

I feel that we must be particularly grateful to Professor Salvadori for undertaking this anything but easy task.

Pier Luigi Nervi

PREFACE TO THIRD EDITION

As stated in the preface to the first edition, this book has been written for those

- who love beautiful buildings and would like to know why they stand up;

- who dream of designing beautiful buildings and would like them to stand up;

- who have designed beautiful buildings and would like to better know why they stand up.

The principles of structure are eternal, but new developments in structural materials, methods of design, and construction techniques constantly change the application of such principles to the building of buildings, and require frequent reassessment of the field of construction.

As one starts revising a book such as this it becomes obvious that virtually every page requires clarifications, additions, and updating. Besides innumerable changes of this nature this edition contains:

- A new chapter on structural failures, a topic of increasing concern in our society.

- A new chapter on structural aesthetics, a subject of growing awareness to architects and engineers, that has interested me for many years.

- A new treatment of space-frames for large roofs that have become, because of their economy and beauty, the most popular structures of our time, the world over.

- The first presentation of new techniques for the erection of membrane roofs unsupported by air pressure.

- A new treatment of earthquakes and methods of earthquake attenuation.

- An updating of structural material properties and construction methods.

- A record of new limits reached in the field of architectural structures.

- Over eighty new or modified figures by the original illustrator, Felix Cooper.

The intuitive and descriptive presentation is unchanged from that used in previous editions: irrespective of background, the book can be understood by anyone interested in why buildings stand up.

The structural concepts presented here were formerly introduced mathematically to graduate students of the School of Architecture at Princeton University and, later, to students of the graduate School of Architecture at Columbia University. The same concepts have been presented without mathematics to freshmen in architecture at Columbia, with the help of the models and motion pictures of my friend Robert Heller. Professor Heller did not participate in the preparation of this edition, and the changes and new material are solely my responsibility.

I hope that my latest efforts will meet with the same favor accorded previous editions throughout the world.

My deep gratitude goes

- to my former collaborator, Dr. Robert Heller, for his help in conceiving the original illustrations and for his constructive suggestions;

- to my teachers at the Faculty of Pure Mathematics of the University of Rome, who made mathematics part of my mental makeup and allowed me to move beyond it;

- to Charles R. Colbert, the former dean of the School of Architecture at Columbia, for encouraging me to try this intuitive approach to structures;

- to Felix Cooper for drawing the illustrations;

- to Tim McEwen of Prentice-Hall for suggesting that I prepare this revised edition;

- to all my friends for their interest and support during the relatively brief but intense period when the thoughts accumulated in years of study became this book;

- to my wife, Carol, who stood by me from the time this book was first conceived to the day I corrected the proofs of this present edition.

New York   *Mario Salvadori*

This book is joyfully dedicated
to my architectural students
who for thirty years taught me
how to teach structures.

FUNDAMENTAL CONCEPTS

Like many disciplines, the knowledge base of structures is rooted in fundamental concepts that apply at all levels of understanding. The first five chapters of this book introduce those essential principles upon which all the later chapters are developed. Chapter 1 discusses the basic idea of structures and the relationship between architects and engineers, while Chapter 2 describes the types of forces (loads) that structures must resist, and the relationship to building codes that prescribe them. Chapters 3 and 4 present the basic properties of materials used in construction and the basic conditions required for structures to exist, while Chapter 5 illustrates the essential types of behavior that structural elements are subjected to.

The Soccer Stadium in Braga Portugal (see Figure 2.1)

Photo courtesy of Deborah Oakley

ONE

STRUCTURE IN ARCHITECTURE

1.1 WHAT IS STRUCTURE?

It can be argued that the essence of a building is structure, for no physical object, whether built or natural, can exist but for the structure that gives it form (Figure 1.1). Without the structural armature of our bones, we would be like jellyfish or octopi, slithering on the ground going about our daily business. It is the structure of its wood fibers that enables the tree to stand, just as it is the structure of the bridge that enables it to span a river. The difference between the two is merely that one developed from nature, the other by the will of a human creator. Over the centuries, humanity has come

FIGURE 1.1 The Burj Al Arab hotel in Dubai, UAE, is a stunning example of expressed structure in architecture. The steel exoskeleton provides for lateral stability in the region's high winds and perfectly compliments the sail-like shape of the design.

Photo courtesy of Joceclyn Hidalgo

to understand many of the secrets of nature. We have learned how to employ that understanding to a desired end in the creation of structures that meet our specific needs for shelter, commerce, worship, recreation, and transportation.

The purpose of this book is to take the reader on a journey into both becoming aware of the wide variety of built structures in the world and developing an understanding for the key principles that underlie them. The complete engineering design of a structure is normally a complex undertaking, especially for larger structures. It requires an ability to mathematically model forces that exist in response to the loads that the structure may experience during its useful life, and the proportioning of materials to resist those forces. Nevertheless, it is entirely possible to arrive at a very strong intuitive understanding of structural behavior and materials with little or no math. This text presumes no formal training in advanced mathematics, and so it is well suited for beginning students of architecture, as well as practitioners who need a refresher, or the serious lay student seeking to understand more about the subject. As such it can serve as a good preparation for the undertaking of more advanced study in the principles of engineering structures.

1.2 STRUCTURE IN NATURE

The place where we first and most directly encounter structure is not in architecture but within nature herself. Every living thing, from the smallest cell to the tallest tree, has a structural form that is shaped in direct response to the forces of its environment, such as gravity, water pressure, and wind. Other natural structures serve the needs of their builders. The spider's web is built of the arachnid's own secretion, The bee's geometrically precise honeycomb and the beaver's dam could not be better constructed by humans.

Each of us therefore has an innate understanding of structures at a very subliminal level because our very bodies *are* structure. We physically sense the pull of gravity and intuitively know to widen our stance and lean into a strong wind, for example. Capitalizing on this, we can use our own bodily sense and a growing awareness of structural forms in nature to help understand how built structures are designed and constructed.

The shape and proportions of a structural form are significantly a matter of scale (Figure 1.2a–c). We can observe that the branching pattern of a dandelion stalk is far more

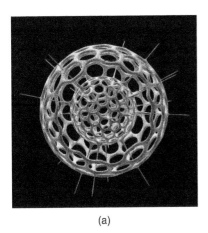
(a)

(b)

(c)

FIGURE 1.2 Size matters: a microscopic radiolarian skeleton (a type of oceanic zooplankton), a dandelion, and a tree branch. A dandelion seed (b), light and made to catch in the wind, is a distinctly different pattern than that of a tree (c), and both far different than the radiolarian (a). In fact, scaled up to the size of a tree, the dandelion branches could not support their own weight. Every structure in nature has a unique shape that is largely determined by both its size and environmental influences.

(a) Photo: Alfred Pasieka/Stockbyte/Getty Images; (b) Photo: Achim Prill/123rf; (c) Photo: Potapov Alexander/Shutterstock

slender than if the plant were enlarged to the size of even a small bush, never mind a large tree. This is due to the fact that the amount of material increases by the cube of its size and yet the pull of gravity is essentially constant. This is to say that the doubling in size of an object or organism increases its volume by a factor of not two or four, but by *eight* times. An ant is known to be an organism that can carry a load many times its own body weight. Enlarge it to the size of an elephant, though, and the spindly legs of the ant, even if proportionally enlarged, would no longer be sturdy enough to even support itself despite how appropriate the form may be at its natural size (so much for 1950s science fiction films!).

We can therefore look to nature as an aid in our quest to understanding the behavior of human-created structures. Consider a tree branch (Figure 1.2c). It is essentially a cantilever, which is a beamlike element that is supported only at one end (see Chapter 7). Notice how the tree branch is thickest where it meets the tree trunk…this is also the place where the internal stresses that the wood fibers must resist are at their highest value for the branch. As a consequence, more material is needed here to resist these stresses. We can similarly look to the behavior of natural materials, many of which, such as wood and stone, we use in our constructed buildings. Hair is an example of a material that can have a high degree of ductility—it can be pulled and stretched somewhat before it breaks. A blade of grass, though exhibiting some ductility, will snap much more readily when pulled. These are but a few of the *engineering properties* of materials that will be discussed at length in Chapter 3, *Structural Materials*.

1.3 THE ARCHITECT AND THE ENGINEER

Architects interact with numerous specialists in the creation of a building, many of whom are engineers. It is helpful to gain a perspective of the variety of engineering disciplines and understand the roles that these professionals play in relation to architects. At its basis, engineering is the art of creation in the service of a desired end, by employing known principles of science and properties of materials. There are many branches of engineering that architects work with, and even within individual branches there are subspecialties.

Civil engineering is a very broad discipline, which encompasses a wide range of subspecialties. These include individuals responsible for site design (addressing land surveying, site grading, drainage, and parking), transportation engineers who design highways and other transit systems, environmental engineers who focus on the treatment processes to provide clean water and dispose of waste, fire protection engineers who focus on the safety of structures against fire hazard, and geotechnical engineers who specialize in the analysis of the soil and rock that buildings are built upon.

Electrical engineers are responsible for the design of systems to electrically power buildings, as well as the internal distribution of power to lighting, electrical outlets, and machinery. There are also electrical engineers with whom architects typically have little interaction, including those who are responsible for the large-scale production of power and distribution through the "power grid" on a regional scale, and electronics and computer engineers who design the "high-tech" systems of the modern world.

Mechanical engineering is another particularly broad discipline. Some mechanical engineers design automobiles (automotive engineers) or airplanes (aerospace engineers), while others create the many machines that we are all familiar with in our daily lives, such as kitchen appliances and household utilities. The types of mechanical engineers that architects most frequently interact with are those who design the heating, ventilation, and cooling systems in buildings, as well as elevators and escalators.

The most important branch of engineering in relationship to the subject of this book is yet another subspecialty of civil engineering, the structural engineer. Structural engineers, as the name implies, are responsible for the creation of safe structures. This includes those who design bridges and other highway structures, as well as those whose main focus is on the design of building structures. The subject of this book is fundamentally focused on the principles that underlie the profession of structural engineering.

What distinguishes engineers from architects? How do engineers think? The popular stereotypical image of an engineer is the introverted nerd lacking social skills with thick-rimmed glasses and a pocket protector. Although there may be some who fit that description, the truth is actually far from the reality. Engineering is, in fact, a very creative process and, fundamentally, engineers are problem solvers. It is actually rather difficult to lump engineers into one class, because there are so many branches of engineering. Many engineers are the types who enjoy logic puzzles or figuring out how to make something work and other intellectual challenges. A good percentage are tinkers who like to work with their hands. If any generalization can be made, it is that all good engineers excel at rational problem solving.

So how do architects think in contrast to engineers? In many regards, architecture is among the last of the great humanist fields of study. Whereas engineers are most often specialists within their given field, architects are generalists who learn to see the big picture. A good architect must have a basic understanding of each of the many disciplines needed to construct a building. Architecture spans across many levels, from the most sublime sculpting of form and manipulation of such intangibles as light and shadow, to the social responsibility of the project at an urban scale, to the physical realization of building construction. With the increasing complexity of the world, and increasing recognition of the role that buildings play in our environment, a good architect is called on like never before to be conversant in the supporting roles of an ever-growing number of disciplines including,

and going beyond, engineering and, of growing importance, environmental sustainability.

Fundamentally, architecture is a collaborative experience that requires the integration of all disciplines. In this regard, the architect can be compared with an orchestral conductor, the one individual who has an overview of the entire process. By integrating key decisions early on in the architectural design process—especially those that deal with the building structure—the best and most satisfying result is more likely to be achieved.

1.4 HISTORICAL DEVELOPMENT

As noted earlier, structure is an essential component of architecture, and has always been so. No matter whether man built a simple shelter for himself and his family or enclosed large spaces where hundreds could worship, trade, discuss politics, or be entertained, humans had to shape certain materials and use them in certain quantities to make their architectural creations stand up against the gravitational pull of the earth and other powerful forces of nature. Wind, snow, and rainstorms, earthquakes, and fires had to be resisted. If possible, this was accomplished with expenditures of labor and materials that were not unreasonable in relation to their availability and cost. And because from earliest times a sense of beauty has been innate in humans, all constructions by civilized peoples were also conceived according to certain aesthetic tenets. This would often impose on the structure far more stringent requirements than those of strength and economy.

It may be thought, therefore, that structure was always considered important, and, in a sense, dictated architecture. This is simply not so. Magnificent buildings have been created in the past, and are created even today, with a notable disregard for the "correctness of structure." The Parthenon (Figure 1.3), divinely beautiful as it is, translates

FIGURE 1.3 The Parthenon in Athens, Greece. This most majestic of all ancient buildings—an acknowledged architectural masterpiece—is nonetheless "incorrect" from a purely structural viewpoint. Stone as a building material is weak in tension and unsuited for spanning long distances; hence, the column spacing must be quite close to keep the stone spans very short.

Photo: Brent Wong/Shutterstock

structural forms typical of wood construction into marble and is, structurally speaking, "wrong." Since wood is a material capable of withstanding tension and compression, and long horizontal elements require both tensile and compressive resistance, they are well built out of wood but much less so of stone.

Stone withstands compression well, but has very little ability to carry tension. Thus, horizontal elements can be built in stone only by reducing their length and supporting them on heavy vertical elements, such as columns or pillars. Hence, horizontal elements of stone are "incorrect" from a structural point of view. On the other hand, Gothic cathedrals could span up to one hundred feet (30 meters) and cover hundreds of square yards (hundreds of square meters) crowded with worshippers by making use of the arch—a curved structural element in which tension is not developed. Thus, stone is the correct material for a vaulted type of structure, and the beauty of the Gothic cathedrals satisfies both our aesthetic sense and our feeling for structural strength (Figure 1.4). This precept is echoed in the famous statement by architect Louis I. Kahn when he "asked" a brick, *What do you want to be?* And, metaphorically, the brick replied, *"I like an arch,"* which is a pure compressive structure. Like stone, brick is a material weak in tension, and so structures in which it always remains in compression are the most appropriate form for such materials.

It has been argued by some architectural historians, as well as by some structural engineers, that a deep concern for structure will unavoidably lead to beauty. It is undeniable that a "correct" structure satisfies the eye of even the most unknowledgeable layman, and that a "wrong" structure is often offensively ugly. But it would be hard to prove that aesthetics is essentially dependent on structure. It is easy to show, instead, that some "incorrect" structures are lovely, while some "correct" ones are aesthetically unsatisfying. It may perhaps be wiser to say that correctness of structure is, most of the time, a necessary condition of beauty, but is not sufficient to guarantee beauty. Some contemporary architects and engineers, such as Santiago Calatrava and Christian Menn, or their equally famous predecessors, such as Felix Candela and Pier Luigi Nervi, are so imbued with artistic sense that their structures are beautiful (Figures 1.5a and 1.5b). But some grandiose buildings, recently erected by the use of daring engineering techniques, undeniably lack beauty.

We may thus conclude that knowledge of structures on the part of the architect is highly desirable, and that correctness of structure cannot but add to the beauty of architecture. But considerations of beauty aside, no architecture can be effectively constructed without consideration of structure, and so the better an architect understands the principles of structure, the more empowered will he or she be to make a positive impact on the design from the earliest stages. Final engineering will then be a confirmation of early design decisions, as opposed to a determination of conflicts that must be resolved in order to ensure structural strength and stability, potentially with negative consequences to the original design intent.

FIGURE 1.4 The groin vaults of the magnificent Rouen cathedral in northern France are an expression of a "correct" structure. Here, stone is used to its best ability in compression, with little or no tensile stresses being developed. The stone arches of the groin vault roof effectively channel the heavy load of the stone roof out and down to the exterior walls, where further structures on the exterior known as flying buttresses counteract the outward push.

Photo courtesy of Terri Meyer Boake

1.5 THE PRESENT INTEREST IN ARCHITECTURE

In the recent as well as the ancient past, the figure of the architect was unique: He or she was both an artist and a technologist, a designer, and a builder. Michelangelo could be a painter, a sculptor, an architect, and a master builder: The Vatican in Rome bears his imprint in all four fields. During the last century, however, specialization of knowledge has taken over the field of architecture, and the various functions—once entrusted to the same individual—are now frequently exercised by several different specialists. At least two key persons are essential in the construction team of any important building: the architect and the structural engineer. Today, no architect would dare design a building of even modest size without consulting a structural engineer. The roots of this dependence are to be found in the increasing importance of economic factors, in the technological direction of our culture, and, above all, in our mass civilization's need for an increasing number of all types of structures.

(a)

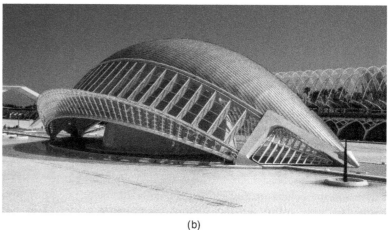
(b)

FIGURE 1.5 The works of engineer-architects such as Felix Candela (a, Our Lady of the Miraculous Medal Church in Mexico City) and Santiago Calatrava (b, Planetarium at the Science Center, Valencia, Spain) exemplify the harmony possible between architecture and engineering. As with a Gothic cathedral, here structure *becomes* the architecture. In the case of Candela, structural efficiency shapes the structure in a series of tilted hyperbolic paraboloids (see Chapter 12), and in the case of Calatrava, high-tech kinetic architecture with operable components translates the abstract form of a human eye into a structure.

(a) Photo courtesy of Benjamin Ibarra-Sevilla; (b) Photo courtesy of Deborah Oakley

As the number of human beings multiplied at an increasing rate during the last few centuries so as to create a "population explosion," civilized societies have also given each human more services, sharply increasing the "psychological density" of the population. Each one of us requires and is given more schooling, more travel, more medical care, more entertainment. Large numbers of people gather under the same roof for all the gregarious activities so typical of our era. Large stations and airports, large stadia, large theaters, large churches, large arenas appear in increasing numbers. Urban agglomerations require the sprouting of taller buildings. The large structure has become a symbol of our culture and a monument to governments, churches, or corporations. In addition, housing the millions and supplying them with schools and hospitals are among the basic goals of civilized societies.

The architect is challenged by these tremendous tasks; the layman becomes aware of the importance of architecture in his own life. Thus, the specialists meet to solve new, difficult problems in a climate of public interest. The general public whose monies are often used for these large projects takes a personal interest in their construction. This interaction between the specialists and the public may lead to better, and more correct, architecture, provided the layman understands the basic problems of the specialist, and the specialists themselves have a common bond of mutual understanding. This is the central theme of contemporary architectural education, including both the education of the architect and the popularization of architecture.

1.6 STRUCTURES AND INTUITION

It is obvious that only the most serious training in mathematics and the physical sciences will allow a designer to analyze a complex structure to the degree of refinement required by modern technology. Today's structural engineer is a specialist among specialists, a subgroup among civil engineers. As new technologies develop, even structuralists specialize: At present some structural engineers specialize in reinforced concrete, others in reinforced concrete roofs only, and some in roofs of only one particular shape or even another material such as high strength fabric. One goes to these specialists for advice on a particular type of structure as one would go to a medical specialist for advice on a rare type of disease.

But it is just as obvious that, once the basic principles of structural analysis have been established, it does not take a specialist to understand them on a purely physical basis. As previously noted, we all have some familiarity with structures in our daily lives: we know at what angle to set a ladder so that it will carry our weight without sliding on the floor; we can have a good sense of how thick a board must be to function as a bookshelf between two supports. We instinctively lean into the wind and widen our stance on a gusty day. It is a fairly easy step to capitalize on these experiences, to systematize such knowledge, and to reach a basic understanding of how and why a modern structure works.

While the layman may find this inquiry fascinating, the architect should find it mandatory: Without it he or she will soon be out of the field of contemporary architecture. For the interested public, it may be one more hobby; for the architectural student and the practicing architect, it is one of the basic requirements of the profession.

Once he or she has grasped the fundamentals, the architect must become conversant with the more refined points of the theory of structures. This will allow the intelligent application of a wealth of new ideas and methods unavailable until a few years ago even to the greatest architects (Figure 1.6). Architecture at its finest incorporates an understanding of structure from the earliest planning stages rather than something that comes after the architectural design is complete. This is only truly accomplished in close collaboration with skilled engineers who understand and share a common vision with the architect.

There is an obvious danger in this new availability and freedom. Art is enhanced by limitations, and freedom may easily lead to anarchy. Since, today, almost any structure can be built, the important question is: "*Should* it be built?" instead of: "Can it be built'?" The architect is less hampered by technological difficulties and may be led astray into the world to the most unjustifiable structures. It is true that the average contemporary architect can aspire to greater achievements in the field of structures than even those of the exceptional practitioner of only a hundred years ago, but such achievements, the fruit of technology, are also obtained through blood, sweat, and tears. In the early decades of the twenty-first century, technological advances have enabled increasingly daring structures (Figure 1.7) that make even the tremendous technological leaps of a few decades ago seem pale (Figure 1.8).

What follows in the subsequent chapters of this book is an attempt to introduce the reader to the field of structures without appealing to a formal knowledge of mathematics or physics. This does not imply that structures will be treated in an elementary, incomplete, or simplified manner. On the contrary, some of the structural concepts presented in the last chapters of this book are refined and complex. Nevertheless, they can be grasped by the reader and recognized in general architectural constructions on a purely intuitive basis. It is hoped that this better knowledge of structural action may lead the interested student to a deeper understanding of the finer points of structural design, and architects to a better facility in embracing structure as a fundamental concept of architectural planning and aesthetic opportunity.

FIGURE 1.6 Contemporary structural analysis software enables visualization of forces in a structure in ways not previously possible. Such technologies enable both the practicing structural engineer and aspiring architect better understand the behavior of increasingly complex structures. The colors in the image are a visual depiction of deformations or stress levels in a structure under load.

Photo courtesy of Autodesk, Inc. © 2012

FIGURE 1.7 The Central China Television headquarters tower (CCTV) in Beijing. A stunning example of a gravity-defying structure impossible to construct even a few decades ago, but made possible through contemporary developments in computerized structural analysis and advancement in materials and fabrication capabilities. But is it an example of structure built simply because we can?

Photo: yxm2008/Shutterstock

FIGURE 1.8 The John Hancock Tower in Chicago, Illinois, regarded as an exemplary model of structural efficiency and simple elegance, was designed in an era before advanced computational methods were widely available.

Photo courtesy of Deborah Oakley

KEY IDEAS DEVELOPED IN THIS CHAPTER

- Structure is the external or internal armature that gives physical objects form and resistance to external forces.
- Structure may be human-made or natural.
- Built structures frequently imitate nature.

- Cooperation between architects and engineers is essential for successful structures; they complement each other. Multiple engineering disciplines are needed in a building project.
- Architecture evolved and changed with the development of building materials from stone and wood to steel, concrete, and composites.
- The advent of computers has simplified the design of complicated and daring forms.

QUESTIONS AND EXERCISES

1. Look around you at the world of your immediate experience. Everything you can see and touch is some form of structure: from the smallest mineral crystal to the largest high-rise building or the longest spanning bridge. Notice what types of materials they are made of, and the different patterns they take. Begin to develop a questioning mind of how and why a structure is made the way that it is. Take notes and draw sketches. Keep a record of these observations.
2. You've been living in and around built structures all your life, but have you ever stepped back to really look at the variety of systems comprising this built world? How many different systems and materials do you note in the buildings and structures you interact with every day? Do you see patterns in the types of systems? Are some of them more supportive of the architecture while others seem more utilitarian? The first step in learning structures is to begin to develop this type of awareness.

FURTHER READING

Rice, Peter. *An Engineer Imagines.* Ellipsis London Press Ltd. 1998.

Saint, Andrew. *Architect and Engineer: A Study in Sibling Rivalry.* Yale University Press. 2008.

Millais, Malcom. *Building Structures: From Concept to Design,* 2nd Edition. Spon Press. 2005.

Ching, Francis, D. K., Onouye, Barry S., and Douglas, Zuberbuhler. *Building Structures Illustrated: Patterns, Systems, and Design,* 2nd Edition. Wiley. 2013. (Chapter 1)

Wentworth, Thompson, D'Arcy. *On Growth and Form: The Complete Revised Edition.* Dover Publications. 1992.

Gordon, James. E. *Structures: Or Why Things Don't Fall Down.* Da Capo Press. 2003.

Moore, Fuller. *Understanding Structures.* McGraw-Hill. 1998.

BUILDING LOADS AND CODES

2.1 THE PURPOSE OF STRUCTURE

Structures are always built for a definite purpose. This utilitarian element is one of the essential differences between structure and sculpture: There is no structure for structure's sake.

Structure's main purpose is to enclose and define a space, although, at times, a structure is only built to connect two points, as in the case of bridges and elevators, or to withstand the action of natural forces, as in the case of dams or retaining walls.

Architectural structures, in particular, enclose and define a space in order to make it useful for a particular function. Their usefulness stems, generally, from the total or partial separation of the defined space from the weather and may not require its complete enclosure: A suspended or cantilevered roof of a stadium stand protects the spectators from the weather without enclosing them in a space (Figure 2.1).

The enclosed space may serve many different purposes: the protection of the family, the manufacture of industrial products, the worship of deity, the entertainment of citizens, the gathering of lawmakers. Different purposes, served by different spaces, require different structures, but all structures, by the simple fact of their existence, are submitted to and must resist a variety of loads. Only in rare cases is

resistance to loads the primary purpose of a structure: loads are, usually, an unavoidable design consideration in the creation of a structure.

2.2 BUILDING LOADS AND CODES

All structures, particularly buildings, must conform to a variety of regulations called building codes and zoning laws. Some of these are regulated by local authorities, such as minimum "setbacks" from roads and neighbors, maximum height and type of structure. On the other hand, safety-related concerns, load-carrying capacity, the strength of materials used, and fire resistance of materials are governed by the *International Building Code* (IBC), developed by the International Code Council (ICC). First published in 2000, the code is updated on a regular basis. Despite the title, it is predominantly a U.S. code, replacing the three previous codes used in different regions of the country.

The determination of the loads acting on a structure is a complex problem. The nature of loads varies with the design, the location of the structure, and the materials used. Loading conditions may vary with time, change of occupancy, and applied loads. The most important loads on an architectural structure change slowly with time: They are called static. These include the weight of the building, furniture, and so on, thermal expansion and contraction, snow, as well as

Figure 2.1 Partial protection from weather

The soccer stadium in Braga, Portugal, designed by the architect Eduardo Soto de Moura, utilizes suspended cables to support a roof that partially covers the seating areas. The cables supporting the roof are attached to a second seating area mirroring this one on the opposite side of the playing field.

Photo courtesy of Deborah Oakley

lateral earth and water pressures below grade. Rapidly applied loads such as violent winds, earthquakes, reciprocating machines, and explosions are considered to be dynamic loads. The IBC divides loads into various classifications, such as Dead, Live, Wind, Snow, Seismic loads, Soil and Water Pressures, and so on.

The load to be carried by the floor of a building varies so much, depending on the occupant of the floor, the distribution of furniture, the weight of machines, or the storage of goods, that codes substitute for it an equivalent load. This equivalent load is derived, on the basis of statistical evidence, for each type of building, and is modified from time to time as new conditions or knowledge arise (see Section 2.4).

Code loads are conventional loads: A floor load may be assumed to be a constant number of pounds on each square foot (PSF) of floor (or kiloNewtons per square meter (kN/m^2) in SI units), even though in practice no floor is ever loaded uniformly. Similarly, the pressure exerted on a building by the wind may be assumed to be constant in time and distributed uniformly over its surface. The wind, instead, blows in gusts, and wind pressure varies from point to point of a building. Here again, the code simplifies the design procedure by taking the wind variations into account statistically and suggesting "safe" conventional wind pressures.

Whenever the loads on a building are not considered by codes, and when they present characteristics that may endanger a structure's life, they must be accurately determined through experiments or by mathematical calculations. The effect of hurricane winds on a skyscraper may have to be found by means of aerodynamic tests on a model, conducted in a wind tunnel.

It is not always sufficient for the designer to consider only code loads, since the responsibility for the strength of the design rests with him and not with the code authorities; this is particularly true in circumstances where code regulations do not apply. It is therefore essential for the architect to acquire an awareness of loads.

2.3 DEAD LOADS

To determine the required structural strength, the loads to be applied to the components are dictated by the IBC. Because the loads carried by the floor of a building vary, depending on the occupants of the structure, location of furniture, the storage of goods, and so on, the code substitutes an "equivalent load" under which the floor will not fail or deflect so much as to become unusable. This code load must therefore be a multiple of the load that would produce failure or unacceptable deflection. This multiple is denoted as the "safety factor" (SF). The SF accounts for uncertainties in load estimation, as well as irregularities in materials and workmanship. It should be recognized that it is impossible to know these facts with 100 percent certainty.

The magnitude of the SF depends on the usage of the structure as well as on economics. A large SF may require larger components or stronger materials to make a structure much safer, but may make it more expensive or even dysfunctional. An airplane, for instance, may not be able to fly if it is too heavy when oversized parts are used.

New techniques based on probability methods have been developed that compare the statistical variations of loads and of the structural strength of the materials used in order to determine an optimal safety factor. A further discussion of these techniques follows the section on live loads. A more detailed analysis of "Load and Resistance Factor Design" (LRFD) is, however, beyond the scope of this book.

The unavoidable weight of the structure itself and the weight of all loads permanently on it constitute its dead load (Figure 2.2). It is one of the paradoxes of structural design that one must know the dead load beforehand in order to design the structure, while it simultaneously cannot be determined until the structure is designed. The dimensions of a structural element depend essentially on the loads acting on it, and one of these loads is the dead load, which in turn depends on the dimensions of the element. The designer is compelled to start the calculation of a structure by making an

Figure 2.2 Live and dead loads

Live loads are characterized by their transient nature. It is likely obvious to the reader that people and easily movable furnishings like chairs are live loads. But so, too, are even more stationary furniture and shelving that is rearranged less frequently, but which is otherwise not permanently attached to or supported by a floor or wall.

Dead loads are those elements that are essentially permanent, such as the building structure, floor and wall surfaces and finishes, as well as mechanical and electrical equipment. The self-weights of typical building components determine these loads. Dead loads, however, also comprise built-in furnishings, such as shelving that is more or less permanently mounted to walls, as opposed to free-standing units that can easily be repositioned.

educated guess as to its dimensions and, hence, its dead load, which also depends on the construction material (e.g., concrete vs. steel). He or she then adds to it all the other loads, checks its strength, and finds out at the very end of the calculation whether the guess was correct. Long practice alone will prevent innumerable wrong guesses in structural design. The checking of the strength of a given structural element for given loads, called structural analysis, is a fairly routine operation. The initial educated guess, called structural design, must come from experience and is often the result of an almost artistic intuition rather than of scientific calculations.

The dead load is, in many cases, the most important load on a structure. It may greatly outweigh all other loads, particularly in large structures and those built of heavy materials. In bridges, in roofs over wide unencumbered areas (e.g., halls, churches, theaters), and in stone and masonry structures of a massive type (columns, buttresses, walls), the dead load often dictates the dimensions of the resisting elements. In certain cases, the dead load is not only important but also useful or even essential—as, for example, in a gravity dam (Figure 2.3), where it is used to resist the horizontal pressure of water.

Modern structural materials, such as high-strength steel, prestressed concrete, composites, or aluminum, in some cases reduce the importance of the dead load in relation to the other loads, but in no case can the dead load be ignored. Its main characteristic is its continuous presence: put it is a permanent load.

The dead load is easily computed. Once the dimensions of the structure have been determined on the basis of prior experience, its weight is evaluated by consulting tables of unit weights for structural materials (Table 2.1).

TABLE 2.1	Weights of Some Building Materials	
Material	**Weight (lb/ft³)**	**Weight (kN/m³)**
Aluminum	170	26.7
Wood (Dense)	40	6.3
Wood (Light)	28	4.4
Steel	490	77.0
Brick	120	18.9
Sand	95	15.0
Concrete	144	22.6
Glass	160	25.1

Though there are few uncertainties about the dead load, its calculation is a painstaking and tedious job. It is also a fundamental job, since the amount of material used in a structure is, together with labor, one of the major components of its cost.

2.4 LIVE, SNOW, AND WIND LOADS

Loads such as the weight of people, machines, movable furniture, partitions, nonstructural elements, and movable fixtures are of an uncertain nature and of uncertain location in a structure and hence require safe averages established by the codes. Some occupancy loads prescribed by the IBC are listed in Table 2.2.

The live loads suggested by the codes are usually so conservative that, in order to avoid unrealistically high live loads, the codes allow live load reductions depending on the number of floors of a building and the area supported by a single structural element, such as a column. Live load reductions take into account the negligible chance that every floor in a building, or the entire large area supported by a single structural element, will simultaneously carry the full live load.

The weight of snow depends on the climate and elevation of the region where the building is to be erected: For

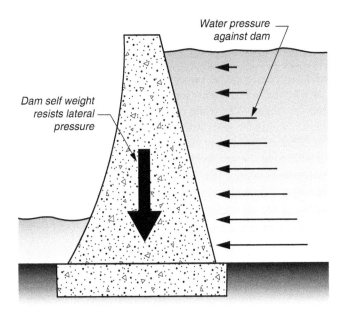

Figure 2.3 A gravity dam
The fluid pressure of water behind a gravity dam is balanced largely by the dead weight of the dam itself. The water pressure tends to slide the dam as well as topple it over, while the weight of the structure acts to stabilize it against these actions. The self-weight for such a structure is thus the most essential factor in its design.

TABLE 2.2	Occupancy Loads	
Type of structure	**Load (lb/ft²)**	**Load (kN/m²)**
Balcony	80	4.8
Garage	50	2.5
Library	150	7.0
Office building	80	4.0
School room	40	2.0
Shop	100	5.0
Theater	60	3.0

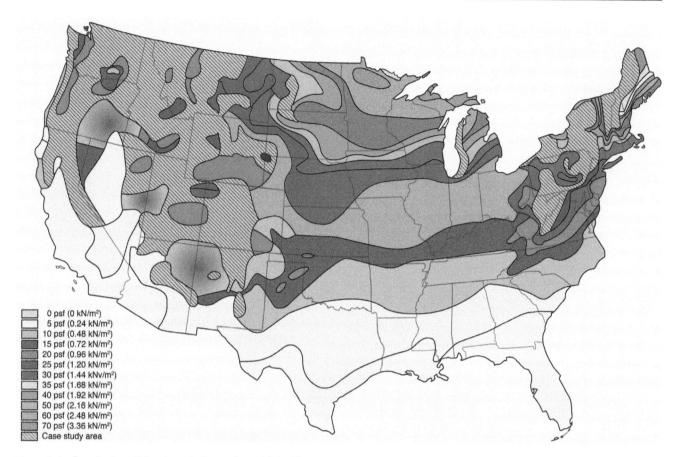

	0 psf (0 kN/m²)
	5 psf (0.24 kN/m²)
	10 psf (0.48 kN/m²)
	15 psf (0.72 kN/m²)
	20 psf (0.96 kN/m²)
	25 psf (1.20 kN/m²)
	30 psf (1.44 kNv/m²)
	35 psf (1.68 kN/m²)
	40 psf (1.92 kN/m²)
	50 psf (2.16 kN/m²)
	60 psf (2.48 kN/m²)
	70 psf (3.36 kN/m²)
	Case study area

Figure 2.4   Snow loads on flat surfaces in the continental United States
The IBC maps out in detail the range of snow loads across the country. The magnitude varies greatly depending on latitude and elevation, with larger values farther north as well as higher in elevation. Many mountainous areas are extremely localized, and case study analysis of the exact location is required, whereas other regions can be generalized. Regions with graduated color are those where loads vary depending on elevation.

instance, according to the IBC, it is equivalent to 30 pounds per square foot (1.4 kN/m²) in New York City, whereas it is 50 psf (2.4 kN/m²) in Minneapolis, Minnesota. The map of Figure 2.4 presents the distribution of basic snow loads in the United States.

Snow load is usually measured on the horizontal projection of the roof. The slope of the roof affects the retention of snow; a steep roof has lesser retention than a flat one. The nearly flat roof of the Blacksburg, Virginia, High School auditorium collapsed under the weight of a 9 inch (23 cm) layer of snow (Figure 2.5). While not an extreme snow level, it contributed to the structure's demise because it coincided with some substandard welds connecting roof trusses to vertical columns. Subsequent examinations indicated that the foundation was also inadequate and uneven.

The wind load on a building is difficult to ascertain with any degree of accuracy, because it depends on the wind velocity and on the shape and surface of the building, as well as on the roughness of the surface terrain (Figure 2.6). Average wind velocities are known with some degree of certainty, but it is difficult to measure the highest instantaneous velocity of a hurricane wind, or to forecast the largest velocity the wind will have in a certain locality. Figure 2.7 illustrates the variation of design wind velocities in the United States.

The dead, live, and the other "gravity" loads due to the pull of the earth are resisted by suitable structural systems. Resistance to wind pressures and suctions and other horizontal loads often requires separate structural systems. Horizontal wind-bracing systems may be seen on the underside of bridges (Figure 2.8), while vertical wind-bracing systems are hidden within the inner walls of most buildings (Figure 2.9).

Framed buildings (see Chapter 8) may be wind braced by stiffening alternate bays at alternate floors with diagonals or panels (Figure 2.10). In the Areva (formerly Fiat) Tower in Paris, engineered by Weidlinger Associates, stiffening panels are set in the frames of the outer walls (Figure 2.11). In some buildings, the outer walls are wind braced by diagonals spanning a number of floors.

The influence of the building itself presents even greater uncertainties with regard to wind load: Its shape may produce pressures on the windward side and suction on the leeward side, and the roughness of its surface may change the value of the local pressures. In any case, codes approximate these fluctuating dynamic loads by prescribing safe, static pressures or suctions due to wind, which also vary with height. These code values are revised from time to time in order to take into account the knowledge continuously accumulated in the field of aerodynamics.

Figure 2.5 Blacksburg, Virginia, High School gymnasium roof after it collapsed under snow load

Under a relatively modest snow load, the flat roof structure collapsed through a combination of substandard construction, as well as inadequacies in the foundation.

Photo courtesy of Michael McDermott

Figure 2.6 Wind load

Wind loads can be quite substantial in coastal and island areas subjected to hurricane-speed winds. In recent years, the IBC has increased the maximum design wind velocities in these regions to better safeguard life and property from devastating storms.

Photo: Meghan Pusey Diaz/123RF

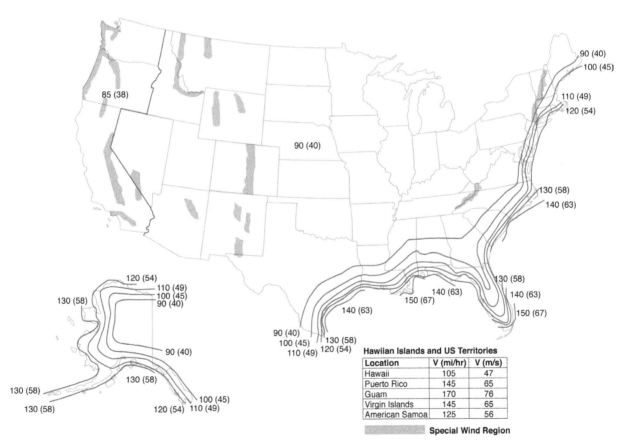

Hawiian Islands and US Territories		
Location	V (mi/hr)	V (m/s)
Hawaii	105	47
Puerto Rico	145	65
Guam	170	76
Virgin Islands	145	65
American Samoa	125	56

▨ Special Wind Region

Maximum Wind Velocity, mph (m/s)

Figure 2.7 Maximum wind velocities in the United States

As with ground snow load values, the IBC provides wind speed maps that specify design velocities. Also similar to snow loads, these values can have great variation in closely separated areas of coastal and mountain regions. The shaded areas represent mountainous regions where local conditions must be considered, and referred to as *"special wind regions"* in the IBC. The vast majority of the interior of the continental United States is designed for a wind speed of 90 mph (40 m/s). The highest wind speeds occur in costal or island areas subjected to hurricane-force winds.

Figure 2.8 Horizontal wind-bracing systems

All structures must transfer lateral forces through a horizontal plane. In bridge structures, the roadway must be stiff enough to perform this function. The underside of the road deck of the Akashi-Kaikyo bridge in Japan uses diagonal chevron bracing in the horizontal plane to stiffen the structure for transfer of the lateral wind forces to the supporting piers. The bracing creates a horizontal truss that functions as a very deep beam spanning horizontally between the supporting piers to resist wind forces.

Photo: Leung Cho Pan/123rf

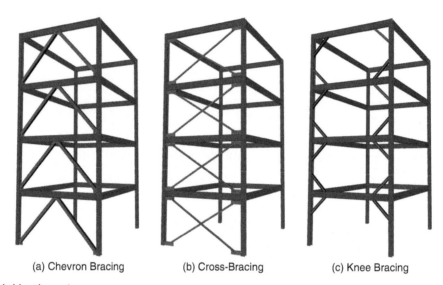

(a) Chevron Bracing   (b) Cross-Bracing   (c) Knee Bracing

Figure 2.9 Vertical wind-bracing systems

A variety of bracing schemes are used in building structures, such as chevron bracing (a), cross-bracing (b), and knee bracing (c). Each serves the function of preventing the structure from displacing laterally under the horizontal loads of wind or seismic forces. Members that serve to resist compression as well as tension forces must be noticeably heavier (a and c) than those that can be sized for tension only (b).

Bracing in only one plane is shown here for clarity, but in reality buildings must be constructed with bracing in multiple directions. Such bracing frames are often located in the core of buildings, as well as around their perimeters. Configurations will vary greatly depending on the size of the building, as well as the type and magnitude of the anticipated lateral forces. See Chapter 5 for a discussion of tension and compression and Chapter 8 for a complete discussion of building frames.

The wind velocities, V (in mph or Km/hr) may be converted into wind pressures, p, by means of code equations that take into consideration not only the speed of wind but also the effects of altitude, terrain roughness and ground obstructions, the shape of the building, as well as the surface on which the wind acts. These equations are derived from the basic physics equation stating that kinetic energy is equivalent to one half the mass of an object times the square of its velocity ($K = \frac{1}{2}MV^2$). The mass in this case is that of air at sea level.

It should be recognized that the depth of snow and wind velocity have statistical variations. Wind may be calm one day and become a gale or a tornado on the next. Similarly, there may be no snowfall today and a major blizzard tomorrow. Such extremes of weather are infrequent occurrences, and designing structures to withstand them with the same factors of safety as required for the more frequently occurring loads may be prohibitively expensive.

The strength of structures is also statistically variable. The strength of concrete, for instance, depends on the size

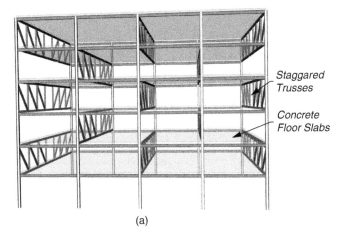

(a)

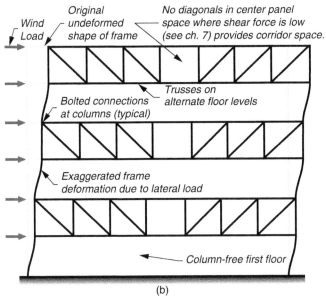

Wind Load

Original undeformed shape of frame

No diagonals in center panel space where shear force is low (see ch. 7) provides corridor space.

Bolted connections at columns (typical)

Trusses on alternate floor levels

Exaggerated frame deformation due to lateral load

Column-free first floor

(b)

Figure 2.10 A staggered truss framing system

The staggered truss system is a unique approach to accommodate both gravity and lateral loads, while simultaneously increasing the usable uninterrupted floor space. It is particularly well suited for residential units. The floor structure is carried by the bottom chord of one truss and the top chord of another truss, thereby creating a two-bay column-free interior space. For lateral loads, the trusses connected to columns at their top and bottom chords, effectively creating rigid frames (see Chapter 8) that use simple and less expensive bolted connections compared to heavy welded moment connections. The entire ground floor is also completely free of interior columns.

and strength of the stones in it, the amount of water used, and the length of cure time. To account for these variations of loads and strength, a "Stress-Strength Interference" method may be used. Loads are converted into stresses, and their frequencies of occurrence are plotted together with the frequencies of structural strength.

As seen in Figure 2.12, the two diagrams have an overlapping region. The size of this region indicates the chances that an infrequent extreme stress coincides with an equally

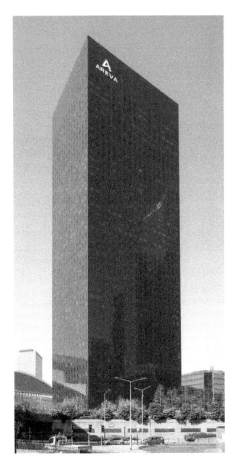

Figure 2.11 The Areva Tower in Paris

The wind bracing of the Areva (formerly Fiat) Tower, at the La Défense complex in Paris by Skidmore, Owings and Merrill and engineered by Weidlinger Associates, consists of stiffening panels set into the frames of the outer wall. These are pierced by window openings that decrease in area from the top to the bottom of the building. This increases the strength of the panels as the total wind force accumulates from the top to bottom of the tower.

Photo courtesy of Stéphane Renou

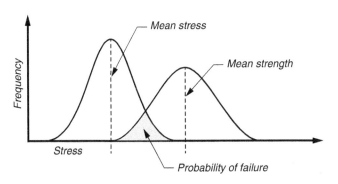

Figure 2.12 Stress (S)-Strength (R) Interference

Stress (S)-Strength (R) Interference diagram. The small shaded area represents the probability of failure of the structure.

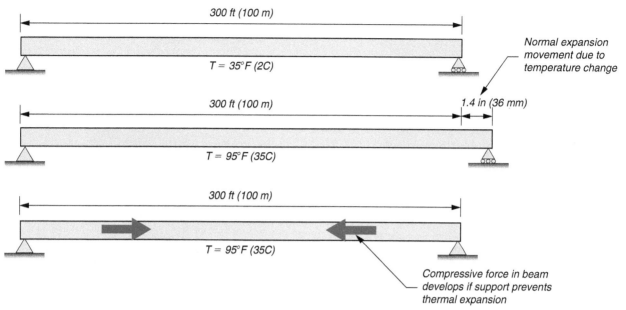

Figure 2.13   Thermal load

A bridge subjected to a temperature increase from when it was constructed will undergo an expansion in length. If this expansion is prevented, a compressive load will be developed in the member, which can be a sizable force.

infrequent understrength component and may destroy the structure. This had happened to the Blacksburg High School (Figure 2.5).

The designer can reduce this "Probability of Failure" by moving the two diagrams further apart. Either the size of the component may be increased, that is, by reducing stresses and thereby moving the stress diagram to the left, or by using stronger materials and moving the strength diagram to the right. Statistical variability will be further discussed in Chapter 13, "Structural Failures."

2.5 THERMAL AND SETTLEMENT LOADS

All structures are exposed to temperature changes, and change shape and dimensions during the cycle of day and night temperatures, and the longer summer and winter cycles. If restrained against movement, the effects of changes in dimensions due to thermal expansion and contraction are often equivalent to large loads, which may be particularly dangerous because they are invisible. A simple example of this type of loading condition will suffice to indicate its nature and importance.

A long steel bridge spanning 300 ft (~100 meters) over a river was built in winter, when the average temperature was 35°F (~2 degrees Celsius) (Figure 2.13). On a hot summer day, the air temperature may reach 95°F (35°C) and the bridge expands because it acquires the temperature of the surrounding air. The increase in length

of the bridge may be calculated with the aid of the "thermal coefficient of expansion," which for steel is 6.5×10^{-6} in/in/degree Fahrenheit (1.17×10^{-5} mm/mm/ degree Celsius). This is a physical property of the material itself that describes the amount of change in length it undergoes for a given change in temperature. Each material has its own characteristic coefficient of thermal expansion. Hence, the length change becomes the length multiplied by the temperature change and the thermal coefficient, or about 1.4 inches (36 mm). This change in length is small compared to its original length. But, if the bridge piers make it impossible for this expansion to take place, they develop in the bridge a horizontal compressive load capable of reducing the length to its winter value. Steel is very stiff in compression; it takes a large compressive load to reduce the length of the bridge by 1.4 inches (39 mm). This load is so high that it would use up half of the strength of the steel, leaving only 50 percent of its original strength to carry the loads for which the bridge was designed. The obvious way of eliminating such an overload is to allow the bridge to change its length with varying temperatures. This is usually done by supporting one of the bridge ends on a roller or "rocker" bearing (Figure 2.14).

The length change of the cables of suspension bridges due to temperature variations does not produce large stresses; it just raises or lowers the roadway. For example, the middle of the George Washington Bridge in New York changes its elevation by as much as 10 feet (3 meters) between winter and summer. Due to thermal conditions,

Figure 2.14 **Rocker support of the New River Gorge Bridge in West Virginia**

To permit free structural movement from thermal expansion and contraction, bearings are used to accommodate this displacement. The bearings either roll (as in the case of a roller bearing) or tilt (as in a rocker bearing, illustrated here). The gear teeth prevent slippage on this critical connection point. The bridge is the longest steel arch bridge in North America (Also see Figure 8.38 for an overall view of the bridge).

Photo courtesy of West Virginia Division of Culture and History

framed structures of high-rise, air-conditioned buildings may also develop stresses (see Section 7.4).

A similar condition of thermal movements, with different but equally dangerous consequences, is encountered in large domes. When the external temperature increases or decreases, the dome tends to expand or contract. Since it is usually prevented from so doing by its underground foundations, which remain at a constant temperature, it will move mostly up or down: The dome "breathes" (Figure 2.15). The top of a dome, covering a hall with glass walls and spanning 200 ft (~61 meters), may breathe up and down as much as 3 inches (~80 mm) due to air temperature changes, and if the

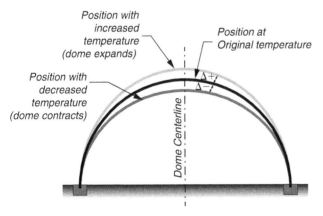

Figure 2.15 **Thermal movements of a dome**

As ambient air temperature rises and falls from the temperature at the time of original construction, a dome will expand and contract somewhat in response. The dome in effect "breathes" with the change in air temperature, although the actual amount of movement may be very small relative to the size of the dome.

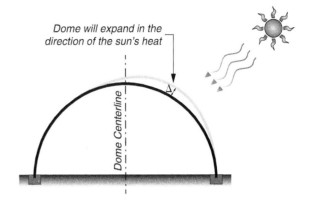

Figure 2.16 **Asymmetric thermal distortion of a dome**

As a dome is subjected to temperature change, it will expand upward or contract downward. When subjected to uneven heating, the expansion will be unequal, causing a distortion of the shape.

walls follow this movement, the glass panes break. A special type of sliding support eliminates this danger.

More complicated thermal loads are developed in a dome during the daily thermal cycle, when one of its sides is heated more than the other. The dome changes shape in an unsymmetrical fashion and becomes distorted (Figure 2.16). The stresses due to this distortion may be complex and high.

These simple examples show that a structure is particularly sensitive to thermal changes if by the nature of its shape, support conditions, and materials it tends to restrain the changes in dimensions due to temperature. On the other hand, acceptable deformations under loads require a structure to be stiff. Hence, the requirements for stiffness and for thermal deformations are opposite. Whenever a structure is to withstand heavy loads and small temperature changes, it may be made quite stiff; but if it must withstand large temperature changes and relatively small loads, it must be made flexible in order to accommodate such changes: The structure successfully resists this loading condition by giving in rather than by fighting it.

Another condition, producing effects equivalent to those of high loads, may stem from the uneven settlements of the building's foundations. A soil of uneven resistance, loaded by the weight of a building, may subside more in a specific portion of the foundation than in others. The soil deflections reduce the support of the foundation in certain areas and the structure may tilt. A prime example of such conditions is the Leaning Tower of Pisa (Figure 2.17).

The tower started leaning soon after construction started. The architects tried to straighten the upper floors but the structure continued to lean and was in danger of collapse. At the turn of the twenty-first century, the Italian government was finally able to stabilize the tower through a complicated process of soil extraction and heavy lead weights, which restored the tower's lean to the point at which it was in the early 1800s.

Figure 2.17   Foundation settlement
The famous Leaning Tower of Pisa, which began to experience uneven foundation settlement even before construction was completed in the twelfth century. Progressive uneven settlement threatened complete collapse of the tower until it was finally stabilized in 2001.

Photo: Lukiyanova Natalia/frenta/Shutterstock

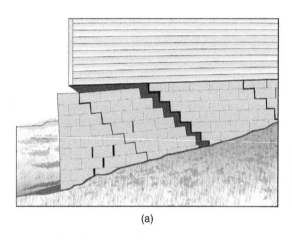

(a)

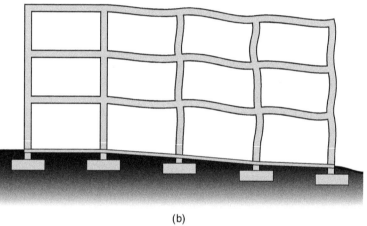

(b)

Figure 2.18   Uneven settlements of foundation
Uneven support settlements causing either a shearing off a portion of the building (a) or deformation of a frame (b).

A portion of a building above an inadequate foundation might either shear off from the rest of the building (Figure 2.18a) or hang partially from it (Figure 2.18b). No additional load has been applied to the building by the uneven settlement, but the supported portion of the building carries more load, and of a different type, than the load it was designed for. The unsupported section of the building is also under strain, as can be seen from the deflection of its beams (Figure 2.18b). Blacksburg High School's demise (Figure 2.5) is partially attributed to support settlement.

Thermal and settlement stresses are examples of stresses due to deformations rather than loads. An example of a stressed but unloaded structure may be built by cutting a piece out of a steel ring and welding the ring again so that it looks untampered with. If cut, this ring snaps open, demonstrating the stresses "locked" in it. A similar condition may be found in some rolled steel beams. If one of these beams is cut down the middle with a saw, its two halves open up and curve in opposite directions, indicating a certain amount of stress locked in by rolling process the (Figure 2.19). One of these

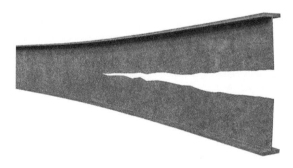

Figure 2.19   Locked-in stresses
Stresses locked into a steel beam due to the rolling manufacturing process may be released if the beam is cut down the center of its length, resulting in an outward curling of the sections.

beams might fail under load because the stresses produced by the loads are superimposed on the "locked-in" stresses and exceed safe values. Locked-in stresses are put to good use in prestressed and posttensioned reinforced-concrete structures (see Section 3.3) and other prestressed structures.

2.6 DYNAMIC LOADS

All the loads considered in the preceding sections were assumed to be stationary or vary slowly with time: They are considered to act statically. In codes, even the wind is assumed to act "statically" (for low and moderate height buildings), although it obviously does not.

Loads that change value or location rapidly, or are applied suddenly, are called dynamic loads. They can be exceptionally dangerous if their dynamic character is ignored.

The common experience of driving a nail with a hammer blow indicates that the sudden application of the moving hammer's weight achieves results unobtainable by the slow application of the same weight.

In earlier times, in order to ring a heavy bell, a church sexton (keeper) would pull its rope rhythmically; the bell would then swing progressively further until eventually it rang. The sexton could not achieve this result by exerting a sudden hard pull on the rope: He must, instead, "yank in step" with the bell's oscillations for a while until it rotated enough to ring. A similar but more common experience is that of being pushed on a swing (Figure 2.20). If one pushes with a sudden force, inertia is difficult to overcome and much force is needed to begin swinging. On the other hand, pushing gently and in rhythm with the swinging causes the person on the swing to go ever higher with each oscillation.

The force exerted by a hammer blow is called an impact load; the rhythmic push to someone on a swing is a particular case of a periodic force, referred to as a resonant load. In the first instance, the higher the velocity of the hammer or the shorter the time for the load to reach its peak value, the greater the effect: An instantaneous blow produces an extremely high force with possibly shattering results. In the second, a relatively small force applied "in step" for a long time produces increasing effects. In the case of the church bell, a small "in-step" force of a few pounds may eventually swing a bell weighing several tons into ringing.

Dynamic loads are exerted on structures in a variety of ways. A high-velocity wind gust may be similar to a hammer blow. A company of soldiers marching "in step" exerts a resonant load on a bridge when their step is in rhythm with the oscillations of the bridge. Workers on the Brooklyn Bridge in the late 1800s were admonished to "break step" to avoid such harmonic oscillations of the construction walkway (Figure 2.21). Although not in danger of overstress, the London Millennium footbridge took engineers by surprise with unexpected lateral motion created by pedestrian traffic. The amount of sway was severe enough to cause discomfort to some pedestrians. The bridge was closed shortly thereafter and renovated with viscous fluid dampers (similar to shock absorbers in cars) that eliminated the problem (Figure 2.22). Some small bridges are said to have collapsed under such resonant loads.

An impact load is characterized by a very short time of load increase and a resonant load by its rhythmic application. But when is an impact time "short?" And when is a varying load "in resonance" with a structure? All structures are, to a certain extent, elastic. They have the property of deflecting

Figure 2.20 Resonant load

Being pushed on a swing is a form of resonance. A person on a swing will sway with a natural period of vibration (the time it takes to swing from one position to the opposite side and back) that is related to the length of cable or rope on the swing. Physics dictates that weight does not affect the period of vibration; however, the greater a person's weight, the more inertial resistance must be overcome. A single sharp push may require significant force to overcome inertia. By pushing in synchronization with the period of vibration, though, only a very small amount of effort can have a large effect, over time sending a person ever higher on the swing with each oscillation. Pushing out of sync, on the other hand, will tend to dampen or cancel the motion, requiring a much higher amount of force.

Photo: Shotsstudio/Fotolia

Figure 2.21 The Brooklyn Bridge under construction, circa 1883

Notice the warning sign about not only the maximum number of workers on the temporary construction bridge but also the admonishment to *break step*, lest the resonant frequency of the temporary bridge be reached, thereby possibly overloading it to failure.

Photo courtesy of New York Historical Society

Figure 2.22 London Millennium Footbridge

When inaugurated in the year 2000, the Millennium footbridge immediately suffered from severe harmonic oscillations simply by pedestrians crossing it. The lateral back and forth sway was controlled after the bridge was closed and fluid viscous damping mechanisms installed.

Photo courtesy of Alfredo Fernández-González

under loads, and of returning to their initial position once the loads are removed. As a consequence of their "elasticity," structures have a tendency to oscillate: a skyscraper swings after the passing of a wind gust, and a railroad bridge oscillates up and down after the passage of a train.

The time required for a structure to go freely through a complete swing, or up-and-down oscillation, depends on its stiffness and its weight, and is called its fundamental period (seconds per cycle), while the number of complete swings in a unit of time is referred to as the frequency of motion (cycles per second). When the frequency of load application coincides with the frequency of oscillation of the structure, the load is said to be in resonance with the structure. In order to go through a complete swing, a modern building will take anywhere from one-tenth of a second to 10 or more seconds. A stiff structure oscillates rapidly; a low, rigid building has a short period. A flexible structure oscillates slowly; a tall steel skyscraper may take over 10 seconds for a complete swing. It has a longer fundamental period. The fundamental period of a structure is, in fact, a good measure of its stiffness.

The time of application of a load is measured against the fundamental period of the structure: If the time is short in comparison with the fundamental period, the load has dynamic effects; if long, the load has only static effects. It is thus seen that the same varying load may be static for a given structure and dynamic for another. A low-velocity long wind gust on a stiff building of short period will have the same effect as a constant static pressure. A high-velocity short wind gust on a flexible building of long period may strain the structure to a much greater extent than its static pressure would lead one to predict.

In a variety of practical cases, the dynamic effects of a load are 100 percent larger than its static effects. If a two and one half-pound (11 Newton) brick is put slowly on the plate of a spring scale, the scale hand will stop at the two and one half-pound mark (22 N) on the dial. If the same brick is instead released on the plate suddenly, the hand will move to the five-pound mark (22 N) and then oscillate, stopping eventually at the two and one half-pound mark (Figure 2.23).

Most loads applied to architectural structures do not have impact characteristics, except those due to hurricanes,

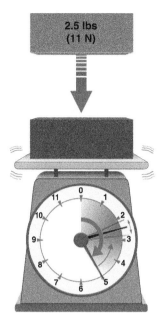

Figure 2.23 Dynamic loading
Effect of a weight dropping on a spring scale: The initial impact load may be as much as twice the weight of the object when applied in a static manner. The scale will bounce to the high point before settling down to the brick's actual weight of 2.5 lbs (11N).

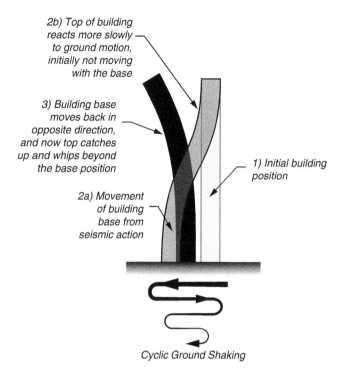

Figure 2.24 Earthquake motions
An earthquake creates complex movements of the ground, with back and forth, up and down, side-to-side, and compressing-extending movements. The greatest forces induced on a building or bridge are the side-to-side and back-and-forth types of movements. In these seismically induced oscillations of the ground, forces are generated in structures due to their own inertia resisting the acceleration of the ground. Structures essentially want to "stay put" while the ground moves below, thus generating shearing forces at the base. The top of taller structures then whip back and forth as the base moves below. The initial ground shaking can be quite intense and then die down over a period of several seconds or, in the case of large earthquakes, several minutes. This two-dimensional illustration is far simplified from the actual complex motions in all directions.

tornados, and earthquakes. Because such extreme winds have dynamic characteristics, the pressure predicted from the wind velocity formula may be doubled.

An earthquake is a sudden motion of the ground. This series of randomly variable jerks, transmitted to a building through its foundations, produces much larger oscillatory motions of the higher building floors (Figure 2.24). An earthquake occurs whenever the locked-in stresses in the earth crust, generated by movements in the molten mantle of the earth, become high enough to suddenly shift a portion of the crust with respect to an adjoining portion, opening up, at times, gaps several feet (meters) wide and hundreds of miles (kilometers) long, over underground fractures called faults. In the United States, faults exist all over the country but are particularly numerous and active in California, where strong earthquakes occur frequently. Figure 2.25 presents a Seismic Hazard Map of the United States, which highlights the most and least seismically-active regions of the country.

The study of earthquakes and the design of earthquake-resistant structures is still a relatively new science. Just as circularly expanding waves are generated by a stone dropped in a lake, the explosion of a small charge embedded in the earth crust generates outward-moving stress waves, whose speed depends on the value of the stresses locked in the crust. An increase in the speed of these waves indicates an increase in the value of the crust stresses and, hence, the danger of an impending earthquake.

Since the dynamic forces due to the jerky motion of the earth crust are mostly horizontal, they can be resisted by the same kinds of bracing systems used against wind. Various devices have also been invented to "isolate" buildings from earthquake vibrations: One of them consists of foundation

piers, made out of alternate layers of elastic materials and steel, which act to effectively decouple the building movement from the ground and allow the soil to move under the building (Figure 2.26).

Inasmuch as the earthquake action on a building depends on the nature of the soil and the structural characteristics of the building itself, earthquake design is a complex chapter of structural theory. It is only in the last few years that enough information on earthquake motions and on dynamic building characteristics has been gathered to allow safe dynamic earthquake resistant design. Tall, more flexible, steel buildings thus correctly designed have survived earthquakes that destroyed smaller, more rigid, masonry buildings.

Wind force that act in resonance with tall buildings can also be the cause of discomfort for occupants, with reports similar to the effects of seasickness when the period of vibration closely matches that of the human digestive tract. Devices called tuned mass dampers function similar to the shock absorbers in automobiles but act horizontally. They reduce the building's motions by means of a large mass of concrete, steel, or lead. The mass tends to stay in place due

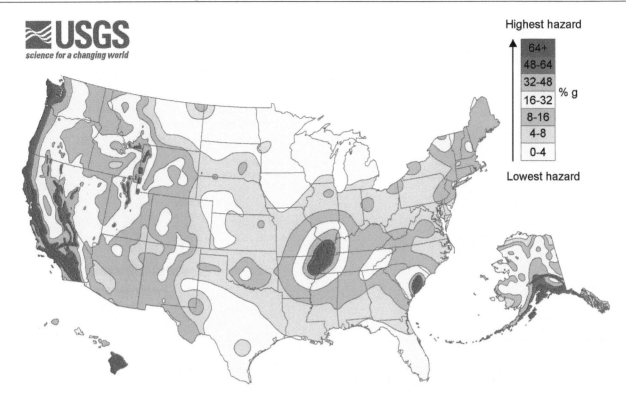

Figure 2.25   Seismic zone map of the United States

Seismic hazard map of the United States, produced by the U.S. Geological Survey. Similar to ground snow loads and wind velocity maps, the IBC has very detailed maps that indicate peak seismic accelerations throughout the country, which translates into a greater potential for structural damage. A complex interaction between the natural period of vibration of the building and the soil can greatly amplify these accelerations and consequent hazard.

Image: From Seismic Hazard Maps and Data - National Seismic Hazard Maps

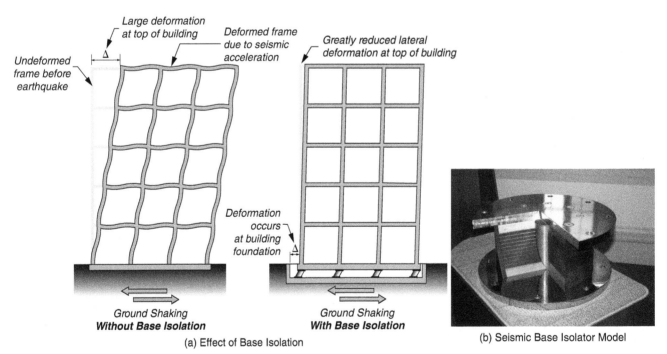

(a) Effect of Base Isolation   (b) Seismic Base Isolator Model

Figure 2.26   Seismic base isolation

Base isolation is an effective and increasingly popular strategy to reduce the impact of seismic forces on structures. Base isolation separates the building or bridge structure from the ground motion with deformable foundation structures (a). There are a number of variations in the construction of these devices. The style illustrated in the 2/3 scale cutaway model (b), referred to as a *lead-rubber bearing*, is comprised of layers of steel plates and flexible elastic material, with a solid lead core to absorb energy. The isolator is capable of both carrying very high-gravity forces while allowing for lateral displacements of up to two feet (0.6m), owing to the layered plate structure. The effect is to reduce the fundamental vibrational period of the building such that it is not in phase with that of the soil on which it is founded.

(b) Photo courtesy of Deborah Oakley

to its inertia while the building moves, and the attachment to the buildings with fluid-filled dampers and springs absorbs the wind energy (Figure 2.27).

Such structures are frequently comprised of large blocks on near-frictionless bearings riding on a thin layer of oil (Figure 2.28). Some systems are actively computer controlled: Whenever the building moves under the impact of a wind gust, an electromechanical feedback system pushes the mass in the opposite direction, extending the springs on one side and compressing those on the other side, and these respectively pull and push back the building toward its original position.

Resonant vibrations may also be attenuated by *passive*-tuned dynamic dampers, in which the period of the mass-springs system is made equal to that of the building. The mass, whose motion relative to the building is damped by

friction devices, passively (i.e., without an external control device) vibrates in resonance with the oscillations of the building but in opposite direction, thus keeping the building much closer to stationary through the action of the connecting dampers. A 730 ton (660 metric ton) sphere of steel hangs suspended near the top of the Taipei 101 tower in Taiwan to help dampen wind-induced oscillations (Figure 2.29). Unlike most dampers, the architects chose to make this a feature of the building viewable by patrons from several observation decks.

Resonant loads occur at times in architectural structures for other reasons as well. A piece of heavy machinery may vibrate because of the motion of its parts. If this vibration has a period equal to that of the supporting structure, the vibrations of the machine will be transmitted to the structure, which will swing with increasing oscillations. Floors,

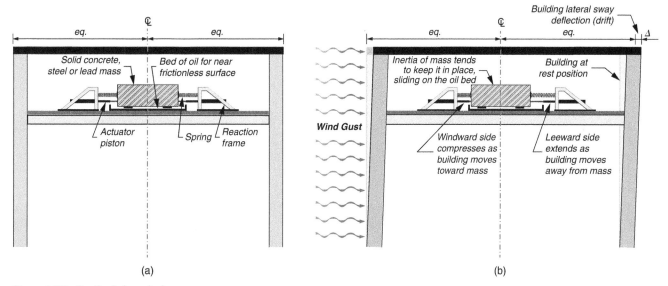

(a)   (b)

Figure 2.27   Feedback dynamic damper

Tuned mass dampers placed at the top of buildings are carefully calibrated to match the building's natural period of vibration. At rest, the damper remains centered (a). Under sway forces, the damper effectively shifts in the opposite direction due to its own inertia maintaining it in position while the building moves around it (b), thus canceling (dampening) much of the sway. Dampers may be actively controlled by a computer mechanism, or passively act by gravity, or a combination

Figure 2.28   Semi-active tuned mass damper

Atop the John Hancock Tower in Boston, Massachusetts, two 300 ton (2.7 MN) semi-active dynamic dampers help to counter the sway due to wind load. Each of the masses (made of steel boxes filled with lead) sits on a near frictionless bed of oil and is attached to a series of hydraulic pistons, springs, and control mechanisms. When a large wind gust occurs, computer controls and actuators facilitate moving the large mass in the exact opposite direction of the building motion, thus canceling out the effects of the lateral load effects. In actuality, the mass is tending to stay in place while the building moves around it. Wind energy is thus dissipated as heat through the hydraulic pistons acting as giant shock absorbers for the mass.

Photo courtesy of LeMessurier Consultants

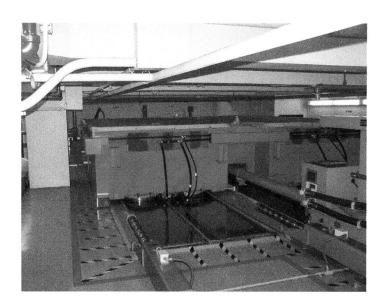

Figure 2.29 Passive tuned mass damper

Taipei 101 passive-tuned mass damper. The suspended sphere has the same fundamental period of the building itself, and so counters the sway by its own inertia. Hydraulic pistons attached to the sphere limit its travel as well as absorb energy and dissipate it as heat. Unlike most buildings where such structures are not visible to the public, the architects chose to make the sphere a prominent part of the architectural design. The damper is on public view at three levels on the restaurant and building observation decks.

Photo: Alvinku/Shutterstock

Figure 2.30 The Tacoma Narrows Bridge ("Galloping Gertie")

The original Tacoma Narrows bridge in Washington developed wild oscillations under light wind loadings. It was nicknamed "Galloping Gertie" by locals due to its tendency to sway in the wind. In 1940, after only four months in use, the bridge dramatically failed in a moderate wind of 42 mph. The bridge buckled up and down and twisted with increasingly more violent swings until it finally gave way and the center section collapsed to the river below.

Photo courtesy of Library of Congress Prints and Photographs Division [LC-USZ62-46682]

foundations, and entire buildings may thus be endangered by rather modest loads with a resonant period.

Whatever their origin, building vibrations can also be damped by friction devices applied at the intersection of any two members free to slide one with respect to the other. In steel frames, friction dampers are located at the intersection of diagonals that are not bolted or welded together.

Even more complicated dynamic phenomena are due to the wind. If a scarf is held out of the window of a moving car, it oscillates rapidly up and down. This "flutter," produced by the constant rush of wind on the scarf, is called an aerodynamic oscillation. The reader may produce such an oscillation by blowing against the edge of a thin piece of paper. Aerodynamic oscillations were produced by a wind of constant and fairly low velocity blowing for 45 minutes, against the Tacoma Narrows suspension bridge at Tacoma, Washington, in 1940;

these increased steadily in magnitude, twisting and bending the bridge, until it collapsed (Figure 2.30).

Dynamic wind loads on a structure are also created by its deflection of the wind stream. This explains, for example, the dynamic overpressures measured on the windward side of a building, which may blow windowpanes in. Underpressures or suctions, on the leeward sides of buildings, may suck windowpanes out. All dynamic phenomena are complex. The designer must be aware of their action and utilize with circumspection even the "equivalent static loads" prescribed by codes.

A closer examination of the maps of snow loads, wind loads, and earthquakes (Figures 2.4, 2.7, and 2.30) indicates that all three dangerous conditions seldom occur simultaneously in the same localities and a structure does not need to be designed to withstand a combination of all three such loads.

KEY IDEAS DEVELOPED IN THIS CHAPTER

- Structure is the load-carrying part of physical objects.
- Loads are categorized as applied and hidden loads.
- Applied loads are the ones the structure must support. These are as follows: Dead loads (the weights of building materials and of all permanent fixtures); live loads (occupancy loads and movable fixtures); wind, snow, and earthquake loads; as well as others, such as dynamic loads produced by machinery.
- Hidden loads are produced by thermal environments and support settlements.
- Building codes dictate the design values for each load category depending on the purpose of the building. The codes may also prescribe heights, and type of construction depending on locality.

QUESTIONS AND EXERCISES

1. Look around the places where you live and work or go to school: What objects would be classified as dead loads and which ones as live loads? Are there any types of loads that possibly fall into either category? Why?

2. What types of structural elements can you observe in buildings you are familiar with that function to resist lateral forces? What makes them different from that part of the structure that resists gravity (i.e., primarily vertical) loads, and how can you tell the function?

3. A developer wishes to build a skyscraper hotel in Miami, Florida. What type of loading should be considered? If the same structure is to be built in Maine, should the loading be different? How about building it in San Francisco?

4. Why do houses located on the Palisades on the California Coast require special foundations? What type of foundations should be used?

FURTHER READING

Charleson, Andrew. *Seismic Design for Architects: Outwitting the Quake*. Oxford, England. Elsevier Press, 2008.

Millais, Malcom. *Building Structures: From Concept to Design*, 2nd Edition. Spon Press. 2005. (Chapter 1)

Sandaker, Bjorn N., Eggen, Arne P., and Cruvellier, Mark R. *The Structural Basis of Architecture*, 2nd Edition. Routledge. 2011. (Chapter 3)

STRUCTURAL MATERIALS

3.1 THE ESSENTIAL PROPERTIES OF STRUCTURAL MATERIALS

A wide variety of materials are used in architectural structures: stone and masonry, wood, steel, aluminum, reinforced and prestressed concrete, plastics, and newer "high-tech" materials such as carbon fiber. They all have in common certain essential properties that make them suitable to withstand loads.

Whether the loads act on a structure permanently, intermittently, or only briefly, the deformation of the structure (a) must not increase indefinitely, and (b) must disappear after the action of the loads ends. The distribution (average) of load (or *force*) over an area is referred to in engineering terms as *stress*, which is the same as the more conventional term *pressure*, measured as force per unit area. In U.S. customary units, this is expressed in pounds per square inch (lb/in², or psi). In S.I. (*Le Système International d'Unités*, the world's most widely adopted unit system), the units are Newtons per square meter (N/m²), which is expressed as the *Pascal* (Pa)—that is, 1 Pa = 1 N/m².

A material put into tension will experience tensile stress and stretch, and conversely a material placed into compression will experience compressive stress and contract. Stress is one of the most elemental and important concepts in structural mechanics. It will therefore frequently be referred to for the remainder of this book.

Understanding the concept of stress in one's daily life experience is intuitive. For example, the blow of a hammer on a nail will drive the nail into a piece of wood, but the blow of the same hammer *directly* on the wood will not similarly embed it into the wood; this is because the force of the hammerhead is distributed over a much larger area. In other words, even with the same force, the stress is much lower on the hammerhead itself than on the tip of a nail. The familiar "bed of nails" demonstration also illustrates this principle (Figure 3.1). Although the nail points are individually very sharp and can easily pierce skin, when the entire body weight of the performer is distributed over hundreds of nails, the force becomes so low on each individual nail that one can safely lie upon the bed without injury—no superhuman powers are required. Another example is the snowshoe, which enables one to walk on soft snow that would otherwise not be able to bear a hiker's weight (Figure 3.2).

A material whose deformation vanishes rapidly with the disappearance of the loads is said to behave *elastically* (Figure 3.3). All structural materials are elastic to a certain extent. If they were not, and a residual deformation were present in the structure after unloading, new loadings would

Figure 3.1   A bed of nails

The concept of stress illustrated by the "bed of nails" performance. Although a single nail pressed on by one's full body weight would easily pierce skin, when the same weight is distributed over many hundreds of nails the force becomes so low that one can safely rest on a bed of nails without injury. For example, if a man weighing 200 pounds (~890 N) has his weight distributed over 400 nails, the resulting force is only 200 pounds ÷ 400 nails, or one-half pound per nail...only eight ounces (~2.2 N)—about the weight of a small glass of water.

Photo courtesy of Chris Beckett

Figure 3.2 A snowshoe

As with the bed of nails, a snowshoe distributes force over a large area; the pressure is then reduced to the point that one may walk on soft snow without penetrating deeply through the surface.

Photo: Karolina Vyskocilova/123RF

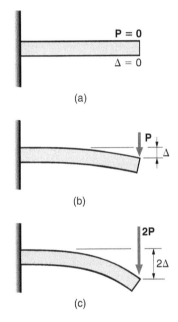

(a)

(b)

(c)

Figure 3.3 Linear elastic behavior

For a material that is linearly elastic, the deformation is directly proportional to the load. In a simple cantilever beam supporting a concentrated load "P" on its free end, the corresponding relative deflection is shown in 3.3b. A doubling of the load will lead to a doubling of the deformation (3.3c), and if the stress on the loaded member is kept within the elastic region, it will return again to its undeformed shape when the load is removed (3.3a).

gradually increase this residual deformation and the structure would eventually become useless. On the other hand, no structural material is perfectly elastic: depending on the type of structure and the magnitude of the loads, permanent deformations are unavoidable whenever the loads exceed certain values. Hence, the loads must be limited to values that will not produce appreciable permanent deformations: Structural materials are usually stressed within their so-called elastic range.

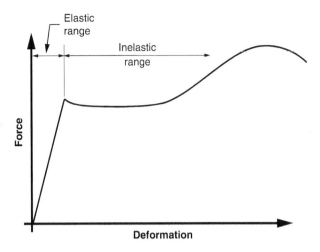

Figure 3.4 Force-deformation diagram

The graph illustrates the response of an elastic material to a force applied to it. The line plots the relative amount of deformation corresponding to the applied force. The sloping straight-line portion of the graph demonstrates that there is a direct proportionality between the amount of deformation (stretch) and the amount of force. For the type of material represented by this graph (such as steel and some plastics), this is true up to a limit, beyond which the proportionality breaks down and a large amount of inelastic deformation occurs with little increase in force.

Most structural materials are not only elastic but, within limits, are *linearly* elastic: This means that their deformation is directly proportional to the load. Thus, if a linearly elastic beam deflects one-tenth of an inch (2.5 mm) under a vertical load of 10 tons (88.9 KN), it will deflect two-tenths of an inch (5.1 mm) under a 20-ton (177.9 KN) load (Figure 3.3c). Most structural materials are used almost exclusively in their linearly elastic range.

These concepts may be visualized by examining a "Force-Deformation Diagram" (Figure 3.4), which illustrates the relative effects between load (plotted on the vertical axis) and deformation (plotted on the horizontal axis) for a structure in tension. When a force is applied to a structural component, it changes its shape: It stretches due to the internal material response of the external load. As the load is increased, the deformation (stretch) will similarly increase. In the elastic range, the initial straight-line portion of the diagram illustrates the direct proportionality of load and deformation.

While the force-deformation diagram provides an easy visualization for the structure in tension, the amount of deformation is dependent on the magnitude of the force and size of the member. In order to make the diagram independent of these values, they are normally plotted as a *stress-strain* diagram (Figure 3.5), which then illustrates a characteristic property for a given material. In terms of stress, if a steel rod with a 1 in^2 (645 mm^2) cross-section is pulled by a 1 pound (4.45 N) force (i.e., a 1 psi (6.9 KPa) stress), it stretches. The corresponding *percentage change* in its length is denoted by the term "strain." This is the change in length of the material relative to its original length, measured in units of inches per inch, millimeters per millimeter, and so on. The applied stress and the resulting strain are

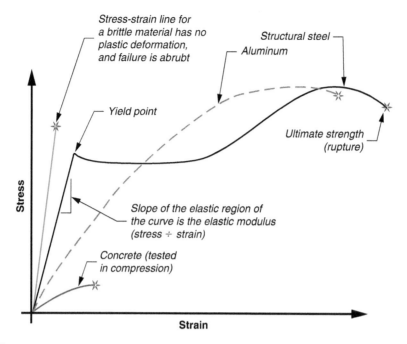

Figure 3.5   Stress-strain diagrams

By dividing the applied force by the cross-sectional area, the vertical axis becomes a plot of stress values, and by dividing the amount of deformation by the original length, the horizontal axis becomes a plot of strain. This diagram then represents a characteristic property for a given material, independent of the size and force magnitude, and provides important insight into the behavior of specific materials.

plotted, with stress on the vertical axis and strain on the horizontal axis, similar to the force-deformation diagram.

The stress-strain diagram for some typical structural materials is presented in Figure 3.5. Steel is an "ideal" material to explore in more depth because it has such a well-defined behavior over an extensive range. Because steel is a linearly elastic material up to a certain amount of strain, the initial portion is a straight line: For any increase in stress, there will be a direct increase in strain, such that doubling the stress will double the strain, and so on. Furthermore, elastic behavior means that removal of the force will return the material to both a zero-stress and zero strain condition.

This behavior is true up to the "yield point" stress (F_y), at which point the steel undergoes a permanent deformation that will remain even after the load is removed. The load at which a material starts behaving in a clearly plastic fashion is called its yield stress or yield point. Further sustained loading beyond this point will cause strains to increase rapidly. This continued deformation with little change in load is referred to as *plastic* behavior (see Chapter 9 for a further discussion on plastic behavior).

Once the steel has completely yielded, which involves considerable deformation, it enters a new phase called *strain hardening*. The material exhibits very little change in strain while the stress level increases. The sample begins to exhibit a pronounced "necking down" in size, and an internal molecular transformation has caused it to become a very tough material. Finally, with continued tension the stress level exceeds the capacity of the steel's molecular bonds and it breaks at a stress called the *Ultimate Strength* (F_u) of the

material. The photograph of Figure 3.6 shows a steel test sample after it has been loaded to failure.

Materials such as structural steel behave plastically above their elastic range, and have a clear and well-defined yield point. Other common building materials such as aluminum and concrete do not have such a well-defined behavior, or even a clear yield point. For such materials, a yield stress is empirically defined based on a specific strain amount (see Section 3.2).

Although large permanent deformations are to be avoided, plastic behavior above the elastic range does not actually make a material unsuitable for structural purposes: in fact, the opposite is true. For example, while the material is in the linearly elastic range, its deformations increase at the same rate as the loads. Above the yield point, deformations increase more rapidly than the loads (Figure 3.7), and eventually keep increasing even if the loads are not increased. This "flow" or yield under constant loads is thus the clearest sign, and a good warning, that failure is imminent. There are other structural advantages of plastic behavior as well, which are discussed in Section 9.3.

Materials to be used for structural purposes are chosen so as to behave elastically in the environmental conditions and under the type of loads to be expected during the life of the structure. The importance of this requirement can be illustrated by a simple example. If a wax candle is improperly inserted in a candlestick, the seemingly stiff candle will slowly bend out of shape at a warm temperature, and after a few weeks or months (depending on the temperature and type of wax) may be completely bent over (Figure 3.8). This deformation would occur much more rapidly in an oven, while in a refrigerator it

Figure 3.6 Failure of steel test sample

Figure 3.6a illustrates a machine performing a tensile test on a material sample. The device is capable of generating substantial forces, great enough to break samples while simultaneously carefully recording the magnitude of force and deformation. The test sample enlarged in 3.6b is a steel rod loaded to failure, as evidenced by the thin crack at its center. Notice the pronounced "necking-down" of the specimen near the center. When such necking occurs, the steel has gone well beyond its yield point, the location on the graph of figure 3.5 where diagram sharply changes direction.

Photo courtesy of MTS Systems Corporation

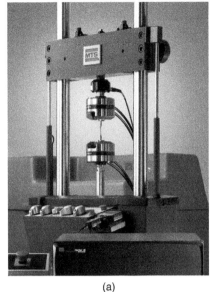

(a)

(b)

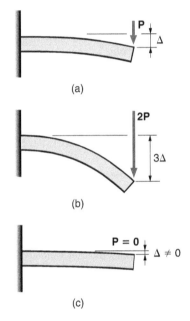

(a)

(b)

(c)

Figure 3.7 Plastic behavior

Unlike elastic behavior, the response of the cantilever beam is not proportional to the applied load (i.e., a doubling of the load can lead to *greater* than doubling of the deflection (Figure 3.7b)). Furthermore, when the load is removed, there remains a permanent deformation (or *set*) in the beam and it does not return to its original position. The beam has deformed plastically (Figure 3.7c).

Figure 3.8 Candle deformed

An ordinary wax candle will plastically deform under its own weight, given sufficient time in a warm environment. With an increase in temperature, the plastic flow will become more pronounced, whereas with a decrease in temperature, the candle becomes a brittle material.

Photo: Laurie Fish/123rf

might be indefinitely limited. Under arctic conditions, therefore, wax is "more structural" than in the tropics.

While given a sufficiently long time, a candle will deform under its own weight alone, if one tries to bend it quickly, it breaks; it becomes brittle. All structural materials exhibit a similar behavior: at low temperatures and under fast load applications they are elastic and brittle; at high temperatures and under long-time loads they "flow." The temperature and time required for flow differ for different materials.

Materials that are linearly elastic up to failure, such as ordinary annealed glass (the type used in a picture frame, for example) and some plastics (note the misnomer), are highly unsuitable for structural purposes. They cannot give warning of approaching failure and often shatter under impact (Figure 3.9). Recent developments, however, that combine the flexibility of plastics and the strength of glass have lead to exciting architectural developments. *Laminated structural glass* is comprised of sheets of high strength tempered glass combined with resilient plastic film fused into

Figure 3.9   Brittle failure

Some materials, such as glass, are brittle by nature. They have little elastic behavior; instead of deforming gradually, internal stresses are released suddenly and powerfully, rupturing the material itself.

Photo: Cosma/Fotolia

Figure 3.11   Brittle fracture of a steel beam

On an unusually cold winter night in 1985, a roof girder for the open-air theater at Wolf Trap Center for the Performing Arts in Vienna, Virginia, suffered a brittle fracture. A complex combination of brittle steel and minute fabrication flaws combined with the bitter cold was determined as the cause. Fortunately, enough redundancy was present that the roof did not collapse and the building was subsequently strengthened and has been successfully functioning ever since.

Photo courtesy of Deborah Oakley

Figure 3.10   Laminated structural glass

Widely popularized in Apple computer stores, laminated structural glass has made it possible to design architectural elements such as this circular stairway in Boston, Massachusetts. The inherent brittleness of glass is overcome by laminating multiple layers of glass with plastic material. All primary load-bearing elements are comprised of glass with stainless steel connectors, where the highest stresses need to be transferred between individual glass elements.

Photo courtesy of Deborah Oakley

multiple layers. Its strength and flexibility is such that it can be used in applications such as beams and floor slabs that have previously been impossible. Dramatic multistory facades, stairways, and floors made of structural glass combined with steel tensile elements have now become commonplace (Figure 3.10). Structural glass and other materials such as reinforced concrete (which combines plain concrete

with structural steel; see section 3.3) are referred to as *composite* materials. They combine the best properties of two or more materials into one new material that has unique properties all its own.

Steel at normal temperatures has a useful linearly elastic range up to the yield point, followed by a long plastic range. It will, however, become suddenly brittle at a temperature of −30° Fahrenheit (−34°C). Some unexpected failures of steel bridges in Canada in the 1950s were traced to this sudden transition from elastic-plastic to brittle behavior at low temperature. A similar brittle fracture failure occurred on an unusually cold winter night in 1985 to one of the main roof girders of the open-air roof pavilion of the Filene Center Theater at the Wolf Trap Performing Arts Center in Vienna, Virginia (Figure 3.11).

At high temperature, even steel—one of the strongest structural materials—loses most of its strength: It keeps deforming more and more, even under constant loads. At not much above 1000° Fahrenheit (538°C) it loses almost all bending resistance and becomes limp like cooked spaghetti (though it does not melt until much more than twice that temperature) (Figure 3.12). Hence, if steel is to be used safely in a building, it must be protected, or *fire-retarded*, so that it will not reach high temperatures at least for a few hours, during which time the building can be safely evacuated. When a material is fire retarded for an indefinite time, it is said to be fireproof: Reinforced concrete is practically fireproof, provided the reinforcing steel is sufficiently protected by a cover of concrete. Building codes have specific requirements for fire retardation.

Some materials have a relatively limited elastic range and behave plastically under low loads. Certain plastics (thus correctly named in this case) flow under almost any load. The yielding behavior of these plastics, and the brittle

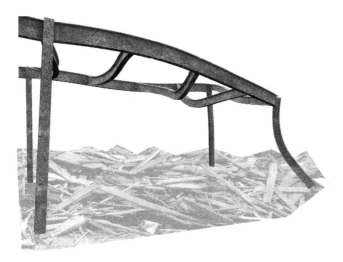

Figure 3.12   Warping of steel due to fire

While unprotected steel will not burn in a fire, at around 1100°F (538C) it substantially loses stiffness and will warp like plastic under its own self weight. It is notable that this loss of stiffness occurs more than 1500°F (816C) below its melting point.

behavior of others, makes them unsuitable for structural purposes. But reinforced plastics, such as fiberglass, present acceptable structural characteristics, and their increased use is easily foreseeable. Materials such as these are specifically useful in certain applications such as the corrosive environments of chemical manufacturing plants and wastewater treatment facilities where unprotected steel would be quickly damaged.

Modern structural materials, such as steel, are what are known as *isotropic* (from the Greek *isos*, meaning "equal," and *tropos*, meaning "direction") and homogeneous, that is, their properties and behavior under load are the same throughout the structural component. Because of this, their resistance does not depend on the direction in which they are stressed. Wood, on the other hand, is *anisotropic* (from the Greek *an*, meaning "not"), since it has different strengths in the direction of the grain and at right angles to it. This drawback is remedied by joining, with plastic glues, sheets or pieces of wood with grain in different directions. The *engineered wood* thus obtained presents more homogeneous strength characteristics and moreover can be made weather resistant. Many varieties of engineered lumber are now widely available such as plywood, oriented-strand board (OSB), particle board, medium-density fiberboard (MDF), glued-laminated timber (GluLam), parallel-strand lumber (PSL), and others. Each of these not only improves upon the properties of natural wood (which is subject to the variations that accompany any organic material), but of increasing importance allows the use of small pieces of "leftover" materials that would formerly have been waste, thus making better use of resources.

Structural materials can also be classified according to the kind of basic stresses they can withstand: tension, compression, and shear. These basic forces are discussed in depth in Chapter 5. In brief, tension is the kind of stress that pulls apart the particles of the material and lengthens elements (Figure 3.13a); compression pushes the particles one against the other and shortens elements (Figure 3.13b); shear, however, tends to make the particles slide, as they do

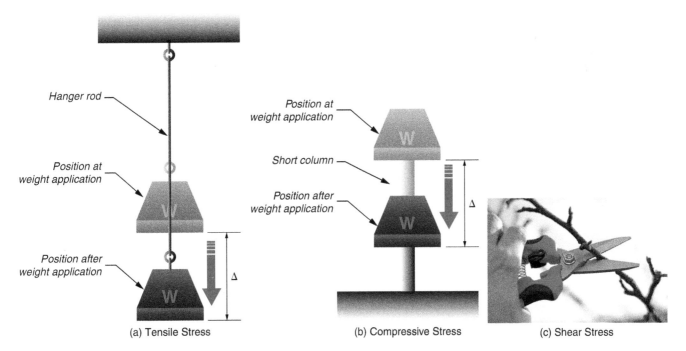

(a) Tensile Stress   (b) Compressive Stress   (c) Shear Stress

Figure 3.13   Basic types of stress

An element that is pulled is in tension (a), one that is pushed or squeezed is in compression (b), and one that experiences a slicing or transverse force (such as a pair of scissors or a branch pruner) experiences shear (c). It should be noted, that longer members in compression are subject to buckling behavior (see Chapter Five). All deformations in these diagrams are exaggerated for clarity.

(c) Photo: Zbigniew Guzowski/123rf

in in a pair of wood pieces glued together under either tension or compression, or as they do in a wire or branches sheared by a pair of cutting pliers (Figure 3.13c).

All structural materials can withstand compressive stresses. Some, such as steel, resist compressive and tensile stresses equally well. Others, such as stone, brick, unreinforced concrete, or wood loaded perpendicular to its grain, do not: Their tensile strength is about one-tenth of their compressive strength. Their use is necessarily limited to loads and shapes that will not develop significant tensile stresses. Materials capable of resisting tensile stresses usually resist shearing stresses also; materials that are essentially useful in only compressive loads, on the other hand, do not have high shear strength either (see Section 5.3).

3.2 MATERIAL CONSTANTS AND SAFETY FACTORS

For structural purposes, materials are mostly used within their linearly elastic range, and this implies that their deformations are proportional to the loads upon them. But different materials deform differently under the same loads. If a steel wire 5 feet long (1.5 m), 1/16 of an inch (1.6 mm) in diameter, is loaded by a weight of 1000 pounds (4.45 kN), it lengthens by 2/3 of an inch (17 mm); the same wire made of aluminum stretches three times as much, that is, 2 inches (51 mm) (Figure 3.14).

Steel is stiffer in tension than aluminum. The measure of this stiffness is a property of each structural material, called its tensile *elastic modulus*. The higher a material's elastic modulus, the stiffer it is—meaning that it will stretch or deform less than a material having a lower elastic modulus. The elastic modulus is the force theoretically capable of stretching elastically a wire, 1 square inch (645 mm^2) in area, to twice its original length (theoretically, because in

actuality the wire will break before stretching that much). Another concept, toughness, is the energy required to tear a piece of material apart. Materials with a long plastic range are tougher then brittle materials.

The elastic modulus (E) of a material may be determined from the stress-strain diagram. For highly elastic materials; it is the slope of the initial, straight-line portion of the diagram, that is, the ratio of a stress level and the corresponding strain. For the structural steel of Figure 3.5 at a stress of 30,000 psi (207 MPa), the strain is 0.001 in/in (0.025 mm/mm). Consequently, $E = 30,000,000$ psi (207 GPa). The slope of the stress-strain curve for aluminum is flatter. The strain at 30,000 psi (207 MPa) is 0.003 in/in (0.076 mm/mm) and $E = 10,000,000$ psi (69 GPa).

It is seen on the diagram that aluminum does not have a well-defined yield point. A yield stress is defined for such materials by drawing a line parallel to the initial straight portion of the diagram at a strain of 0.2 percent. The stress level where the line intersects the curve is the yield stress. For the aluminum of Figure 3.5, $F_y = 36,000$ psi (248 MPa).

The modulus in compression differs, in general, from the modulus in tension. The compressive modulus of concrete varies with its composition and has an average value of 4 million pounds per square inch (27.6 GPa); its tensile modulus has little significance, since its tensile strength is negligible. The compressive modulus of wood varies between 1 and 2 million pounds per square inch (6.9 GPa and 13.8 GPa) in the direction of the fibers, and is approximately 3/4 million pounds per square inch (5.2 GPa) at right angles to them. Since they are isotropic and homogeneous, steel and aluminum have the same moduli in tension and compression.

For purposes of safety, knowledge of the stress at which a material will start yielding is of the utmost importance. The yield stress or yield strength in tension or compression for common structural steel (and for aluminum) varies

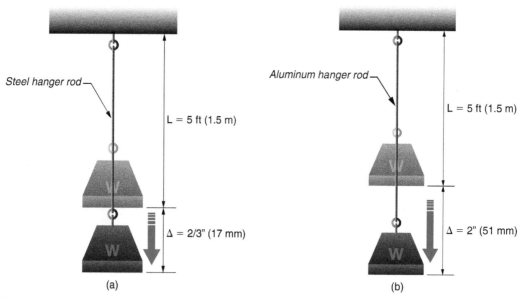

Figure 3.14 Tensile deformations
A rod of aluminum will stretch three times as much as a steel rod supporting the same load, because the elastic modulus of aluminum is 1/3 that of steel.

between 36 thousand and 50 thousand pounds per square inch (248 to 345 MPa). Since structures cannot be allowed to yield under loads lest their permanent deformations accumulate in time, safe stresses are usually assumed as a fraction of the yield point of their material (usually 60 percent, or a safety factor of about 1.7). Thus, steel and aluminum may be safely stressed in tension or compression to about 20 to 30 thousand pounds per square inch (138 to 207 MPa), and concrete in compression to approximately 1 to 5 thousand pounds per square inch (7 to 34 MPa).

The safety factors thus introduced depend on a variety of conditions: the uniformity of the material, its yield and strength properties, the type of stress developed, the permanency and certainty of the loads, and the purpose of the building. This last factor is of great importance from a social viewpoint: The safety of a large hall is more critical than the safety of a one-family house—and most certainly more critical than a storage shed—and so must be evaluated more conservatively. Building codes (Chapter 2) define appropriate safety factors for most commonly used structures. Factors of safety for buildings of exceptional dimensions, to be occupied simultaneously by large numbers of people, are established today so as to make it highly improbable that even a predetermined, small number of persons would be killed by a structural failure during the life of the structure. The calculation of such safety factors involves the use of probability theory and leads to results gauged in terms of human lives. (See Sections 2.3 and 13.5.)

Safety factors cannot be established on the basis of the yield point whenever the material does not present a well-defined yield point, or does not exhibit a distinct yielding behavior. The first case occurs with concrete, which does not have a clear transition from elastic to plastic behavior; the second with brittle materials, which behave linearly up to failure. In these cases, safety must be measured directly against failure, as far as the material is concerned. It is thus important to realize that steel will break in tension at a stress of 50 to 300 thousand pounds per square inch (345 MPa to 2.1 GPa), and concrete in compression at a stress of 3 to 12 thousand pounds per square inch (21 MPa to 83 MPa). These stresses are called the *ultimate strength, F_u,* of the material.

The load, or combination of loads, inducing stresses equal to the ultimate strength of an element, is called the *ultimate load* for that element. When the design of a reinforced concrete structure is based on ultimate strength (an approach called *strength design*), codes specify that certain load combinations should not reach values higher than the ultimate load, U. For example, indicating by "D" the dead load, by "L" the code live load, and by "W" the code wind load (up until recently when the factors have become more complex), the American Concrete Institute (ACI) code required that U be less than $0.75 \times (1.2D + 1.6L + 1.6W)$, or less than $1.2D + 1.6L$ in the absence of wind. Similar ultimate design formulas are also becoming the preferred approach in steel and wood design as well.

Some advantage may be realized in establishing safety factors on the basis of failure, even when the yield point is clearly defined. These safety factors give direct information on the overload the structure can support before it collapses, rather than on the overload that will make the structure unusable because of excessive deformations. Knowledge of both overloads may be useful in establishing higher safety factors for permanent and semi-permanent loads on the basis of yield, and lower safety factors for exceptional loads (hurricane winds, earthquakes) on the basis of failure. This criterion is accepted by most codes: A combination of normal gravity loads and of exceptional lateral loads (such as full wind or earthquake forces that have a minimal chance of occurring simultaneously) is often allowed to stress the structure 33 per cent above the stresses due to gravity loads acting alone.

3.3 MODERN STRUCTURAL MATERIALS

Iron has been used as a structural material for thousands of years, mainly in combination with other materials. Since the tensile properties of iron and the compressive properties of wood were well known, a combination of wood struts in compression and iron tie-rods in tension was often used in trusses spanning the naves of medieval churches, for example (Figure 3.15). It should be noted, however, that iron has

Figure 3.15   Wood and iron truss
Heavy timber structures often employ iron (and today, steel) tie rods and sag rods. The function of the sag rod is simply to prevent the tie rod from deflecting under its own weight. The sag rod serves no other structural function, unless a large object such as a chandelier is suspended below it, in which case it transfers that weight up to the top timber members. The purpose of the pilasters is to provide a good bearing surface for the truss and to strengthen the walls.

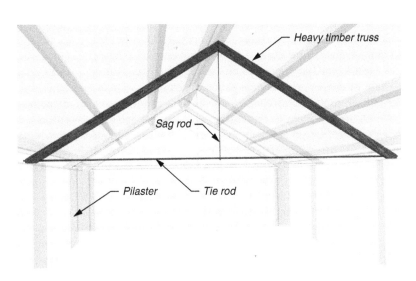

a much smaller elastic range than structural steel. Cast iron is actually brittle—it does not exhibit any plastic behavior, thereby limiting its use. Steel, with its greatly improved properties, could only be produced in small quantities until the Industrial Revolution. Steel was thus limited to small objects and weaponry like swords until the middle of the nineteenth century, when large quantities of steel suitable for construction projects could be produced.

Increased knowledge in metallurgy, chemistry, and physics has substantially improved the properties of structural materials during the last 150 years. Stainless steel has all but eliminated the danger of rust in most applications, while rust-resistant steel, such as U.S. Steel's "Corten" (also known as "weathering steel"), stops rusting after two or three years and does not require painting. This tough rust coating acts similarly to the oxide coating that aluminum develops, both of which form a protective layer that inhibits continued corrosion under ordinary usage. Ultimate tensile stresses of steel alloys have reached values as high as 300 thousand pounds per square inch (2.1 GPa), and those of minute steel "whiskers," millions of pounds per square inch (7 GPa and higher), forecasting revolutionary developments in steel design. Similarly, some new aluminum alloys have the strength of structural steel and are only one-third as heavy. Plastic glues have transformed wood into a more permanent, practically isotropic material. Glass reinforced plastics combine the strength of glass with the nonbrittle behavior of plastics.

Possibly the most interesting (and one of the most important) modern structural materials is reinforced concrete, which combines the compressive strength of concrete with the tensile strength of steel. The use of reinforced concrete in contemporary construction is so pervasive that it may surprise the reader to learn it was only developed in the mid-1850s, and did not achieve wide usage until the early part of the twentieth century. Although the Romans used a form of concrete made with volcanic ash (known as *pozollana*) for many centuries, it was never combined with steel to create reinforced concrete. The art of creating the pozollana disappeared with the fall of the Empire. Its modern incarnation, *Portland cement*, was developed in 1824 by the Englishman Joseph Aspdin.

Concrete is actually a mixture of Portland cement, stone aggregates, and water. When activated by water, a chemical reaction is initiated in the cement causing it to set and harden to make the structure a single, monolithic entity. Concrete has been likened to "liquid stone" and can be poured in a variety of shapes with a wide range of textures, so as to adapt itself to the architecture of the building and the loads on it. A new freedom in the design of structures has thus been realized, far greater than that inherent in assembling beams and columns of standard shapes like those of rolled steel sections or sawn lumber.

Concrete is very strong under compressive stresses. The continued use of ancient Roman structures such as the Pantheon (to this day the largest unreinforced concrete dome in the world) give testament to its durability when used in a compressive manner. Under tensile stresses, however, concrete is very weak and breaks easily. The reinforcing steel therefore acts like a web of tensile elements, which pervades and holds together the mass of concrete.

The properties of steel depend on the careful proportions of ore mixtures, furnace and quenching temperatures, and minute amounts of alloyed chemicals. The properties of concrete depend on the quality and amounts of components in the mixture. Concrete is so sensitive to variations in mixture that it must be carefully and scientifically "designed" in specialized laboratories when used in large construction projects. The grain size and distribution of the sand and gravel aggregates, the quality of the cement, and the quantity of water used, all substantially affect the concrete strength and its hardening time.

It should be noted that concrete does not become hard by "drying out." In fact, moisture is a critical component of the hardening process, which is actually a chemical reaction known as *hydration*. During hydration (or as more commonly referred to, the curing process), heat is produced, and unless the concrete is properly protected from both low and high temperatures, it may set improperly and crack, or have an ultimate strength lower than its design strength. Cracks due to drying shrinkage may permit humidity and water to reach the reinforcing steel, which then rusts and eventually disintegrates the entire reinforced concrete mass. It is therefore a critical part of the process that each batch of concrete placed in a building, bridge or other construction project be tested to guarantee that its strength is in agreement with the laboratory design, and be properly cured in a moist and temperature-controlled environment, lest it need to be ripped out and replaced—a time-consuming and expensive process which has resulted in numerous lawsuits over the decades.

The manufacture of concrete is thus seen to be a delicate process, requiring as much care as that of any chemical. The concrete for important construction jobs is seldom mixed at the site; it is manufactured off site in concrete plants and transported to the site. On the other hand, because concrete properties vary so much with composition and methods of mixture, many different types of concrete can be obtained, each suited for a specific purpose. Concrete with a compressive strength as high as 12,000 pounds per square inch (6.9 MPa), rapidly hardening concrete, and concrete with lightweight aggregates, are constantly being introduced. In addition, *chemical admixtures* of many types (such as those improving weather resistance in freeze-thaw cycles, those yielding a speeding up or retarding of curing time, and those making the mixture more flowable and easily pumped up many stories) are also under continual development. The reader will appreciate the potentialities of this new material after becoming acquainted with some of its applications to structural systems.

As previously noted, the function of the reinforcement in concrete is to carry tensile stresses in a member, and there are several methods of employing the tensile reinforcement: one that is "passive" and two that are "active." Fundamentally (as will be further discussed in Chapters 5 and 7), a concrete beam on two supports that is loaded by downward forces has its upper surface in compression while the lower portion is in tension and would crack if not reinforced by steel (Figure 3.16a). The oldest and still most common method of reinforcing a beam (referred to as *conventionally reinforced*) provides steel that acts in a passive manner. When a steel rod is inserted in the lower portion of a beam, it is in the region of tensile force on the beam and tightly bonded to the concrete. The reinforcing steel (colloquially referred to as "rebar") is stamped with ribbed patterns to ensure a good bond between the concrete and steel—an essential transfer of force if the steel is to carry loads from the concrete. In bending under load, the concrete engages this tensile steel and prevents cracking as illustrated in Figure 3.16a.

Introduced in the mid-twentieth century, prestressed concrete further advanced the basic concept of a passive tensile reinforcement for a nontensile material by actively precompressing the concrete member. There are two methods of creating prestressed concrete. The difference between the two techniques has to do with *when* the pretensioning occurs: One type is referred to as *precast prestressed* (or more commonly just *prestressed*), and the other as *post-tensioned*. Prestressed concrete is created in factories for the prefabrication of structural elements. Here, *tendons* of exceptionally strong wire are pulled against the ends of steel forms in which the concrete is poured. Once the hardened concrete grips the tendons,

these are cut from the formwork ends and their tension puts the concrete in compression as the tendons naturally attempt to return to their initial unstressed state. In post-tensioned concrete, slack tendons are threaded through hollow sleeves (known as *ducts*) set in the concrete forms and, once the concrete reaches a sufficient strength, are pulled against the hardened concrete by outside jacks thereby compressing it (Figure 3.16b). Post-tensioning is used mostly on site and, in large structures, may proceed in stages as required by the stresses in an element, which increase under the increasing dead load.

An unloaded beam of prestressed concrete may appear from the outside to be unstressed, but "locked-in" stresses compress the concrete and tense the steel. The tensile stresses developed in a beam by the loads reduce, or at most wipe out, the initial compressive stresses due to prestressing, so that the concrete is never in tension. With no appreciable tension, it has less of a tendency to crack (see also see Chapter 7 for a more thorough presentation of beams). Early investigators of reinforced concrete first proposed the prestressing of concrete members in the mid-nineteenth century. Unfortunately, the steel available at the time had a low yield point and could not be stressed high enough to allow the tendons to remain substantially tensed following the plastic flow of the concrete under the prestressing compressive stresses. Only after high-strength steel suitable for this

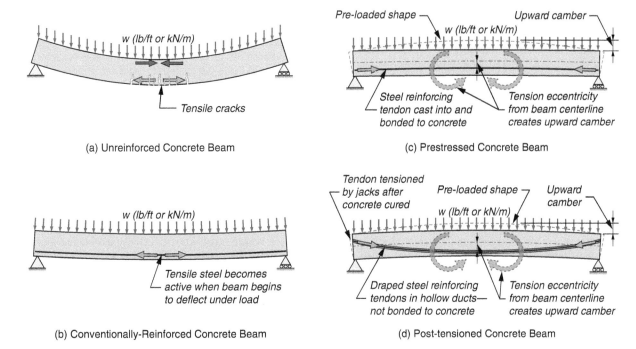

(a) Unreinforced Concrete Beam

(b) Conventionally-Reinforced Concrete Beam

(c) Prestressed Concrete Beam

(d) Post-tensioned Concrete Beam

Figure 3.16 Concrete beams: Unreinforced and reinforced
A concrete beam with no reinforcement has little tensile capacity. With downward loads, the bottom surface will be put into tension, while the upper surface is in compression (see Chapter 5) and thus crack and fail at a modest level of load (a). Conventional reinforcement bonds to the concrete to form a composite element where concrete carries compressive stresses (which it is well suited for) and steel carries tensile stresses (b). Conventional reinforcement is *passive*: no stress is placed on the steel until the beam is placed under load. An *active* form of reinforcing is to either pre- or posttension the steel with powerful hydraulic jacks (c and d). When the tension is released, it places the bottom surface of the concrete into compression, thus counterbalancing loads that would put it into compression. A pre- or posttensioned beam requires less reinforcement than a conventionally reinforced beam, although in complete buildings both pre- and posttensioned reinforcement is still augmented with conventional reinforcement.

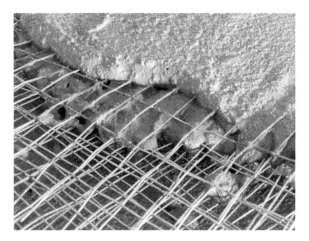

Figure 3.17   Ferrocement

Rather than using heavy reinforcing bars or high-strength strands, Ferrocement uses multiple layers of fine mesh placed in random layers to distribute reinforcement throughout a thin structural member. Ferrocement has been used for boat hulls and larger projects, and is a technique favored in regions where labor is plentiful but materials are expensive. It can be molded into virtually any shape.

Photo courtesy of Milinkovic Company

Figure 3.18   Cathedral of St. Mary of the Assumption in San Francisco, California

One of the last projects in his life by famed Italian engineer Pier Luigi Nervi, the roof of the cathedral is comprised of triangular "pans" of Ferrocement tiles following the profile of the hyperbolic paraboloid roof (see Chapter Twelve). The tiles are both formwork for the concrete cast on the outer roof surface, as well as creating the final exposed interior surface, thereby eliminating additional formwork. The edges of the pans link together to form thin skewed beams (see Chapter 10) on the interior surface. Nervi is renowned for his modern reinvention of this material first explored in the mid-1800s. He completed a number of prominent projects with this technique, most located in his native Italy. See also figure 14.16b for an exterior photograph.

Photo courtesy of Deborah Oakley

purpose was manufactured at reasonable prices did prestressed concrete become an economic reality in the mid-twentieth century.

"Ferrocement" (iron-concrete), a structural material first successfully used by Italian engineer Pier Luigi Nervi, is a combination of dense steel mesh and cement mortar. In Ferrocement, a number of layers of wire mesh, with square holes less than half an inch on a side, are packed randomly one on top of the other across the thickness of thin elements and embedded in a mortar of cement, water and sand with no coarse aggregate (Figure 3.17). The resulting material has the compressive capacity of an excellent concrete and the tensile strength of steel, since the steel is so thoroughly distributed through the mortar as to hold its particles together even when tension is applied to it. One of the most successful applications of Ferrocement is to the construction of small-boat hulls, which Nervi started in the 1920s. Nervi went on to create a number of spectacular architectural projects using this material (Figure 3.18).

In a more recent development, steel reinforcing bars or mesh have been replaced by short glass or metal fibers mixed with the concrete, thus allowing the creation of structural elements capable of resisting tension homogeneously in all three dimensions. Furthermore, new *Ultra-High Performance Fiber Reinforced Concrete* mixtures (such as the proprietary mixture known as *Ductal® Concrete* by the French company, Lafarge) produce concrete with previously unobtainable tensile strength. Delicate tracery and shapes formerly possible only with cast steel can now be formed with this high-tech wonder concrete (Figure. 3.19).

The high compressive strength of ceramic tile may similarly be combined with the tensile strength of steel and the cohesive properties of concrete to create a material similar in behavior to reinforced concrete. Italian engineering pioneer of reinforced concrete, Eduardo Torroja, among others, showed how imaginatively one could use such material in countries where steel is prohibitively expensive and labor costs are low (see Section 10.6).

Plastics, synthetic materials using organic substances as binders, are finding widening applications in the field of structures because of their varied properties, but, generally, are not yet widely used as structural materials because of their relatively high costs.

Most structural plastics are glass reinforced. Glass fibers with a tensile strength of up to 600 thousand pounds per square inch (4.1 GPa) are commonly used to reinforce epoxy resins molded into airplane parts and other structural elements. Glass fabrics give tensile strength to epoxy and polyester resins used to manufacture structural panels, and to polyvinyl and other plastic materials constituting

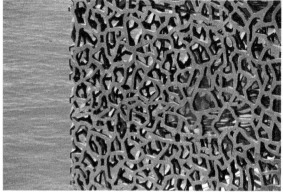

Figure 3.19 Ultra High-Performance Fiber-Reinforced Concrete

Ductal® is a proprietary fiber-concrete mixture developed by the French company Lafarge. Elements created using this material can be far thinner than conventionally-reinforced concrete, and the high tensile capacity enables it to be used in applications such as thin tracery where previously only cast steel could have been employed. The entire southeast and soutwest-facing facade (as well as much of the roof) of the Museum of European and Mediterranean Civilizations (MuCEM) in Marseille, France, by architect Rudy Ricciottti, is wrapped in a bris soleil web of Ductal® concrete.

Photo courtesy of Rudy Ricciotti

the membranes of large air-supported roofs (see Section 11.3). Urethane and polystyrene foams, with low compressive and tensile strength but sufficient shear strength, are used as cores of sandwich panels of plywood, gypsum, aluminum, steel, or plastics, with excellent weight-to-strength ratios and thermal-insulating properties (see Section 10.6).

Perhaps one of the most exciting new materials is graphite carbon fiber reinforced plastic. Carbon fiber combines an extremely high tensile strength with very low weight. Filaments of carbon fiber can be woven into textile sheets like any conventional fabric. When combined with epoxy or other resins, they can be formed into virtually any shape imaginable to form an extremely tough yet flexible and strong but lightweight structural element. Presently used extensively in the manufacture of sporting goods as well as in the aerospace industry (in the famous *Stealth* fighter plane and the Boeing "Dreamliner," for example), its use in construction projects is still limited to specialized applications where its superior "strength to weight ratio" is important. Because their manufacture is a highly laborious process, carbon epoxy structures are expensive. New research in this area, however, portends the day when large-scale application in buildings and bridges will be commonplace. It is easy to forecast that the ever-growing developments in the field of plastics will make these materials more and more competitive with classical structural materials.

This chapter has briefly introduced some of the most important characteristics of the various materials used in structures; it is really the tip of the iceberg of an entire field known as *materials science*. Nevertheless, this basic understanding can create a framework of understanding to carry the reader through an appreciation of much of the field of structures. It should also be seen that the proper use of structural materials is essential to correct design, since the availability of materials limits the choice of structural systems in any important structure.

KEY IDEAS DEVELOPED IN THIS CHAPTER

- Materials of construction must carry loads without permanent deformation. They may be (1) elastic, within a certain range; (2) elastic-plastic; and (3) brittle.
- Some materials are able to carry tensile and compressive loads equally well, while others can only be used with compressive loads.
- Stress is defined as the force per unit area and strain as elongation per unit length. There are specific characteristic properties of yield point, yield stress, and ultimate strength for any given material.
- Rigidity/flexibility of a material is quantified by the modulus of elasticity, defined as the ratio of stress to strain.

QUESTIONS AND EXERCISES

1. Many common plastics used in household products possess a remarkably similar behavior to structural steel, including heavy-duty trash bags. You can experiment with this by taking a cutting from a trash bag, approximately 1/2 inches (13 mm) wide by 6 inches (152 mm) long, and pulling on it. Depending on the plastic, there will be some stretch initially, but it will return to the initial state after the pulling force is removed. Pull hard enough (but not so hard as to break it) and a permanent set will be made in the plastic. It has thus reached its yield point. Some plastics are highly anisotropic due to their manufacturing process. This means that strips cut 'across the grain' (so to speak) will exhibit a pronounced necking down, whereas strips cut 'parallel to the grain' will not. Try both ways to see the difference!

   If the cuts are clean (no notches or defects on the sides), continued pulling will result in a surprising amount of stretch of this material. This is the plastic range of the material. Finally, it will become very "tough" and resistant to further stretching. Continued pulling will not stretch it further, but one can notice the increase in force that the material will sustain. This is the point of strain hardening for this material. Ultimately, when the force is high enough, it will suddenly break. It has reached its ultimate stress level. This characteristic behavior is, in miniature, exactly the way structural steel behaves.

2. Make a beam out of foam rubber, about 1 in square cross-section and a foot long (25 mm wide and 300 mm long). Hold one end in your hand and put a small weight on, or simply press on, the free end. Observe the deflection. Now cover the top side with transparent tape and repeat the loading. Observe the reduced deflection: You have created a reinforced beam!

3. Turn over the reinforced beam and apply the load again. Notice the wrinkling of the tape. Why is it wrinkled?

FURTHER READING

Millais, Malcom. *Building Structures: From Concept to Design*, 2nd Edition. Spon Press. 2005. (Chapter 5)

Sandaker, Bjorn N., Eggen, Arne P., and Cruvellier, Mark R. *The Structural Basis of Architecture*, 2nd Edition. Routledge. 2011. (Chapter 4)

Allen, Edward and Zalewski, Waclaw. *Form and Forces: Designing Efficient Form and Forces: Designing Efficient, Expressive Structures*. John Wiley & Sons, Inc. 2009. (Chapter 13)

STRUCTURAL REQUIREMENTS

4.1 BASIC REQUIREMENTS

All structures must conform to certain physical and human constraints. Modern developments in materials production, construction techniques, and methods of structural analysis have introduced new freedoms in architectural design that were inconceivable even in the recent past, considerably widening its scope.

These new freedoms, however, do not exempt modern structures from satisfying certain basic requirements that have always been the foundations of good architecture. We may list them under the following headings: *equilibrium, stability, strength, functionality, economy*, and *aesthetics*. The sections of this chapter will address each of those conditions in turn.

4.2 EQUILIBRIUM

The fundamental requirement of equilibrium is concerned with the guarantee that a building, or any of its parts, will not move. More technically this is *static* equilibrium versus *dynamic* equilibrium, which deals with objects in motion. For simplicity in this text it will be simply referred to as equilibrium. Obviously, this requirement cannot be interpreted strictly since some motion is both unavoidable and necessary (*thermal expansion, etc.*). The displacements allowable in a building, though, are usually so small compared to its dimensions that the building appears immovable and undeformed to the naked eye.

The principles governing the motion of bodies, published by Sir Isaac Newton in 1687, are called Newton's laws. The particular cases of these laws governing equilibrium (i.e., the lack of motion) are of basic importance in structural theory because they apply to all structures and are sufficient for the design of many of them. Such structures, called statically determinate, support loads by developing reaction forces whose values do not depend on the material used. These reaction forces can be determined by the two simple equations of linear and rotational equilibrium, stating that the numerical sum of all forces and of all rotational actions must equal zero for static equilibrium to exist.

Structures that cannot be designed on the basis of Newton's laws alone and which require a knowledge of material properties are called *statically indeterminate*, meaning that the external forces supporting them cannot be determined solely by the basic equations of statics. Many modern structures are statically indeterminate, and possess certain advantages such as using less material, as well as the ability to redistribute loads, making them more resilient. Modern computing methods have made the design of statically indeterminate structures far easier than in in the past when calculations were performed with manual techniques.

Newton's three Laws of Motion can be concisely stated as follows: (1) Inertia: An object at rest will remain at rest, or an object in motion will remain in motion, unless acted on by an external force; (2) Force is directly proportional to the mass of a body multiplied by its acceleration; and (3) Equilibrium: Every force action has an equal and opposite reaction.

The first and third laws together form the engineering field of study referred to as "statics," and adding the second refers to the engineering field of "dynamics." Dynamics is all about our modern world, from the cars and airplanes we travel in to the machinery in factories that make the "stuff" of our daily lives. Strikingly, a basic understanding of statics, coupled with knowledge of material properties and their internal response to forces, is sufficient for the design of the vast majority of all building structures. Very tall buildings, long span bridges, or structures located in regions of high seismic activity are notable exceptions, however.

As noted above, there are essentially only two basic conditions of equilibrium: linear and rotational. Linear equilibrium states that for an object to stay at rest, a straight push or a pull on the object in any direction in space must be balanced by a net equal force in exactly the opposite direction (this opposite force is referred to as an *equilibrant*). Rotational equilibrium refers to a similar balance of forces. In this case, however, a force that causes an object to *rotate* about a point must be balanced by an equal and opposite rotational tendency. To look at these conditions more closely, it is easiest to first visualize linear equilibrium.

4.2.1 Linear Equilibrium

Certain elementary conditions of linear equilibrium in simple structures can be easily visualized. An elevator hanging from a cable (Figure 4.1a) is supported by the pull of the cables; the cables, in turn, hang from the pulley at the top of the building.

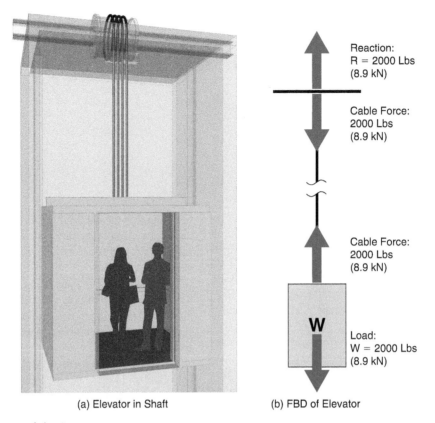

(a) Elevator in Shaft   (b) FBD of Elevator

Figure 4.1   Free body diagram of elevator

A schematic representation of equilibrium called a *Free Body Diagram* (FBD) illustrates forces as vector arrows, typically with lengths proportional to the force magnitude. The FBD (4.1b) illustrates the equilibrium between acting and reacting forces on an object. In this illustration, the acting forces include the weight of the elevator itself as well as the live load of the occupants (*W*). The reacting force on the cables (*R*) must be equal and opposite for static equilibrium to exist.

 Multiple cables support a traction elevator to both reduce the cable diameter and provide for greater safety. The four cables in this example would each carry ¼ of the total weight, or 500 pounds (2.2 kN) each. Notice that the cable forces appear to be *pushing* toward the center, which may at first appear to be compression. With respect to the load however, they point in the opposite direction *pulling away* from the elevator cab, and thus in tension. In a static condition, force actions and reactions always balance linearly.

If the elevator and its occupants weigh two thousand pounds (8.9 kN) and is at rest, the cable exerts on the elevator a pull of two thousand pounds (8.9 kN). The weight (i.e., *the downward force of gravity acting on an object*) of the elevator and the upward pull of the cable are equal and "balance out": the elevator is "in linear equilibrium" (Figure 4.1b). The simplified abstraction of this balance of forces, which uses vector arrows to represent the forces, is referred to as a *free body diagram*, (FBD), and is one of the most essential conceptual tools in engineering mechanics.

 In another example, children pulling on a rope with total equal and opposite forces do not move: The rope is in horizontal linear equilibrium (Figure 4.2). But if one or more individual exerts a greater pull than the others, this will yank the opponents off their stands, and all individuals and the rope will move: Equilibrium is lost. Similarly, if a sculpture weighing 1000 pounds (4.5 kN) is set on a pedestal (Figure 4.3), the pedestal exerts an upward push of 1000 pounds (4.5 kN) on the sculpture. The pedestal weight of 2500 pounds (11.2 kN) *plus* the sculpture weight then apply a combined total load of 3500 pounds (15.7 kN) to the ground, distributed across the entire surface of the pedestal base. If the soil is soft and exerts a smaller

upward push, the sculpture would move down, and there would be lack of linear equilibrium, and the principles of dynamic structures would come into play.

 These elementary examples show that a body does not move in a certain direction if the forces applied to it in that direction balance out: A force exerted in a given direction must be opposed by an equal force exerted in the opposite direction. Whenever this happens, there is linear (or *translational*) equilibrium in that direction.

4.2.2 Rotational Equilibrium

Rotational equilibrium is an everyday experience. It is readily visualized by the familiar experience of a seesaw, which the reader may have personally had as a child. Two children of identical weight sitting at the end of a seesaw *at equal distances from the fulcrum* (i.e., pivot point) will place the seesaw into rotational equilibrium. The upward force at the pivot of the seesaw is in linear equilibrium with the combined weights of the two children and the board of the seesaw (though the forces are not along the same line, they are all vertical). We thus say that it "equilibrates vertically" the weights of the two children and "reacts" with an

Figure 4.2   A Tug-of-war

The rope being pulled on by the children is in linear equilibrium. So long as the *total force* exerted is the same magnitude but in opposite directions on each side, there will be linear equilibrium even when the children individually pull with different force magnitudes.

Photo: Ilike/Shutterstock

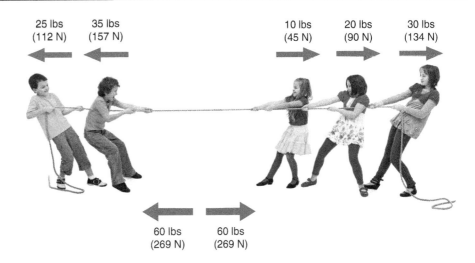

Figure 4.3   Equilibrium of vertical forces

(a) Statue on a Pedestal   (b) FBD Equilibrium of Forces

Figure 4.3a shows a statue as it appears before us. Unseen are the forces at play internally. In the FBD of 4.3b, we see that the total weight of the statue (W_s) is supported by a reaction force at the statue pedestal (R_P) that is equal and opposite to that of the statue weight. This reaction force can then be considered as a load (P_P) of the same magnitude acting in the opposite direction on the pedestal itself. The weight of the pedestal (W_P) is then added to the load on the pedestal, the total reaction on the ground then being the sum of all components. This reaction at the bottom of the pedestal (R_{Total}) is distributed across the entire surface of the pedestal base. In reality, the weight of the sculpture on the top of the pedestal is also a distributed force, but shown here as a concentrated force for clarity.

upward push equal to their combined weight and that of the board. The supporting force is therefore known as a *reaction* (Figure 4.4 a).

The two children will also be balanced rotationally about the pivot point, and thus the board stays level. Rotational equilibrium breaks down, however, when the children sit at different distances from the point of support: The seesaw rotates in the direction of the child sitting farther away from the pivot (Figure 4.4 b). Such distances from points of support are called "lever arms." In order to guarantee "equilibrium in rotation" when the two children have equal weights, their lever arms must be equal. If the two children

have different weights, equilibrium in rotation can still be obtained by giving the lighter child a larger lever arm and the heavier child a smaller lever arm.

Equilibrium in rotation requires that the weight times the length of the lever arm of each child have the same value (Figure 4.4 c) and tend to rotate the board in opposite directions. The product of a force multiplied by its lever arm is referred to as the ***moment of the force*** (Figure 4.5).

Such simple equilibrium principles apply to all structures. Equal and opposite forces guarantee linear equilibrium in a given direction, and equal and opposite moments (i.e., the products of forces times lever arms) guarantee equilibrium in

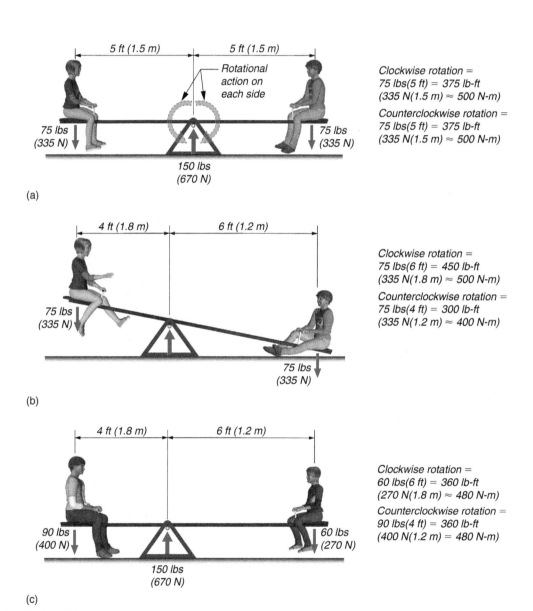

Figure 4.4 Rotational equilibrium

Two children on a see-saw can be seen as a system of rotational equilibrium—more conventionally known as *balance*. If the children weigh the same, then at equal distances they will have the same rotational tendency and thus balance out (4.4a). If one child sits farther away, rotational balance will be lost (4.4b). On the other hand, if an older/heavier child sits on the end, to achieve balance each child must sit at a different distance from the support, or the entire board must shift by the same amount to relocate the pivot off center (4.4c).

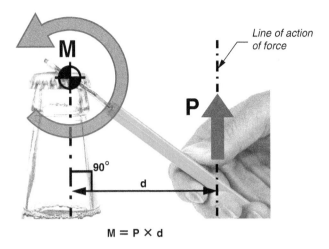

Line of action of force

M

P

90°

d

M = P × d

Figure 4.5 Principle of moment

By definition, moment (known as *torque* in physics) is the action of a force removed a distance from an object. The result of its action relative to that point is to generate rotation. Its technical definition is *force multiplied by perpendicular distance*. The perpendicular distance is measured between the line of action of a force and a line parallel to this that passes through the object.

We experience moment primarily in one of two ways. One is balance: The seesaw is an example of balance. The second is force multiplication: All manner of household tools make use of force multiplication in objects ranging from bottle openers to lever handles on doors, and tools such as wrenches. It is force multiplication that governs the behavior of elements in bending. A tree branch, for example, gets thicker toward the trunk where the internal forces become larger due to greater rotational action. See Chapter 7 for a complete discussion of bending behavior.

Photo: Li Xuejun/123RF

rotation. In Figure 4.4c, the seesaw is in rotational equilibrium because the clockwise moment about the pivot from the right child's weight of 360 pound-feet (480 N-m) equals the counterclockwise moment of the left child's weight of 360 pound-feet (480 N-m). One can also see that *the mathematical units for moment are always force times distance* (e.g., pound-feet (lb-ft), Newton-meters (N-m), etc.).

An apparently more complicated situation is presented by the equilibrium in the vertical direction of a 150 pound (670 N) man walking across a 10 foot (3 m) long plank. It will be shown however, that this apparent complication is simply rooted in misperception.

In this case, the plank rests on supports at its two ends and (for clarity in this example) the weight of the plank itself will be neglected. When the man stands at midspan, each support carries half of his weight: We say the weight of the man is "equilibrated" by two equal "support reactions," each being equal to half of his weight (Figure 4.6a). When the man is standing closer toward the left end, his weight is supported by a greater reaction of the left support (Figure 4.6b). As he moves across the plank, the reaction of the left support decreases and the reaction of the right support increases, until they become equal again when he is at midspan. From then on, the right support reaction increases and the left decreases (Figure 4.6c) until the right support carries the entire weight of the man as he moves to step off the plank. Whatever his location, the plank transfers to the supports the entirety of his weight. To have vertical equilibrium, the two reactions must always add up to this weight, but the reactions differ in value depending on the weight's location.

With a little reflection even a child can easily recognize the truth of the foregoing paragraph in terms of vertical forces. What may not be immediately apparent, though, is that this is *also an example of rotational equilibrium*. In comparing Figures 4.4c and 4.6b, one can visualize the man on the plank as an *inverted diagram* of the children with differing weights on the seesaw.

Numerically, the forces in the beam of 4.6b are identical to those of the seesaw of 4.4c, except in mirror image (Figure 4.7). The upward support reaction on the seesaw becomes the downward load of the man on the plank, and the downward weights of the children on the seesaw become upward plank reactions. The rotational balance is therefore the same (but in reverse direction), and thus beam reactions can be seen as a form of balance. The moment with respect to the 150 pound (670 N) force due to the forces on either end is the same, regardless of whether the 150 pound (670 N) force is looked at as the reacting support for the seesaw or as the man's weight acting as a load on the plank.

This conceptual understanding of linear and rotational equilibrium is a foundational study: The principle of rotational equilibrium is among the most essential and fundamental concepts in all of structures. But this is merely an introduction; the interested reader would do well to study the topic in greater depth from the various references listed at the end of this chapter.

Next, application of the concept of equilibrium will be considered in the study of the second structural requirement: stability.

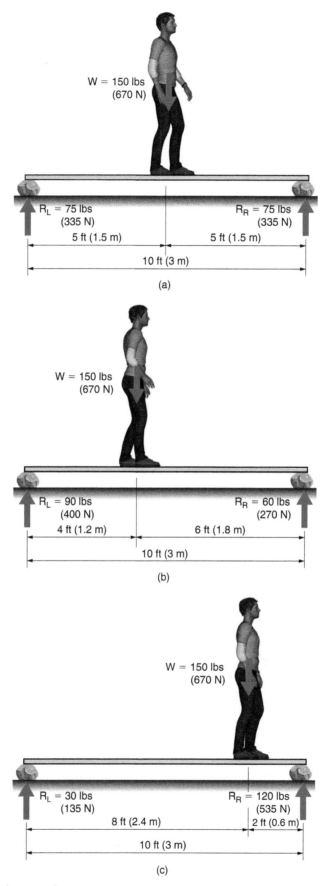

Figure 4.6   Rotational equilibrium in beam reactions

Beam reactions can be seen to be a form of rotational balance. A load on the center of a span will have symmetric reactions on the left and right sides. When located toward the left, however, the left side will have a greater reaction, and when located toward the right, the right side will have a greater reaction.

Figure 4.7 FBD Equivalence between seesaw and beam

Combining the FBD of 4.4c (top of diagram) with that of 4.6b (bottom of diagram), it can be observed that they are inverted numerical equivalents. In this way, beam reactions can be seen also as a system of rotational balance, just like the seesaw.

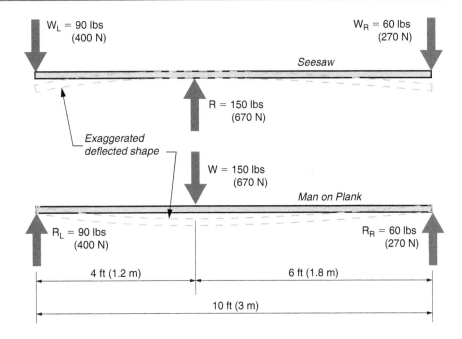

4.3 STABILITY

The requirement of "rigid-body" stability is concerned with the danger of unacceptable motions of the building as a whole. When a tall building is acted upon by a hurricane wind and is not properly anchored in the ground or balanced by its own weight, it may topple over without disintegrating. The building is unstable in rotation. This is particularly true of tall narrow buildings, as one may prove by blowing on a slim cardboard box, resting on a rough surface to prevent it from sliding (Figure 4.8).

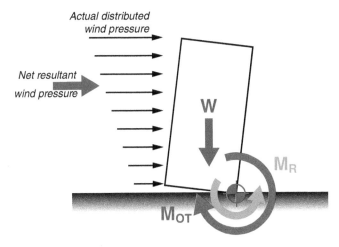

Figure 4.8 Instability due to wind

The pressure of the wind (which increases with height) causes a resultant clockwise *overturning moment* (M_{OT}) with respect to the front edge of the structure. For rotational stability, this moment must be balanced by at least as great of a *resisting moment* (M_R). The resisting moment may be due to the self-weight of the structure (W), which with respect to the front edge of the structure creates a counter-clockwise resisting moment. One can observe this behavior by taking an empty cereal box and simply blowing on the long face of it.

The danger of rotational instability is also present when a building is not "well balanced" or is supported on a soil of uneven resistance. If the soil under the building settles unevenly, the building may rotate, as the Leaning Tower of Pisa still does, and only relatively recently was stabilized from completely toppling over (Figure 4.9).

A catastrophic example of foundation instability is seen in the dramatic failure of the foundation support in the apartment blocks of Niigata, Japan, in the 1964 earthquake (Figure 4.10). Here, the ground experienced a phenomenon known as *liquefaction*, in which the seismic motions caused groundwater to become pressurized and thereby turn solid soil into a mud. The shaking did not substantially damage the buildings, but the ground beneath them simply gave way and the entire apartment block sank into it.

A building erected on the side of a steep hill may, by its own weight, have a tendency to slide down the slope of the hill. This may happen either because the building slides on the soil (Figure 4.11a) or because a layer of soil adhering to the foundations slides on an adjoining lower layer (Figure 4.11b). The second occurrence is not uncommon in clay soils when water seeps through the ground, transforming the clay into a slippery material. There have been numerous cases of such hillside failures, particularly in areas such as the mountainous West coast of the United States and other similar regions of the world.

The equilibrium FBD of a building on a slope is illustrated in Figure 4.11c. The weight of the structure acting vertically is decomposed into two *component vectors*; these are two sides of a right triangle, whose hypotenuse is proportional to the weight. The downhill component is responsible for the tendency of siding and is resisted by the friction force between the ground and the foundation, and also between different layers of soil. The component perpendicular to the ground exerts pressure on the ground and is equilibrated by the soil reaction.

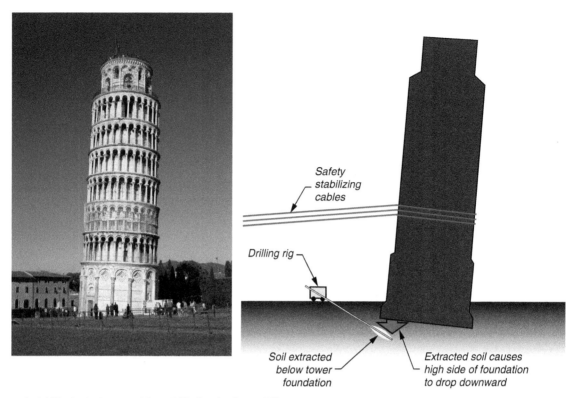

Figure 4.9   Instability due to uneven settlement: The Leaning Tower of Pisa

Almost immediately after construction was started in 1173, the tower began to lean due to uneven and highly compressible soils. During construction, attempts were made to straighten the tower by making walls taller on one side, but it continued to develop an increasing tilt over the centuries. By the mid-twentieth century, the tower was sloping nearly 6° from vertical and was in danger of toppling over. After years of study and discussion, a technique of soil extraction was employed that removed soil below the tower foundation, which caused it to subside and tilt more upright. The tower has been stabilized with a safe tilt angle that is estimated to last at least two hundred years into the future.

(a) Photo: 4745052183/Shutterstock

Figure 4.10   Niigata Earthquake apartment block failure

A magnitude 7.4 earthquake struck Niigata, Japan, in 1964. Due to soil liquefaction some apartment buildings simply tilted over. With one side sinking below the other, the building's weight was no longer concentric on the foundation, and thus an overturning moment developed that caused the failure through rotational instability. Other buildings that did not tilt simply sank down a full story or more into the soil.

Photo courtesy of USGS

All these cases of instability are related to the soil and to the building's foundations. From the viewpoint of economy and usage, foundations are a "necessary evil"; moreover, they are out of sight so that the layman is seldom aware of their importance and cost, and yet they are of fundamental significance. As in the case of the Niigata apartment block failure, even a well-built structure is only as strong as the foundation upon which it sits.

Some foundations of heavy structures erected on loose sand permeated by water must allow the building to "float" on such a soil: They are built by means of "rafts" which are similar to the hull of a ship (Figure 4.12). A raft or a mat

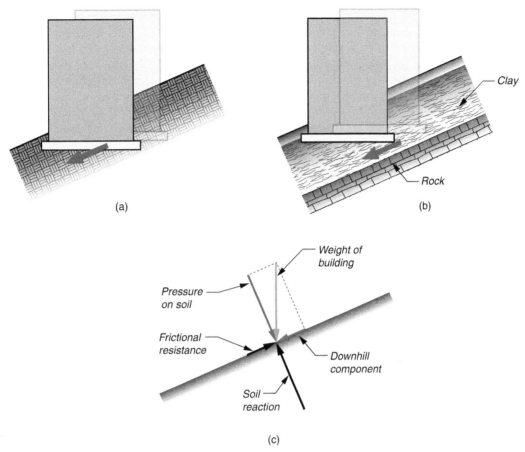

Figure 4.11　Instability due to sliding

A building built on a steep hillside will have a tendency to slide simply due to its own weight (4.11a), or by deeper soil layers sliding against one another (4.11b). Figure 4.11c is an FBD of the forces on the slope. Any single vector can be broken into two or more smaller vectors that have the same net result on the structure, referred to as *vector components*. In the case of the building on the hillside, it is the downhill component of the building's weight parallel to the slope that causes sliding to occur. As the slope increases in steepness, the magnitude of the downhill component will also increase. For equilibrium, the downhill component is resisted by the equal and opposite friction force, while the component of the weight perpendicular to the incline is equilibrated by soil reaction.

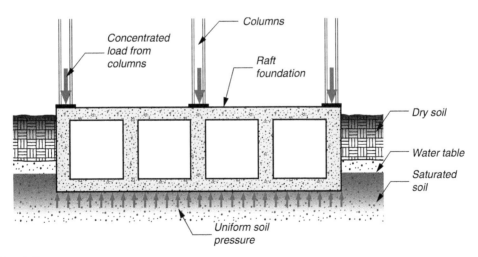

Figure 4.12　Raft foundation

A raft foundation is a cellular concrete box forming a unified foundation for an entire building structure. A raft behaves something like a boat in water, displacing a large amount of soil while also frequently holding back groundwater. Raft foundations are often designed so that the weight of soil displaced is close to the weight of the building itself, resulting in little or no net increase in pressure on the soil at the base of the raft, and thus limiting the amount of settlement experienced by the building.

foundation allows the weight of the building to be spread over a larger soil surface generally equal to the footprint of the building itself, thus reducing the pressure on soils of low bearing capacity. With soils of average bearing capacity and buildings of moderate scale, the weight of the building is usually supported on isolated rectangular spread footings of reinforced concrete.

Elaborate precautions against failures must often be taken to guarantee the stability of structures. Wood, concrete, or steel piles can be driven into the soil to depths which permit the building to be supported either by the friction between the surface of the piles and the soil (Figure 4.13a) or by a deeper layer of solid rock (Figure 4.13b). The piles may be rammed into the soil or may be made to slide into it by rapid vibrations. Soils may also be consolidated by chemical means.

The design of proper foundations is based on thorough soil investigations. *Soil mechanics,* however, is a complex science and is its own subspecialty of civil engineering known as *geotechnical* engineering. To this day, most of the damage to buildings comes from faulty foundations, even though their cost may reach 10 per cent or more of the total cost of the building.

4.4 STRENGTH

The requirement of strength is concerned with the integrity of the structure and of each of its individual parts under any and all expected loads; in other words, no portion of it will physically fail in use. A common expression is that a chain is only as strong as its weakest link. Similarly, in order for the entire structure to have the required strength, each of its constituent parts must be made to resist all forces that it may encounter during its useful life.

The design process of a structure for strength is a very rational and systematic one: The structural system is first chosen and the expected loads on it are established, members are selected initially as a trial, and the state of stress is then determined at significant points within the structure and compared with the kind and amount of stress the material can safely withstand. If necessary, these "trial members" are revised with new selections and checked again until the requirements are met. In this iterative process, safety factors of varying magnitude are used to take into account uncertainties in loading conditions and material properties (see Section 3.2).

Strength should not be confused with rigidity: Two structures may be equally strong, even though one deflects more than the other under the same loads (see Figure 3.13). Although it is often a measure of strength against loads, rigidity may be a sign of weakness in a structure subjected to temperature changes, uneven settlements, or dynamic loads (see Sections 2.5 and 2.6).

Certain structural weaknesses may lead to modest damage, while others may produce the collapse of the building. Hence, the designer must check strength under a variety of loading conditions to obtain the worst stress situation at significant points in the structure. The structural optimist is inclined to believe that a structure collapses only if faulty design is compounded with faulty construction, and helped by an act of God; the cautious pessimist fears, instead, that a structure may collapse at the slightest provocation. In practice, completed structures do collapse, although in relatively small numbers; moreover, owing to the plastic behavior of structural materials (see Section 3.1), most collapses do not occur suddenly and seldom take human lives. Chapter 13 addresses the structural failures more extensively.

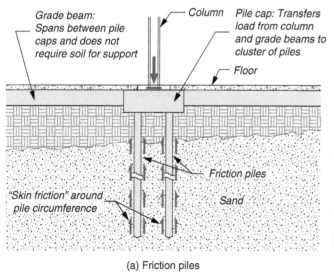

(a) Friction piles   (b) Load-bearing piles

Figure 4.13   Foundation on piles
If the supporting soil is weak, or if it is subjected to heavy building loads, or if strong surface soils are underlain by a weak and/or compressible layer of soil, pile foundations are frequently used to carry the building loads. In the case of friction piles (4.13a), the building weight is carried by friction between the piles and surrounding soil. Load-bearing piles (4.12b) act like long slender columns (braced against buckling by the surrounding soil) that carry the loads to firm bearing strata or bedrock below the surface. Piles may be constructed of wood (typical of houses and piers on a waterfront), steel, or concrete. Some steel piles may be as long as several hundred feet.

(a) Column load testing

(b) Completed interior

Figure 4.14 Columns of the Johnson Wax building

The Johnson Wax building by the architect Frank Lloyd Wright has columns that, per the building codes in place at the time, were deemed too slender relative to their height. Wright argued that their construction using ferrocement would be more than adequate and produced a test column to prove it. The column was demonstrated to carry more than five times the required load, and thus Wright was allowed to proceed with the building, which is now an icon of modern architecture.

(a) Photo: Hedrich Blessing Collection/Chicago History Museum/Getty Images; (b) Photo courtesy of James S. Russell, FAIA

The strength of a structure is evaluated according to the rules and regulations of building codes. These procedures are usually safe, but can be uneconomical when they ignore newly developed systems and materials, and their conservative nature can at times stifle aesthetic expression. A famous example of this is the Johnson Wax Building designed by Frank Lloyd Wright in the mid-1930s (Figure 4.14), whose columns were considered too slender by the building code of the day and officials were concerned that they may be inadequate. But Wright persisted and subsequently, using a full-scale mockup, proved the strength of the columns far beyond their required capacity. His tenacity led to the creation of a building regarded as a masterpiece of modern architecture, which likely would never have achieved that status by strictly following the code.

Most structures are, however, less dramatic, and it is the role of building codes to ensure public safety since full-scale structural testing is impractical in most all cases. But code provisions should not be applied blindly, that is, without careful consideration of their original purpose. Building codes are typically provided with "code commentaries" to clarify these purposes and aid designers in their correct application.

The responsibility for strength rests squarely on the shoulders of the structural engineer. Every day, this job is made more complex—but also safer—by improved theoretical understanding and the improved tools available. In the past few decades, advanced mathematical theories that were first introduced in the mid-twentieth century (but which were largely impractical to carry out manually) have been made routinely possible with the power of computers. Both the simple but arduously repetitive calculations and the more complex ones are now handled through computer software. The computer has done more than simply speed up calculations: It has liberated the structural engineer to become more creative and powerful at his or her job.

The possibility of repeating a long calculation in a matter of seconds or minutes allows the consideration of a variety of combinations of shapes, sizes, and materials, and the choice of "the best" among a number of designs. Computers have thus brought about the "partial optimization" of structural solutions, which until relatively recently was no more than theoretically possible.

Moreover, the computer permits a more accurate analysis of problems of exceptional complexity—for example, those connected with earthquake design—by this "mathematical modeling" of the structure. Before the advent of the computer, the rigorous analysis of such problems cost so much in both time and money that approximate solutions were mandatory, involving wasteful use of materials. Today large savings and greater safety are obtained by the more detailed and realistic studies computers make possible.

Computers, however, still have their limits. When the analysis of an exceptionally difficult problem cannot realistically be carried out through mathematical modeling, the designer may solve the problem by means of a test on a structural model of the building. The construction of a reduced-scale model requires a thorough knowledge of

(a) Sculpture

(b) Maquette of Sculpture

Figure 4.15   Chicago Picasso Sculpture

The massive sculpture by the artist Pablo Picasso (4.15a) stands 50 ft (15.2 m) tall and weighs over 160 tons (1.4 MN). Due to its enormous size and weight, the structural concerns of simply enabling it to stand were paramount. Testing on a 1/30th scale model (4.15b) was conducted for wind and vibration to ensure it could be safely constructed at full scale.

(a) Photo courtesy of Deborah Oakley; (b) Photo courtesy of Robert Heller

material properties, of appropriate scale ratios for lengths and thicknesses (which usually differ), and of correct scale ratios for static and dynamic loads. A number of laboratories, properly staffed and equipped for structural model testing, exist in the world today: Dams, bridges, high-rise buildings, and other exceptional structures are sometimes designed on the basis of these tests.

An example of such model testing is the "Chicago Picasso" (Figure 4.15) a 50 foot (15.2 m) tall, 160 ton (1.4 MN) steel structure designed by Pablo Picasso. The artist gave Chicago an approximately 20 inch (0.5 m) tall sculpture made of construction paper with instructions to build it into a 50 foot (15.2 m) high steel structure. The City Council contracted the consulting firm of Weidlinger and Salvadori to determine the size and thickness of the various parts in order to withstand the winds of the Windy City. Because the structure was too complicated, a 1/30th scale steel model (Figure 4.15b) was constructed and tested for winds and vibrations at Columbia University by Robert Heller. After modifications indicated by these tests, the sculpture was built and is still standing today.

In the rare cases in which doubts may exist even about the reliability of a model test, the engineer may decide to test a full-scale structural member. This procedure is usually adopted when an important component, appearing repeatedly in a structure, does not lend itself to a clear analysis. It is a costly and time-consuming method, and though it yields incontrovertible information on the behavior of a structural element, it is seldom used (see Figure 4.14).

4.5 FUNCTIONALITY

Structural functionality is concerned with the influence of the adopted structure on the purposes for which the building is erected. For example, long-span floors could be built by giving them upward curvature, as in the dome of a church: Their thickness and their cost might thus be greatly reduced (see Chapters 11 and 12 regarding discussion of optimal structures). But, since floors for the most part need to be level to suit functional requirements, they are constructed horizontally.

Suspension bridges are rather flexible structures. The Golden Gate Bridge in San Francisco sways laterally as much as 11 feet (3.3 m) each side of vertical under strong winds. Such motions obviously must be limited, not only so that fast-traveling cars not be swayed from their paths but also because the pressure of a steady wind produces aerodynamic oscillations capable of destroying a bridge if it is too flexible (see Section 2.6).

The excessive flexibility of a structure may impair its functionality even under static loads. Thus, most codes limit beam deflections to one 360th of their spans in order to avoid cracks in rigid ceiling materials such as plaster. Aluminum, which is three times as flexible as steel, in many cases requires design for deflection rather than for strength. Worse conditions may arise under dynamic loads: A stream of traffic may produce a continuous and uncomfortable vibration throughout a flexible adjoining structure, seriously impairing its usefulness. Buildings over subway or railroad tracks are often supported on isolation pads to avoid such vibrations.

4.6 ECONOMY

In the context of this heading, economy refers to cost-effectiveness. Since economy can also mean structural efficiency, this topic is more directly addressed in Section 4.8. Sometimes economy is not a requirement of architecture. Some structures are erected for monumental or symbolic purposes (e.g., Figure 4.15): possibly to aggrandize the owners in the eye of the public or to enhance spiritual values. Monuments to the state or to "corporate images" fall in the first category; churches belong in the second. Their cost has little relation to their financial value.

But the utilitarian character of structure is so fundamental that even the structural systems of nonutilitarian buildings are influenced by economy. In other words, a strict structural budget must always be contended with unless the structure itself is an advertising display: An aluminum structure may be required, regardless of cost, in order to emphasize the ownership of the building by an aluminum manufacturer.

In the great majority of cases, the structural engineer is expected to make comparative cost studies and, other things being equal, to choose "the most economical structure." In a modern building, other engineering costs, particularly those concerning mechanical systems (heating, air conditioning, electrical, and plumbing), and architectural details far outweigh structural costs. The cost of the structure is usually not more than 20 to 30 percent of the entire cost of a building. Hence, even a substantial reduction in structural costs seldom represents savings of more than a small fraction of total building costs.

It must be always remembered that the adopted structure must produce the greatest economy in the total cost of the building and not in the cost of the structure alone. Deeper beams may reduce somewhat the cost of the structure while increasing substantially the cost of the curtain-wall, the partitions, and the mechanical systems, thus increasing the cost of the building. In structures such as large bridges, sport stadia, electric network and TV towers, however the cost of the structural system is the major expense and may govern the design.

The cost of the structural design itself represents usually less than one percent of total costs of a building. The moneys allotted to structural design are subdivided into a budget for the preliminary design, in which the system is established, and one for the final design, which includes preparation of the working drawings and specifications, a check of the shop drawings prepared by the fabricator, and, sometimes, inspection or complete supervision during construction. Final drawings are elaborate pieces of workmanship and may cost many thousands of dollars apiece. Shop drawings are seldom discussed in the education of architects and engineers, but are critical drawings that must interpret the final plans in explicit detail showing every dimension, cutout, bolt hole and so on.

The two largest cost components of a structure are materials and labor. In this connection, two basic types of economy are encountered in the world today. In the first, usually found in the more advanced industrialized countries, the cost of materials is relatively low and the cost of labor relatively high. In the second, usually encountered in less developed countries, this ratio is reversed. This in turn influences the choice of materials.

The solution of the structural problem is influenced in a fundamental way by this ratio of materials to labor costs. In the first type of economy, all kinds of machinery (cranes, conveyors, excavators, compressors, electrical tools, etc.) are used to reduce labor costs and to speed construction; easily assembled, prefabricated elements are the rule; steel is often the typical material. In the second, labor is used in large amounts for both transportation and construction; small elements are employed to minimize the use of heavy equipment; wood, masonry, bricks, and concrete are typical materials. But different ratios of materials to labor costs may influence the choice of structural system even in different areas of the same country.

Continuous changes in productivity and economic balance introduce a variety of intermediate conditions, depending on location and time. Concrete is becoming more favored and more competitive in countries where steel reigned until a few years ago. Metallic structures are becoming popular in countries where, up until now, concrete had been the most economical material.

The availability of heavy equipment is one of the main limiting factors upon the use of large prefabricated elements in construction. Some of the most interesting work performed in Europe in the recent past was conceived with basic elements not exceeding the capacity of the most readily available cranes, which was of a few tons. On the other hand, elements weighing tens of tons are commonly used in the United States even in construction jobs handled by small contractors.

Availability of skilled labor limits methods of construction in a variety of ways. "Ferrocement," as used by famed engineer Pier Luigi Nervi, is not economical in the United States at the present time, because of the hand labor of placing cement mortar between the superimposed wire meshes (see Section 3.3). Cement guns commonly used to spray concrete cannot be adopted in this case, because the velocity with which the mortar is ejected makes it bounce off the meshes and may displace the wires of the mesh. The use of specially designed guns may solve the problem, although this is a development yet to occur. Most of the delicate stonework typical of medieval buildings is ruled out today by the lack of stonemasons, whose tradition of apprenticeship has vanished. Similarly, a shortage of certified welders may make it impossible to consider an otherwise economical welded steel structure and require bolted steel construction instead. Even a lack of equipment to test the execution of the welded joints may rule out such a solution.

Other more subtle factors may also decisively influence cost. At times, the regulations of local codes tip the economic balance in favor of a specific material by imposing restrictions on another. For example, certain codes limit the thickness of flat concrete slabs to no less than a given value. The application of this regulation to curved slabs, which structurally could be much thinner, may make the construction of a small reinforced-concrete dome uneconomical. The inadmissibility of aluminum as a structural material was typical, in the recent past, due to the limitations imposed by certain codes. Fire regulations may favor concrete because of its fire-resistant properties, and comparative fire insurance costs may just as decisively recommend this material. Ecological and environmental considerations may also impact the choice of materials and will, in turn, affect the cost a structure.

The initial cost of a structure is but one factor in its economy; maintenance is another. The low maintenance costs of concrete and aluminum structures may swing the balance in their favor, when compared with that of steel structures, the latter requiring periodic painting to inhibit rust. Similarly, energy considerations influence the economy of a structure through its life cycle cost.

Speed of construction influences the amount of loan interests to be paid during the financially unproductive building period and is another factor to be considered in the choice of a structural system. Prefabricated elements, whatever their material, allow simultaneous work on foundations and superstructure, and shorten construction time; hence, they are becoming increasingly popular. Governing bodies and labor unions have at times retarded or accelerated the adoption of modern structural systems. Political considerations have had the same effects.

Economy in structure is obtained through the interplay of numerous and varied factors to be weighed carefully in order to develop the most appropriate structural system and method of construction for each set of conditions. This analysis is so complex that, in the case of large buildings, it is entrusted to specialists called construction managers, who advise the architect and engineers during the design phase and the contractor during the construction phase.

4.7 AESTHETICS

The influence of aesthetics on structure cannot be denied: By imposing his or her aesthetic tenets on the engineer, the architect often puts essential limitations on the structural system. In actuality, the architect often suggests the system and material believed best adapted to express the conception of the building, and the engineer is seldom in a position to radically change the architect's proposal.

The best architecture arises from a close collaboration between architect and engineer. In some cases, the architect consults with the engineer from the very beginning of the design, and the engineer participates in the conception of the work, making structure an integral part of architectural expression. The balance of goals and means thus achieved is bound to produce a better structure and a more satisfying architecture. Famed architects such as the English Richard Rogers and Italian Renzo Piano epitomize this type of collaboration. Works by such architects are frequently highly expressive and celebratory of structure, particularly in structures such as terminals and stadiums that require very long spans (Figure 4.16).

The influence of structure on architecture and, in particular, on aesthetics is more debatable. It was remarked in Section 1.4 that a totally sincere and honest structure is conducive to aesthetic results, but that some architects are inclined to ignore structure altogether as a factor in architectural aesthetics. Both schools of thought may be correct in their conclusions, provided their tenets be limited to certain fields of architectural practice. No one can doubt that in the design of a relatively small building the importance of structure is limited, and that aesthetic results may be achieved by forcing the structure in uneconomical and even irrational ways.

At one extreme, the architect will feel free to "sculpt" and thus to create architectural forms which may be inherently weak from a structural viewpoint, although realizable. The sculptural forms of architect Frank Gehry exemplify this. The overall form of his critically acclaimed Guggenheim Museum in Bilbao, Spain, for instance, bears little relation to its structure, as can be seen in his similarly formed band shell in Chicago's Millennium Park (Figure 4.17). From a technical perspective, the structure is

Figure 4.16   London Heathrow Airport, Terminal 5

Designed by the architects Richard Rogers Partnership and engineers Arup, the building is an example of a highly efficient structural system that is expressive and celebratory as well. From the very beginning of the project, a tight collaboration between the architect and engineer was required to achieve this high level of architectural and structural integration.

Photo courtesy of Deborah Oakley

Figure 4.17a & b   Chicago's Millennium Park band shell

Designed by architect Frank O. Gehry, who has become famous for his many sculptural buildings, the project is a good illustration of structure bending to the will of a designer. Although the front-facing curves have a sensuous aesthetic beauty, the structure supporting them from behind is arguably lacking in any such beauty. A web of steel is required to provide structural stability in a manner that is far from an optimal structural form. A more efficient structure would simply not have the shape of the band shell. Such complex structural systems have become possible only since the advent of advanced structural analysis software, as well as computer-controlled manufacturing that is required to achieve incredibly tight construction tolerances.

Photo courtesy of Deborah Oakley

Figure 4.18   Environmental sculpture

The sculptures *Suspended* by Menashe Kadishman and *Adam* by Alexander Liberman stand vigil at the Storm King Art Center in New York State. Such structures do not serve a utilitarian purpose: They are large structures that exist purely for aesthetic enjoyment and to inspire intellectual curiosity and questioning. The gravity-defying *Suspended* sculpture is, from a structural standpoint, an irrational design since the structure is fighting gravity to cantilever in this manner. A column below the vertical portion would be more sensible to enable it to stand, but obviously would completely ruin the sculptural design.

Photo courtesy of Ken McCown

rather inefficient owing to the irregular curvature, and in no way plays a role in the aesthetics of the structure aside from making that form possible. Large environmental sculpture involves an even more extreme case of structural design almost entirely influenced by aesthetics (Figure 4.18).

At the other end of the scale, exceptionally large buildings are so dependent on structure that the structural system itself is the expression of their architecture (Figure 4.19). Here, an incorrect approach to structure, a lack of complete sincerity, or a misuse of materials or construction methods

Figure 4.19   Structural expression in a high-rise building

The 43 story Hong Kong Shanghai Bank (HSBC) building by Foster and Associates is an extreme example of structural expression in a tall building. The set of columns visible at the outer edges of the building are actually a box of four columns each, which work together to create a *mega-column*. There are eight total mega-columns responsible for carrying the entire building load. Spreading the columns widely apart (as one would naturally spread one's feet apart to resist a strong wind gust), helps increase the building's stability against the strong typhoon winds of the region. A further structural feat is that all floors are suspended (seven at a time) from trusses supported by the mega-columns, which also enable the mega-columns to interact as an extra-large *mega-frame* (see chapter 8 for a discussion of building frames). The suspended floors provide for maximum flexibility in the column-free spaces.

Photo courtesy of Vincent Polizatto

may definitely impair the beauty of the finished building. The beginnings of aesthetics of structure itself are being established by semiotics, the science of nonverbal communication (see Chapter 14).

The influence of structure on modern architecture is so prevalent that some architects have wondered whether the engineer may not eventually take over the field of architectural design. The growing importance of technical services and of structure suggests such a danger. And a grave danger it would be, since the engineer, as a technician, is not trained to solve the all-encompassing problems of architectural design. But these fears may, after all, be unjustified: The engineer, while participating creatively in the design process, knows that in a group society his or her role is limited to collaboration with a design team and its leader. This leader is (and, hopefully, will always be) the architect, whose role is both that of creator and, more and more, that of coordinator.

4.8 OPTIMAL STRUCTURES

A discussion of the basic requirements of structure leads naturally to the question of whether one can satisfy all these requirements and obtain "the best structure" for a given building.

To answer the question, one should first clarify *for whom* the structure is to be "best." For the user, it should be the most practical or satisfying. For the owner, it should, probably, be the least expensive. For individual laborers, it should employ the most human-hours. Conversely, for the contractor, it should employ the *least* human-hours to secure the greatest profit. For the supplier of a specific material, the best structure should use that material in large quantities. For the structural engineer, it might be the easiest to analyze, the most interesting to study, or the most daring, depending on whether he or she is more interested in profit, theoretical skill, or personal satisfaction and fame.

From the viewpoint of the basic requirements considered in the previous sections, the best structure may be the most stable, the strongest, the most functional, the most economical, or the most beautiful.

Thus, it is obvious that the question of establishing the "best" structure does not have a simple, single answer. On the other hand, one may strive for the best structure under a number of specific limitations. For example, optimal solutions have been established in aeronautical engineering under the assumption that minimum weight is the only criterion by which structural elements ought to be judged. Similarly, the standard rolled wide flange or W-sections, and the beam I-sections, which are basic elements in all steel structures, have been studied geometrically to approach maximum strength per unit weight when used as beams or as columns (it should be noted that the two shapes are geometrically different, the latter being more uniform in depth and width, the primary criterion being resistance to buckling versus bending—see Chapter 5).

One may establish more general criteria for "the best" column by considering a variety of shapes and materials and by comparing costs. But it soon becomes apparent that the large number of factors in even a simple problem of this kind makes it practically impossible to establish the values of these factors leading to an "optimal" solution. The column, one of the simplest structural elements, may have a variety of shapes (square, round, I-shaped, boxed); each shape may have a variety of sides- or radii-ratios; the thickness of each side may be different; the length of the column may be large or small compared to its lateral dimensions; the column may be supported on a foundation, or be one of a series of superimposed vertical columns; the materials to choose from may be many; the load to be supported by the column may be centered, or off center. It is understandable that the group of structural specialists of the Structural Stability Research Council (formerly, the Column Research Council) has been at work for decades in the United States in order to establish simple criteria of strength and design for columns of steel and aluminum.

A question of common concern is the determination of the "lightest structural system," which supposedly spans the "longest distance" with the "minimum weight" of materials. Even considering a single material, simple studies show that different structural systems do not vary in weight as much as one may believe. The weight saved by the use of certain structural elements is often found to be required in their connections. Sometimes a system appears lighter than others, until a check of its flexibility shows that additional material is needed to stiffen it and make it functional.

The evolution of structural systems is a slow and delicate process. This should not discourage the serious student from investigating new possibilities or the practicing engineer from adopting new techniques. Let them simply be aware that a field as old and tried as structures does not bear new fruits without the lavishment of incomparably more effort than that required by a routine application of established principles.

KEY IDEAS DEVELOPED IN THIS CHAPTER

- Structures must be in equilibrium. All forces acting on the structure must be balanced. All actions must have equal and opposite reactions.
- Equilibrium is required for both linear forces and rotational actions.
- The action of a force at a distance that generates rotation is known as a *moment of a force*. Understanding moment is at the root of all structural knowledge.
- Structures have to be stable. They must not slide or overturn.

- Structural elements should be designed so that stresses do not exceed the yield strength of their materials or stress limits dictated by the appropriate building codes.
- All structures have a purpose, and they have to be designed to function properly to achieve it.
- All structures are to be built within a predetermined budget. They have to satisfy economical requirements, whether a building or a monument.
- Beauty is in the eye of the beholder. A structurally correct building usually satisfies aesthetic tastes.
- With the aid of modern computers and engineering know-how, anything can be designed; structurally optimum buildings do not always satisfy of requirements of economy and aesthetics.

QUESTIONS AND EXERCISES

1. Take two bathroom scales and a piece of wood across them. Applying weight on the beam one can see that the total force reading on the two scales together will always add up to the applied weight. The distribution between those forces, though, will vary depending on the position of the weight on the beam.
2. Take an empty cereal box and stand it on a table with the narrow end facing toward you. If you blow on it, the box will be very difficult if not impossible to blow over. Next, turn the box with the wide face toward you and repeat the experiment. You will notice that the box is very easily toppled. This is due to the shorter moment arm of the resisting weight.

FURTHER READING

Millais, Malcom. *Building Structures: From Concept to Design*, 2nd Edition. Spon Press. 2005. (Chapter 2)

Sandaker, Bjorn N., Eggen, Arne P., and Cruvellier, Mark R. *The Structural Basis of Architecture*, 2nd Edition. Routledge. 2011. (Chapter 2)

Schodek, Daniel and Bechthold, Martin. *Structures*, 7th Edition. Pearson. 2014. (Chapter 2)

Allen, Edward and Zalewski, Waclaw. *Form and Forces: Designing Efficient, Expressive Structures.* John Wiley & Sons, Inc. 2009. (Chapter 5)

BASIC STATES OF STRESS

5.1 INTRODUCTION

When placed under load, all structures undergo an internal response to that load. As introduced in Chapter 3, the materials of construction experience stress internally.

This is intuitive every time one sits down. A flat hard rock is less comfortable than a contoured cushioned chair because the latter more uniformly distributes the body's weight instead of concentrating it on the pelvic bones. Because of this internal response to force, all structures change shape whenever they are loaded. Although these deformations can seldom be seen by the naked eye, the resulting strains and the corresponding stresses have measurable values. Stress patterns may be quite complex; each, however, consists of only two basic states of stress: tension and compression, which is to say, *even complex structural actions can be reduced to the most elementary forces of a pull or a push.*

There are four primary classifications of stress that will be examined in this chapter: Tension, compression, shear, and bending. Although this list is not comprehensive, the vast majority of all structures can be described by these basic actions. A further discussion of more complicated states of stress is found in Chapter 9, *Some Fine Points on Structural Behavior.*

5.2 SIMPLE TENSION

Tension is the state of stress in which the molecular structure of the material tends to be pulled apart. The steel cables lifting or lowering an elevator have their particles pulled apart by the weight of the elevator (see Figure 4.1). Under the pull of the weight the cables become longer: Lengthening is typical of tension. The elongation of a unit length of cable is called its *tensile strain* (Figure 5.1).

Provided the material is not stressed beyond its elastic range (see Section 3.1), the lengthening of the cable depends only on its type of material, its cross section, its length, and the load. The larger the diameter of the cable, the smaller the unit elongation: in the elastic range the tensile strain is proportional to the load carried by each unit area of the cable cross-section; or, in other words, the tensile strain is directly proportional to the tensile stress in the cable (Figure 5.2). This ratio of tensile stress to tensile strain is a characteristic of the material called its "*elastic modulus*" in tension, which was introduced in Section 3.2.

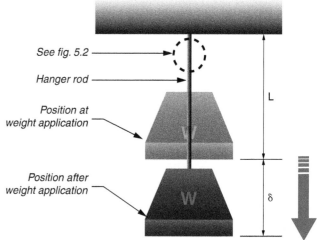

Strain = ε = δ/L = % change in length from original
(Units are inches/inch or mm/mm)

Figure 5.1   Tensile strain
When placed under a tension load (W), a structural element such as a hanger rod will lengthen (deform). The amount of deformation is proportional to the material's elastic modulus. The ratio of the amount of deformation relative to the original length is referred to as the *tensile strain*. Since the amount of deformation and the original length are each measurements of length, strain is a unitless number or a percentage value. It is also frequently shown in units of length/length (i.e., inches/inch or mm/mm).

For a given strain, the lengthening of the cable is proportional to the length of the cable: If the cable elongates by 1/4 inch (6.5 mm) when the elevator is at the top floor of an eight-story building, the cable will elongate eight times that much, or 2 inches (26 mm), when the elevator is at the ground floor (Figure 5.3).

Certain materials, such as concrete, may be easily torn apart by tension with little elongation; others, such as steel, are capable of substantial tensile elongation and are very strong in tension. For example, a high-strength steel cable, 1 square inch (600 mm^2) in area or 1.13 inches (27.6 mm) in diameter can safely carry a load of 100,000 pounds (445 kN) and will break only under a load of 200,000 pounds (890 kN) or more. The reader may get some idea of the magnitude of this force and the strength of steel by considering that this is roughly the equivalent of 70 mid-sized automobiles, each weighing nearly 3000 pounds (13.5 kN)! As an example: A construction crane lifting a 30,000 lbs (135 kN) load on a 100 ft (31 m) long 1 in^2 (600 mm^2)

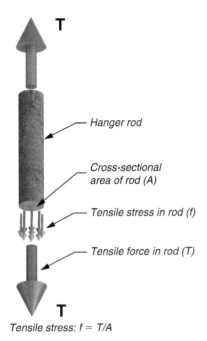

Figure 5.2 Tensile stress
Tensile stress is the most basic type of stress. The entire cross section of the material is acted on with the same magnitude across the member. As introduced in Section 3.1, stress by definition is equal to force divided by cross sectional area. Thus, the higher the force, and/or smaller the cross section, the higher the magnitude of stress will be. In a multi-stranded cable, the stress will further be divided among individual strands within the cable.

cross-sectional area cable experiences a stress of 30,000 psi (207 MPa). With an elastic modulus of 30 million psi (207 GPa), the strain in the cable will be 0.001 ft/ft (m/m) or 0.1 %; therefore, the 100 ft (31 m) cable will elongate 0.1 ft (31 mm). A similar aluminum cable would elongate 0.3 ft (93 mm), because its elastic modulus is 10 million psi (69 GPa).

Using modern ultra-high-tech materials such as carbon nanotubes (which have the highest tensile strength of any known material today), it is actually theorized that one day it may be possible to build a "space elevator" by suspending a cable made from carbon nanotubes from an orbiting platform in space. Instead of rockets, a spacecraft could literally climb the cable into outer space!

Elongation is the most important—but not the only—deformation accompanying simple tension. Careful measurements of the cable before and after the application of the load show that as the load increases and the cable elongates, its diameter decreases. This lateral change in dimension was first discovered by the French physicist Poisson at the beginning of the nineteenth century. The ratio of the lateral to the longitudinal strain is called Poisson's ratio: It is about 0.3 for steel.

5.3 SIMPLE COMPRESSION

Compression is the state of stress in which the particles of the material are pushed one against the other (Figure 3.13b). A column supporting a weight is under compression: It shortens under load. Shortening is typical of compression. In the elastic range, the shortening of a unit length of column, or its

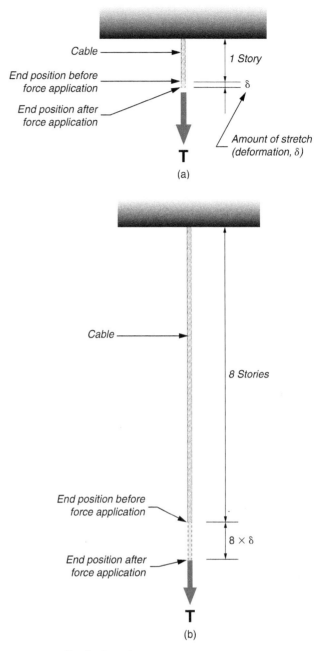

Figure 5.3 Tensile elongation
The amount of elongation that a material undergoes is directly proportional to the length of the tensile element. A doubling of a member's length will result in a doubling of the elongation. If a one-story cable elongates ¼ inch (6.2 mm) under load, an eight-story member will elongate eight times as much, or 2 inch (51 mm).

compressive strain, is proportional to the load per unit of column area. In other words, as with tension, the strain is proportional to the compressive stress: The ratio of compressive stress to compressive strain is the *elastic modulus in compression*.

Deformations in compression are opposite to those in tension: Shortening takes place in the direction of the load, and a corresponding widening occurs simultaneously at right angles to it, due to Poisson's phenomenon. The latter effect is easily visualized by a lump of soft clay: Pushing down on it will not only make it shorter but also cause it to squeeze outward.

Structural elements developing simple compression are very common because, eventually, all loads must be channeled down to earth: They appear equally in modern steel buildings as well as in Greek stone temples.

It is an interesting fact that materials incapable of resisting tension are often strong in compression: Stone, masonry, mortar, and concrete can develop high compressive stresses. A column of marble could be built to a height of 10,000 feet (3000 m) before failing in compression; a concrete column could reach a height of 8,000 feet (2400 m). Modern materials with high compressive strength, such as steel, can be used to build columns much slimmer than those of stone or concrete, but their slenderness introduces an important new type of limitation in the design of compressive elements.

5.3.1 Buckling

The reader may be familiar with the early silent movies of Charlie Chaplin in which he is often seen leaning on a cane made of a slim bamboo rod: Whenever the little fellow leans heavily on his cane, the cane bends outward (Figure 5.4). The same behavior is typical of all long, slender structural elements under compression. As the compression load is slowly increased, a value is reached at which the slender element, instead of just shortening, "buckles out," and usually breaks. This dangerous value is called the buckling load of the element. It becomes a basic design factor whenever the material is strong enough in compression to require only a small cross-sectional area, thus leading to the use of slender elements.

The buckling phenomenon may be usefully visualized from another viewpoint. A slender column shortens when compressed by a weight applied to its top, and, in so doing, lowers the weight's position. The tendency of all weights to lower their position is a basic law of nature. It is another basic law of nature that, whenever there is a choice between different paths, a physical phenomenon will follow the "easiest" path. Confronted with the choice of bending out or shortening, the column finds it "easier" to shorten for relatively small loads and to bend out for relatively large loads. In other words, when the load reaches its "buckling" value the column finds it easier to lower the loads by buckling and bending, rather than by shortening.

It must be realized that, theoretically, the column will bend out even if the load is perfectly centered and the material of the column is perfectly homogeneous. In practice, small imperfections in the centering of the load and/or flaws in the material will facilitate buckling.

It may run counter to expectations, but the danger of buckling is actually *not* related to the stressing of the material above a safe compressive stress or to its flowing under a state of plastic stress. Buckling is actually a phenomenon related to *stiffness*. The value of the compressive load for which a slender column will buckle may produce stresses below the safe values determined on the basis of the compressive strength of the material. Buckling is, in a sense, similar to resonance (see 2.6): If an otherwise safe load oscillates in step with the structure, the structure deflects more and more until it fails; if an otherwise safe compressive load is near the buckling load, the column becomes unstable

Figure 5.4 Buckling under compression
Charlie Chaplin's famous character of the early silent film era, "The Tramp," is often seen carrying a bamboo cane that bows outward when he leans on it. This phenomenon is known as *buckling*, and is a limiting factor on how tall compressive elements may be relative to their diameter or thickness. The scientific explanation and mathematical model of compressive elements in buckling was described in detail in 1757 by the Swiss mathematician and physicist Leonhard Euler. Euler's equation, in fact, proved to be so accurate that to this day it remains the basis for our more complex understanding of slender compressive elements.

and bends out (buckles) more and more until it breaks. Buckling is thus a condition of *structural instability*, not strength.

The stiffness of a column, and therefore its buckling load, depends on its material, its length, the shape of its cross section, and the restraints applied to its ends as well as possible bracing along its height. The buckling load is proportional to the elastic modulus of the material: a column of steel is three times stronger against buckling than an identical column of aluminum. The buckling load is inversely proportional to the square of the column length: A column twice as long as another, and with identical cross section, has a buckling load only one quarter as large (Figure 5.5).

To be strong against buckling and still be efficient, compression members must not be slender and yet must have a small area so as to use a limited amount of material. "Wide-flange sections" (see Chapter 7) (especially those with a thin web and wide flanges approximately as deep as they are wide), boxed sections and, in general, cross sections with most of the material away from the center of the cross-section (large moment of inertia, discussed in Chapter 7.1) are well suited for this purpose.

The buckling load increases with the restraints at the ends of the compressed member. A cantilevered column (Figure 5.6a) buckles as the upper half of a column twice as

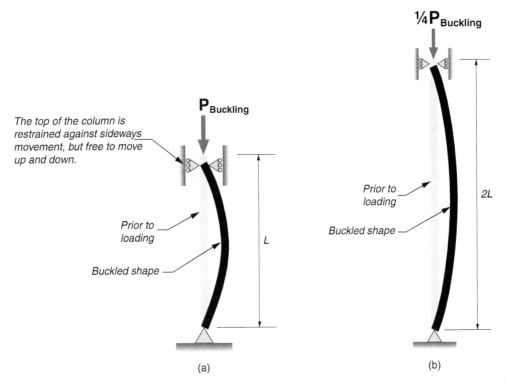

Figure 5.5 Buckling load vs. length
The buckling load of a compression element is inversely proportional to the *square* of its length. Therefore, by doubling the length of the member, its buckling load is only ¼ of the load at the original length.

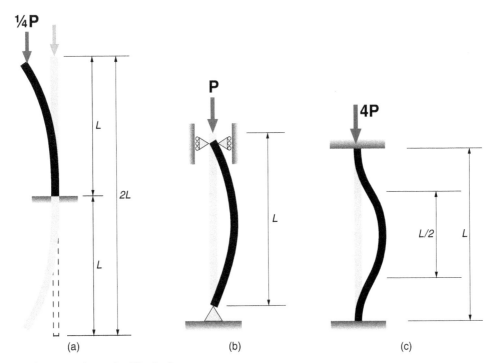

Figure 5.6 Influence of end restraints on buckling load
The manner in which a compression member is supported also influences its buckling load. A compression member that is supported only at one end (a vertical cantilever) behaves like a *simply* supported column that is twice as long (5.6a). It thus has only ¼ the buckling load of a simply supported column the same length as the cantilever column (5.6b). This is evident in the deformed shape, which forms one-half of a c-shaped curve, the other half being a virtual mirror image of the top. On the other hand, if the ends of a compression member are restrained against rotation, it behaves like a simply supported column only half as long, as can be seen by the deformed shape (5.6c). In this case, the column would have four times the buckling capacity of the simply supported column.

long and free to rotate at both supported ends; hence, its buckling load is one-fourth of that for the same "simply supported" column (Figure 5.6b).

5.4 SIMPLE SHEAR

Shear is the state of stress in which the particles of the material slide relative to each other (see Figure 3.13c). Bolts in bolted connections tend to shear (Figure 5.7). A hole puncher uses shear to punch out holes in a sheet of paper since the lever arm of the shear is very small (Figure 5.8). The weight of a short cantilever beam built into a wall tends to shear off the beam from the wall at its support (Figure 5.9).

Shear introduces deformations capable of changing the shape of a rectangular element into a skewed parallelogram (Figure 5.7, 5.8, 5.9). The shear strain is measured by the skew angle of the deformed rectangle (Figure 5.9) rather than by a change in unit length, as in the case of tension or compression.

The forces producing this deformation act on the planes along which the sliding takes place, and when measured on one square inch (one mm^2) of sliding area (Figure 5.10) are called shear stresses. In the elastic range of behavior, the deformation is proportional to the force, and, hence, the shear strain is proportional to the shear stress. The ratio of

shear stress to shear strain is called the "*shear modulus.*" It is a characteristic of the material and is less than half as large as the elastic modulus in tension or compression: Steel, for example, has a shear modulus of 11.5 million pounds per square inch (793,000 MPa).

One of the more interesting and essential characteristics of shear is to produce sliding along not one but *two* planes *simultaneously*, which are always at right angles to each other. If a square element is isolated at the support of a cantilevered beam, it is seen that, due to the action of the beam weight, equal and opposite vertical shearing forces must act on its vertical faces in order to maintain vertical equilibrium (Figure 5.11). These forces also have a tendency to rotate the element, just as the pull and push of a driver's arms tend to rotate the steering wheel of a car. If the isolated element is to be in equilibrium in rotation (see Section 4.2), as it is in the actual beam, two equal and opposite forces must act on its horizontal faces to counteract the rotating action of the vertical forces (Figure 5.12). The horizontal forces required by rotational equilibrium produce a shearing tendency in horizontal planes. Thus, shear stresses in vertical planes necessarily involve shear stresses in horizontal planes, and vice versa. Moreover, to maintain rotational equilibrium the horizontal and vertical shear stresses must have equal magnitudes.

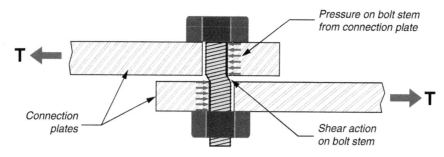

Figure 5.7 Shear in bolts

In steel construction, one of the most common methods to attach members together is by high strength steel bolts. If they are used to transfer tension forces, bolts may be designed to transfer shear forces by direct contact with the material. The shearing action tends to *slice* the bolt in half, and thus develops shear stress across the bolt diameter. To take better advantage of high strength steels, many bolted connections feature so-called *slip critical* connections, where the clamping force is so great pressing the metal pieces together that it is impossible to overcome friction between the parts. In such connections, the bolts do not experience direct shear forces.

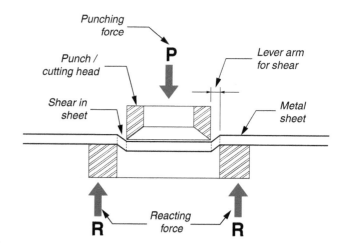

Figure 5.8 Punching shear

Punching shear is a slicing action that occurs when there is a very small distance to create a moment arm between the acting and reacting surfaces of an object. Punching shear is used in everything from paper punches to large metal presses, which can exert many tons of force to pierce thick metal components.

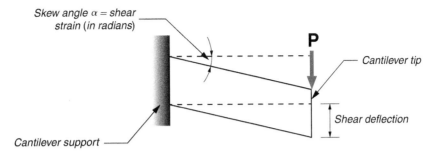

Figure 5.9   Shear strain in cantilevered beam

Shear action in a cantilever beam tends to deform an initially rectangular element into a parallelogram. The angle of deformation is referred to as the *shear strain*, and is analogous to tensile or compressive strain, except measured angularly. Note that shear action in reality cannot be isolated from bending action; these both occur simultaneously in a member. See Chapter 7 for a complete discussion on bending members.

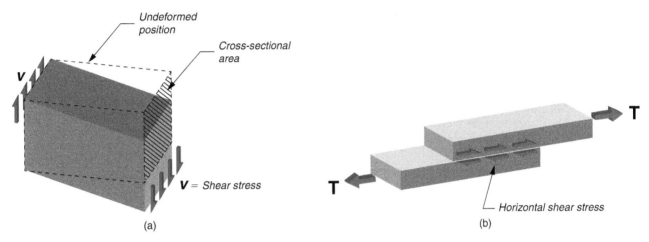

Figure 5.10   Shear stress

Like tensile and compressive stress, shear stress is measured as force divided by cross-sectional area. The area in this case, though, refers to the members' cross-section *parallel* to the force, instead of perpendicular as in the case of tensile or compressive stress. This may occur through the transverse action of a support reaction on a beam (5.10a) or between two plates welded or glued together (5.10b).

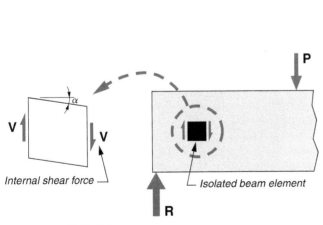

Figure 5.11   Vertical shears

A small element of a beam isolated for inspection can be seen to be in vertical shear due to the (typically) upward reaction of the support and the downward action of the load. The shearing action causes shear strain in the element in the same manner as a cantilever experiences shear deflection (Figure 5.9).

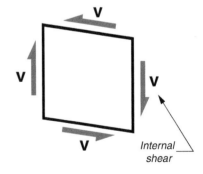

Figure 5.12   Shears required for equilibrium in rotation

A closer observation of the square element in a beam reveals that the vertical shear forces—although balanced in a linear manner—are not balanced rotationally. Therefore, by nature, shear forces exist in two perpendicular planes simultaneously, with the perpendicular shear acting in exactly the opposite rotational direction. This rotational tendency between equal forces separated by a distance, is a special form of a moment, referred to as a *couple*. As will be seen in Chapter 7, couples exist internally in bending members and offer a vehicle for computing internal bending stresses.

This balance of equal and opposite forces is so prevalent in structures that it is given a special name: A pair of equal forces, acting in exactly opposite directions and separated by a distance, is referred to as a *moment couple*, or more commonly just a *couple*. A couple is in linear equilibrium, but will generate rotation, which can only be balanced by an equal and opposite moment couple—precisely the balance we see in looking at the phenomenon of shear.

The existence of the horizontal shearing forces may also be inferred by analyzing the deformation of the square element. The skewing of the element produces a lengthening of one of its diagonals and a shortening of the other. Since lengthening is always accompanied by tension and shortening by compression, the same deformation could be obtained by compressing the element along the shortened diagonal and tensioning it along the other (Figure 5.13). Thus, shear may also be considered as a combination of tension and compression at right angles to each other in directions making an angle of 45 degrees with the shear directions.

The tensile and compressive diagonal forces "result" by combining the horizontal and vertical shears first along the lengthened and then along the shortened diagonals (Figure 5.13). Accordingly, as stated in Section 5.1, the basic states of stress are two, tension and compression, rather than three.

The consideration of shear as equivalent to compression and tension at right angles to each other and at 45° to the shear planes is of great practical importance. A material with low tensile strength cannot be strong in shear, as it will fail in tension in a direction at 45 degrees to the directions of shear (see Section 7.3). Similarly, a thin sheet cannot be strong in shear as it will buckle in the direction of the compressive equivalent stress (see Figure 7.29).

5.4.1 Torsion

The tendency to slide, characteristic of shear, is also found in structural elements *twisted* by applied loads. Consider a bar of circular cross section, on the surface of which is graphed a square mesh of straight lines (Figure 5.14a). If this bar is twisted, so that one end section rotates in relation to the other (as one might twist a screwdriver), the squares on its surface become skewed rectangles (Figure 5.14b). Since this kind of deformation can only be due to shear stress, twisting must produce shear strains and, hence, shear stresses in the cross section of the bar and, for equilibrium, also in the radial planes perpendicular to the cross section. This state of stress resulting from the twist along an axis, although consisting of pure shear, is called torsion.

Since torsion develops shear stresses, it must be equivalent to tension and compression at right angles. This is evident in the act of wringing out a wet rag before hanging it; the rag's surface forms a helix and compression develops perpendicular to the helical lines. The torsion-induced compression squeezes the water out of the rag.

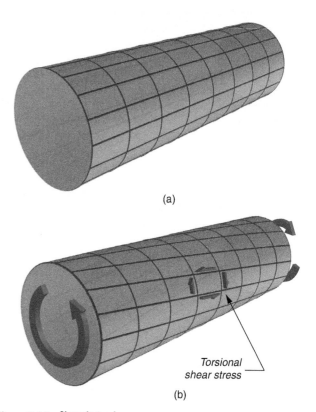

(a)

(b)

Torsional shear stress

Figure 5.14 Shear in torsion
Torsion is the type of stress that occurs in a member that is twisted along an axis. It is frequently encountered in machinery (e.g. in a motor shaft), and also in simple tools such as a handheld screwdriver. Torsion is a type of shear stress, as can be seen in the skew deformations of lines etched onto a cylindrical surface (5.14a), when it is twisted along its length (5.14b).

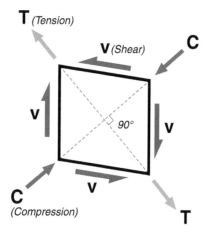

T *(Tension)*

V *(Shear)*

C

V

90°

V

C

V

(Compression)

T

Figure 5.13 Equivalence between shear, and tension, and compression
A study of Figure 5.12 reveals that any pair of shear forces on the elemental block can be looked at as one resultant force, either acting in compression or tension on the block. These resultant stresses (referred to as *principle stresses*) always act at 90° to one another and at 45° to the shear planes. Because there is always a tensile resultant force due to shear action, materials that are weak in tension are also weak in shear. Diagonal cracks in rigid materials such as masonry or concrete are signs of shear cracks in tension. On the other hand, slender elements in shear may experience buckling on a diagonal due to the compression principle stress component. See also Figures 7.28 and 7.29.

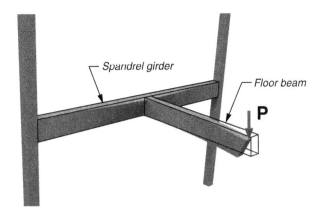

Figure 5.15   Spandrel beam under torsion
Building structures can have elements that require design for torsion, notably on the exterior beams (called *spandrels*) that may receive load from only one side. As a beam bends, the ends will rotate away from vertical. If it is securely attached to the supporting spandrel beam (common in concrete construction), then the end rotation creates a torsional load on the spandrel beam that it must be designed to resist.

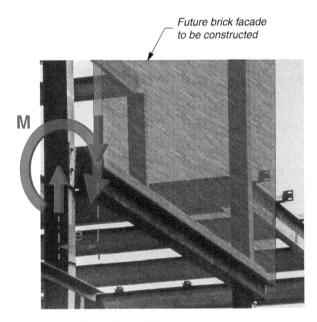

Figure 5.16   Torsion on fascia girder
A closed tubular shape in this steel building under construction enables it to carry the weight of a brick masonry façade. In this case, because the support of the beam at the column is eccentric (removed at a distance) from the load, a couple is created that induces a torsional moment. This type of closed shape is much more effective at carrying torsion than open shapes such as wide flange members.

Photo courtesy of Deborah Oakley

Torsion occurs in a structural element whenever the loads tend to twist it. For example, the eccentric loads transferred by a floor beam to a spandrel beam (a beam located on the perimeter of a building) induce torsional stresses in the spandrel (Figure 5.15). Rigidity against torsion involves the shear modulus. The most efficient cross sections against torsion are circular and hollow, giving the shear stresses the greatest possible lever arm in rotation around the axis of the bar. A closed section in fact is over one thousand times more

rigid in torsion than the same shape with a slit along its length. Round tubes, however, create no flat surface for assembling other structural or architectural elements. Frequently in practice, therefore, closed square or rectangular tubes are used in construction with a slight reduction in efficiency, but with greater ease of assembly (Figure 5.16).

5.5 SIMPLE BENDING

It was stated in Section 5.1 that all complex states of stress are combinations of no more than two basic states of stress: tension and compression. Simultaneous compression and tension in different portions of the same structural element is perhaps the most common of these combinations: it is called bending, and plays an essential role in most structural systems.

Consider a plank supported by stones at each end (Figure 5.17). If two children of equal weight stand at equal distances from the end of the plank, the plank bends downward, and the curve assumed by the plank between the stones can be shown to be an arc of a circle. By drawing evenly spaced vertical lines on a side of a flexible foam beam (Figure 5.18a), it can be observed that upon bending these lines open up at the bottom and crowd at the top (Figure 5.18b). It is seen that the plank's lower fibers lengthen, its upper fibers shorten, and its middle fibers maintain their original length. But the beam has thickness, and all of its fibers must become curved. Hence, the bending down of the beam induces tension in its lower fibers and compression in its upper fibers. Moreover, the tension and compression may be seen to increase in proportion to the distance of the fibers from the middle, or *neutral* fibers. This behavior, carefully described in the sixteenth century by Leonardo da Vinci, was rediscovered only in the nineteenth century by the French physicist Navier. The state of stress in which the stress varies linearly from a maximum tension to an equal maximum compression is called *simple bending*.

For clarity, it should be noted that the use of the term "fiber" is common in engineering mechanics, and is a general term meant to describe any type of material, whether it be truly "fibrous" in nature like wood or solid and uniform like steel which has no fibrous structure. "Fibers" in beams thus more generally refer to "regions" or layers of the cross section, such as the "upper fibers" or "lower fibers."

In Figure 5.17, bending stresses are present along the arc of a circle of the deformed plank, but this deformation is so small in comparison with the length of the plank that the vertical weights of the two children may be said to produce in the plank essentially horizontal stresses. Bending may thus be considered a structural mechanism capable of channeling vertical loads to the supports in a horizontal direction or, more generally, in a direction at right angles to the loads (Figure 5.19). The weights of the children are transferred horizontally to the two stones supporting the plank by bending stresses.

In view of the compressive strength of most structural materials, it is relatively easy to channel loads vertically down to earth. The fundamental structural problem

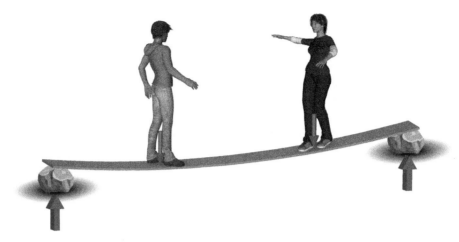

Figure 5.17 Simple bending

Simple bending occurs whenever a horizontal member (supported at either one end as a cantilever or both ends as a simple beam) carries a load *perpendicular* to its spanning direction. The action of bending, as the name implies, is to bend the element into a curve. If a flexible plank is supported on each end and two children stand on it, the board will bend downward into a curve. This curvature induces bending stresses that are greatest at the outer surfaces of the plank.

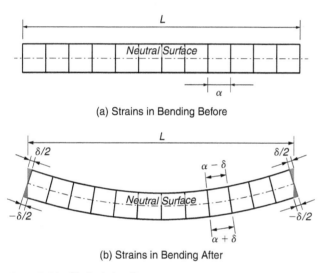

(a) Strains in Bending Before

(b) Strains in Bending After

Figure 5.18 Strains in bending

Bending in a beam can be understood by creating a flexible beam out of foam rubber, and drawing lines vertically at equal spaces and one line in the center along the length (5.18a). With the beam supported at each end, pushing downward on the beam will reveal that the inner surface squeezes together, while the outer surface stretches apart (5.18b). This indicates that the upper portion of the beam is in compression, while the lower portion of the beam is in tension. Between these regions of compression and tension, stresses transition at the middle where there is no stress at all, the so-called *neutral surface* of the beam. Bending (or *flexural*) stress is addressed in detail in Chapter 7.

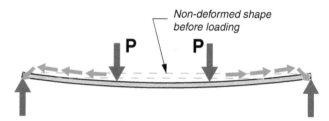

Figure 5.19 Load Channeling

The action of a member carrying a load in simple bending is to effectively transfer or channel those loads to the supports. One could imagine a small mouse crawling from the load and across the beam and then down to the ground. Through internal resistance to bending forces, loads on a member are directed horizontally across a span. The green arrows are meant to represent the direction of load transfer, not actual forces in the member (which are more complex—see Chapter 7).

consists, instead, in transferring vertical loads horizontally in order to span the distance between vertical supports. Bending is thus seen to be of prime importance as a structural mechanism.

A good bending material must have essentially equal tensile and compressive strengths. This explains the historical predominance of wood among natural structural materials, and the long unrivaled role of steel in modern structures. Concrete alone, as previously discussed (see section 3.3) has no useful tensile capacity, and is therefore unsuited as a bending member. *Reinforced concrete*, however, is a composite material with bending properties comparable to those of steel. In this material, the compressive strength of concrete is used in the compressed fibers of the element, and the tensile strength of steel in the tensioned fibers. If the plank considered above were built of reinforced concrete, it would have reinforcing bars near its bottom (Figure 5.20), since this is the region in which tensile stresses develop. The situation is reversed in a cantilevered beam, as shown in Figure 5.21. The tensile surface is now at the top of the beam and the compression at the bottom. The placement of reinforcement must therefore be reversed.

It is worth noting that serious structural failures can result from the improper placement of steel reinforcing bars.

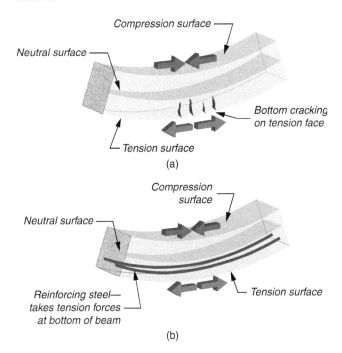

Figure 5.20 Reinforced concrete beam

Since concrete is a material weak in tension, when placed in simple bending there is little resistance to cracking along the bottom surface and failing in bending stress (5.20a). As described in Section 3.3, reinforced concrete is a composite material that combines the tensile strength of steel with the compressive strength of concrete. Working together, the steel carries tensile stresses at the bottom of a simply supported beam, while the compression is carried by concrete in the upper portion of the beam.

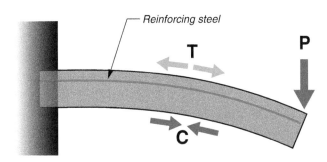

Figure 5.21 Cantilevered concrete beam

If instead of being supported at the ends as a simple beam in Figure 5.20, a concrete *cantilever* beam will experience top surface *tension* instead of compression. The location of steel reinforcement must therefore be on the top of the beam rather than the bottom.

For example in a cantilever, it is of little help placing the tensile steel on the bottom of the beam; the beam will have no more bending strength than an unreinforced concrete beam of the same size. For this reason, careful inspection of reinforcing steel *prior* to the placement of concrete is absolutely critical in construction projects.

KEY IDEAS DEVELOPED IN THIS CHAPTER

- The basic states of stress are tension, compression, bending, and shear. Tensile stresses pull material particles apart, while compression presses them together. Bending and shear change the shape of structural members.
- Tensile stresses produce elongation of components, while compression results in shortening. Both stresses are defined as the applied force divided by the cross-sectional area of the member perpendicular to the force. In each case, the change in length is proportional to the applied stress and to the original length of the member and the strain is the percent change in length.
- Under tension the length increases, and at the same time the thickness of a tensed component decreases. In compression, the shortening of length is accompanied with a widening of thickness. This phenomenon is referred to as the Poisson Effect.
- Most metals are able to carry tensile loads within their elastic ranges. Brittle materials such as concrete, stone, and brick are capable of compressive loads only.
- Tall, slender compressed columns may suddenly buckle as the applied load increases. The buckling load is proportional to the modulus of elasticity and to the distribution of material away from the centroid of the cross section. (Moment of Inertia will be discussed in Chapter 7) It is inversely proportional to the length of the column. Buckling load is also a function of end restraints and of lateral supports.
- Shear produces sliding of materials parallel to the applied force. It distorts a square element into a rhombus. Shear stress is defined as the applied force divided by the cross-sectional area parallel to the force. Shear strain is the change of angle from a right angle to a skewed angle.
- Shear is equivalent to tension and compression at 45 degrees from the shear stress.
- When a cylindrical bar is twisted, squares drawn on its surface deform into rhombuses and indicate the presence of shear. Torsion is a special case of shear.
- Simple bending produces curvature in straight elements. When a component is bent into an arc, the fibers nearer the center of curvature shorten and hence experience compression while the opposite side is elongated and is in tension. A surface between the two sides is unstressed.
- Bending stresses are proportional to the depth of the beam and inversely proportional to the moment of inertia. The farther the amount of material is from the neutral surface the smaller the bending stress.

QUESTIONS AND EXERCISES

1. Take a flexible ruler and place it on two scales, or support one end on some books and the other on a scale and push on it. Experiment with the spacing between these supports and notice how the force varies with the length of the ruler.
2. Push on the same ruler at different points along its length. Observe the variation of the force on the scale.
3. Take a piece of stretch fabric like Lycra® or pantyhose, or a sheet of paper, and tape or glue it to a frame made from a square of cardboard as illustrated. Distort the square and observe the rippling of the fabric. Why does this occur? Can you see the equivalence between shear, and compression and tension?

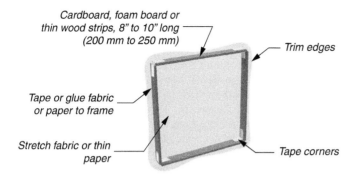

Cardboard, foam board or thin wood strips, 8" to 10" long (200 mm to 250 mm)

Trim edges

Tape or glue fabric or paper to frame

Stretch fabric or thin paper

Tape corners

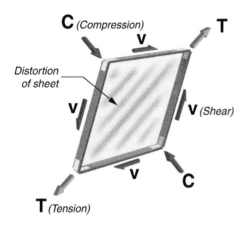

C (Compression)

T

V

Distortion of sheet

V

V (Shear)

V

C

T (Tension)

4. Take a piece of foam, about a foot long (0.3 m) and 2 inches (50 mm) on a side. Draw a grid of vertical and horizontal lines on it with a marker. Place one end on the edge of a table and push down on it in the middle. Observe the distortions of the squares: Why do they become formed in that pattern?

5. Hold a flexible ruler vertically. Place one end on a scale and press straight down on the other end. Observe the force required to buckle the ruler. Now hold the ruler between thumb and forefinger at the middle and again press down on the free end. Observe the new shape of the buckled ruler and see the force required to buckle it. Can you explain the reason for the increased force?

FURTHER READING

Sandaker, Bjorn N., Arne P. Eggen, and Mark R. Cruvellier. *The Structural Basis of Architecture*, 2nd Edition. Routledge. 2011. (Chapters 7)

Schodek, Daniel and Martin Bechthold. *Structures*, 7th Edition. Pearson. 2014. (Chapter 2)

Millais, Malcom. *Building Structures: From Concept to Design*, 2nd Edition. Spon Press. 2005.

Schaeffer, Ronald. *Elementary Structures for Architects and Builders*, 5th Edition, Prentice Hall. 2006.

STRUCTURAL FORMS

Part II of this text illustrates the application of basic principles introduced in Part I. Beginning with the most elementary structural types and moving through complex forms, the reader is introduced to the wide diversity of possible structural types, including tension and compression structures (Chapter 6), bending structures (Chapter 7), frames and arches (Chapter 8), and a more refined look at the nuances of structural behavior (Chapter 9). The fundamental concepts behind those structures, plus illustrative examples, form the basis for this presentation. Since some of the later material of Part III builds on the present material, it is recommended that Chapters 6 through 9 be studied before the later chapters.

The Milleau Viaduct in southern France. See Figure 6.12.
Photo: Igor Plotnikov/123RF

TENSION AND COMPRESSION STRUCTURES

6.1 CABLES

The most elementary of all structural elements, and thus the most easily understood, is the simple tension cable. Cables have been used for thousands of years, beginning with the form of simple ropes made of natural plant fibers and animal hairs, to the present day where the properties of advanced materials have made the long-span bridges of the modern world possible. The high tensile strength of steel, in particular, combined with the efficiency of simple tension, makes a steel cable the ideal structural element to span large distances.

Cables are flexible because of their small lateral dimensions in relation to their length. This flexibility indicates that in a cable there is a limited resistance to bending. As introduced in Chapter 5 (and further discussed in Chapter 7), bending is a complex structural action that involves simultaneous tensile and compressive stresses within a member cross section. In the case of cables, uneven stresses due to bending are prevented by flexibility. The tensile load in a cable is thus evenly divided among the cable's strands, permitting each strand to be loaded to the same safe, allowable stress. This behavior makes cables the singularly most efficient structural form possible. *All* of the material of a cable can be effectively used to carry loads, and there is no

possibility of buckling. Bridges with single spans longer than a full mile are now common due to the combined strength of steel and efficiency of pure tension (Figure 6.1).

As introduced in Chapter 3, it is very easy to visualize the load-carrying action of a simple cable carrying a weight when suspended along its length. This type of behavior is very familiar to almost everyone and clearly produces pure tension in the cable. Loads acting perpendicular to a cable spanning a distance will always produce pure tension; however, the magnitude of the tensile force can vary throughout the cable length, depending on the placement of the loads.

In order to understand the mechanism of how an initially horizontal cable supports vertical loads, consider a cable suspended between two fixed points located at the same level and carrying a single load at midspan (Figure 6.2). Under the action of the load, the cable assumes a symmetrical, triangular shape, and half the load is carried to each support by simple tension along the two halves of the cable.

The characteristic triangular shape acquired by the cable is referred to as the *sag*: the vertical distance between the supports and the lowest point in the cable. Without sag the cable could not carry the load. We see from the vector free body diagram (FBD) of Figure 6.3 that without this sag

Figure 6.1   The Akashi Kaikyo Bridge

With a main span of 6532 ft (1991m) the Akashi Kaikyo bridge currently (2015) has the longest central span of any bridge in the world, demonstrating the efficiency of using high strength steel cables in tension.

Photo: Toshitaka Morita/Getty Images

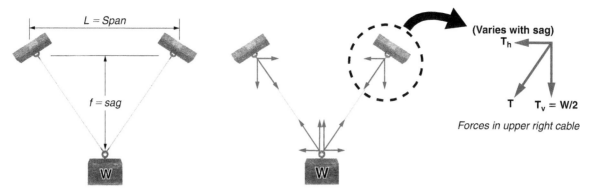

Figure 6.2   Symmetrical load on cable (above left)

A single load, *W*, placed at the midspan of a suspended cable will naturally deform the cable into a triangular shape. The vertical distance of the cable measured from the highest to lowest points along its length is referred to as its *sag*.

Figure 6.3   Internal Cable Force Actions (above right)

This illustration demonstrates the internal forces on the cable of Figure 6.2 as a free body diagram with the upper right joint isolated for clarity.

The tension force on a cable structure can be resolved into horizontal and vertical components. The combination of the two force components, T_v & T_h, have the same net effect on the structure as the actual force, *T*, but facilitate a clearer understanding of the horizontal and vertical equilibrium of the forces. Note that in both horizontal and vertical directions, every force vector is balanced by an equal and opposite vector to maintain static equilibrium— were this not the case, the structure would be in motion.

For a symmetrically loaded cable, it is evident by inspection that ½ of the total vertical load *W* is carried by each support. The vertical components of the forces must therefore each carry ½ of the load *W*. The inclination of the cables generates a horizontal inward pull, T_h, on the supports, each of which must be equal and opposite in direction to maintain horizontal equilibrium. The magnitude of T_h, however, depends entirely on the sag distance *f*, as illustrated in Figures 6.4a and 6.4b.

the cable tension forces would need to be horizontal, but purely horizontal forces cannot balance a vertical load — this would not fulfill the requirement for linear equilibrium in the vertical direction. The inclined pull of the sagging cable on each support may be split into the equivalent of two components: a downward force equal to half the load and a horizontal inward pull. Due to the horizontal action of this inward pull, if the supports were not fixed against horizontal displacements, they would move inward under its action, and the two halves of the cable would become vertical.

A simple experiment can bear this out. If the reader holds a thread in each hand and attaches a weight at the middle, one may sense physically that the string develops no horizontal pull when the fingers are touching, while an increasing pull is developed as the hands are moved apart, thus decreasing the string sag. The pull may be shown to be inversely proportional to the sag: Reducing the sag by half in fact doubles the pull (see Figure 6.4). For vertical equilibrium to be maintained, the vertical pull on the hands is always equal to one half of the load and is independent of

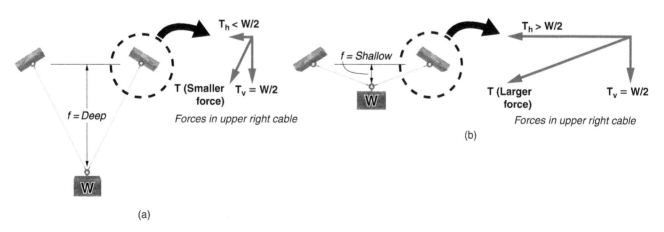

(a)   (b)

Figure 6.4   Variation of Cable Thrust with Sag

The graphic representation of force vectors is drawn in proportion to the force magnitude and direction, so the free body diagram therefore is a visual representation of the force intensity that can be understood at a glance. The vertical components of the inclined cable forces are always equal to one half of the load (when the load is placed at midspan); however, the magnitude of the horizontal force can be seen to decrease as the cable sag is increased (6.4a), and conversely will increase as the sag is decreased (6.4b). There is thus an inverse relationship between the amount of sag and the magnitude of force in the cable.

the sag. Hence, as the sag decreases, the tension the cable exerts on the hands increases because of the increase in the *horizontal*, not vertical pull. If the thread used is weak enough, there comes a point when it snaps, indicating that, as the sag diminishes, the tension eventually becomes larger

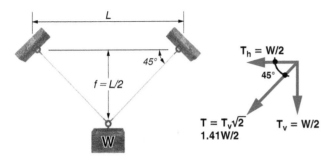

Figure 6.5   Optimal Sag

For a load placed at midspan, the optimal sag is one-half of the span distance, L. This results in a cable inclination of 45 degrees, and so both the horizontal and vertical components of the cable force will be equal to half of the applied load. Although the force magnitude will be higher than a cable with greater sag, the length of the cable will be longer, thereby requiring more total material. A sag distance of one half the span represents an optimal balance between the amount of force in the cable an the total amount of material required in the cable.

than the tensile strength of the thread. In actuality, if we look at the graphic vector equilibrium of the FBD (Figure 6.4b), it is seen that the horizontal component becomes infinitely long as the angle decreases to zero, indicating that it is technically impossible for a *completely horizontal* tension element with no sag to carry a vertical force.

The cable problem just considered raises an interesting question of economy. A larger sag increases the cable length, but reduces the tensile force in the cable and, hence, allows a reduction of its cross section; a smaller sag reduces the cable length, but requires a larger cross section because of the higher tension developed in the cable. Hence, the total *volume* of the cable material, the product of its cross section and its length, is large for both very small and very large sags and must be minimum for some intermediate value of the sag. The optimal or "most economical sag" for a given horizontal distance between the supports turns out to be one-half the span, and corresponds to a symmetrical, 45-degree-triangle cable configuration with a horizontal pull equal to half of the load (Figure 6.5).

If the *load is shifted* from its midspan position, because of its flexibility the *cable changes shape*, and adapts itself to carrying the load in tension by means of straight sides of *different inclinations* (Figures 6.6a). The two supports develop different vertical reactions, but equal horizontal pulls since

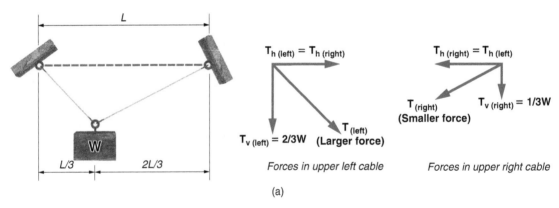

(a)

Figure 6.6a

For a single, asymmetrically placed load, free body diagrams for the upper left and right cable ends clearly illustrate that the force magnitude in the cable is not equal on either side of the load, the left side being greater than the right. For a load positioned at 1/3 the distance from the left support, the vertical reaction components will be 2/3 of the total load on the left side and 1/3 the total load on the right side. Since no other horizontal forces are present, however, the horizontal components on either side will be equal and opposite one another, the actual magnitude varying with the amount of cable sag.

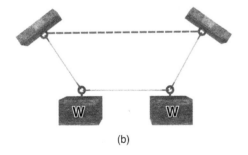

(b)

Figure 6.6b

Two symmetrically placed loads on a cable will reshape the cable into a three-sided polygon.

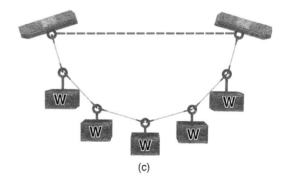

(c)

Figure 6.6c

Additional loads will continue to reshape the cable into a closer approximation of a curve as the straight sides become smaller and smaller.

the cable forces must always be in equilibrium in the horizontal direction. The value of the horizontal pull differs from the value for a centered load, but still varies inversely as the sag. The reader may sense this by shifting the load along the hand-supported thread. Because the vertical components on each side of the cable are different, the *resultant total* tension forces on each side of the load are also no longer equal to one another. This fact may at first be counterintuitive, yet can be clearly demonstrated by a graphic FBD at the upper left and right joints (Figure 6.6 a).

If the single load now is replaced by two equal loads placed on the cable in symmetrical locations, the cable again adapts itself and carries the loads by acquiring a new configuration, this time with three straight sides, the middle of which is horizontal (Figure 6.6b). As the number of loads is increased, with each new loading the cable again acquires a new shape in response to new equilibrium configurations. Each of the new shapes will have straight sides between the loads and changes in direction at the points of application of the loads. This process can continue, and each time the cable will readjust with a new form for each additional load.

It is important to note that for each configuration of loading, there is one and only one shape that the cable will take. This characteristic shape acquired by a cable under concentrated loads is called a *funicular polygon* (Figure 6.6c). The name is derived from the Latin word *funis* for rope, and the Greek words *poly* for many and *gonia* for angle. The funicular form is the natural shape required to carry loads in pure tension or (as will be shown) pure compression and therefore represents the optimum structural form that can be achieved by a unidirectional spanning structure. It is for this reason that the longest-spanning bridges are supported by cables.

As the number of loads on the cable continues to increase, the funicular or *string polygon*, acquires an increasing number of smaller sides and approaches a smooth curve. If one could apply an infinite number of infinitesimally small loads to the cable, the polygon would become a *funicular curve*. For example, the funicular polygon for a large set of equal loads evenly spaced *horizontally* approaches a well-known geometrical curve, the *parabola* (Figure 6.7 a). The optimal sag for a parabolic cable equals three-tenths of the cable span.

If the equal loads are distributed evenly *along the curved length* of the cable, rather than horizontally, the funicular curve differs from a parabola, although it has the same general configuration: It is a *catenary,* the natural shape acquired by a cable of constant cross section or a heavy chain (*catena* in Latin) under its own weight, which is uniformly distributed along its length (Figure 6.7 b). The optimal sag for the catenary is about one-third of the span; for such sag ratio, the catenary and the parabola are very similar curves (Figure 6.7 c).

A cable carrying a combined load of its own self-weight plus a load uniformly distributed horizontally acquires a shape that is intermediate between a catenary and a parabola. This is the shape of the cables in the central span of suspension bridges, which carry their weight and that of the stiffening trusses on which the roadway is laid (Figure 6.8).

A cable would span the *largest* possible distance if it could just carry its weight, but break under the smallest additional load (which is to say that under its own dead load it is stressed to its limit). Assuming an optimal sag-span ratio of one-third to minimize the weight of the cable, it is found that such a steel cable with a strength of 200,000 pounds per square inch (1.38 GPa) could span a distance of 17 miles (27.4 km)! This maximum distance is independent of the cable diameter, since both the weight of the cable and the tension in it are proportional to the area of the cross section. Obviously, actual cable spans are built to carry loads, which are usually much heavier than the cables themselves; hence, cable spans are much shorter than the limit span of 17 miles. The longest North American suspension bridge to date (2015), the Verrazano Narrows Bridge at the entrance of New York harbor, spans 4260 feet (1,298m) between the towers, a length over two-and-a-half times that spanned by the Brooklyn Bridge (which opened in 1883).

The Verrazano Narrows Bridge reigned as the longest in the world for 17 years from 1964 until surpassed in 1981 by the Humber Bridge in England, with a span of 4626 feet (1,410m) between towers. The Akashi-Kaykio Bridge in Japan (fig 6.1), completed in 1998 with a span between towers of 6,532 feet (1,991m) presently has the longest main span in the world. Such spans are rapidly approaching limiting values, beyond which it is not practically conceivable to go with the types of steel available today. Only an improvement in

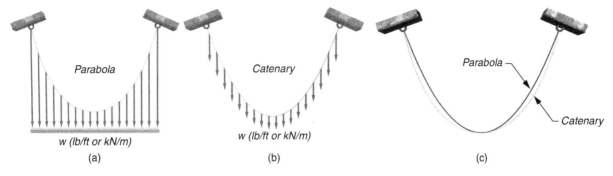

Figure 6.7 Funicular Curves
The variation between a catenary curve and a parabola is illustrated in these series of diagrams with gravity-induced loads. A parabola (6.7a) is the natural result of an equal, uniform load suspended along the length of the cable. A catenary (6.7b) is the result of the cable self-weight, which follows the natural curve of the cable. The two forms are closely similar; however, the catenary has a slightly greater outward bulge as can be seen in the superimposition of 6.7c.

Figure 6.8   The Golden Gate Bridge

One of the most iconic bridges in the world, the Golden Gate Bridge is a classic suspension bridge. Smaller suspender cables are hung from the main suspension cables, which are anchored by massive blocks at either end. The suspender cables support a deep stiffening truss, which carries the weight of the roadway. Without the stiffening truss, the roadway would continually flex with shifting weight of the moving vehicles.

Photo: Martin Molcan/123rf

Figure 6.9   Proposal for a bridge across the Strait of Gibraltar by Prof. T.Y Lin

With a main span of 16,400 ft (5000 m), the bridge would be 2.5 times longer than that of the Akashi Kaikyo bridge of Figure 6.1. This incredible span would be made possible by a hybrid structural system of cable-stayed structures at the masts that in turn support suspension cables, in turn supporting a deck of high strength but light weight glass fiber. At the present time (2015), the bridge remains an unrealized proposal. The obstacles to be overcome, particularly with supporting piers plunging nearly 3300 feet (1 kM) in water, are enormous.

Photo courtesy of Deborah Oakley

cable strength will make substantially larger spans possible with a single suspension cable, although hybrid designs that combined suspension cables with cable-stayed structures can further extend the range, such as the proposed span at the Strait of Gibraltar (Figure 6.9) by Dr. T.Y Lin.

Given the inherent efficiency and economy of steel cables, one may wonder why they are not used more frequently in smaller structures. The limitations in the application of cables stem directly from their adaptability to changing loads: Cables are unstable, and (as introduced in Chapter 4) stability is one of the basic requirements of structural systems. The trusses hanging from the cables of a suspension bridge have the purpose not only of supporting the roadway but also of stiffening the cables against motions due to changing or moving loads.

Stiffening trusses are usually rigid in the direction of the bridge axis, but less so in a transverse direction: Large displacements of suspension bridges caused by lateral winds can be substantial. Moreover, the long roadway and the shallow trusses constitute a thin ribbon, so flexible in the vertical direction that it may develop a tendency to twist and oscillate vertically under steady winds (see Section 2.6). Modern suspension bridges are made safe against such dangers by the introduction of stiffening guy wires (like those used in the Brooklyn Bridge in New York, designed in 1867) and by an increase in the bending and twisting rigidity of their roadway cross section.

In so-called *stayed bridges* of the "harp" or "fan" type (Figures 6.10 and 6.11), the guy wires—or *stays*—have the double role of supporting the deck and of stabilizing it.

Figure 6.10   The Senator William V. Roth Jr. Bridge

Unlike a suspension bridge that uses a main cable from which suspender cables are hung to support the road deck, a cable-stayed bridge supports the deck directly from inclined stay cables that attach to the towers. The cable arrangement of this bridge over the Chesapeake and Delaware Canal is known as a *harp shape* because the parallel arrangement of cables resembles the strings of a harp instrument. In the background can be seen the older arch bridge which the newer one replaced with a much wider road deck, capable of carrying more traffic, as well as longer span that eliminated piers in the waterway.

Photo courtesy of Scott Gore

Figure 6.11   The Clark Bridge

Crossing the Mississippi river near Alton Illinois, the Clark bridge represents another form of the cable-stayed type with the cables arranged in a radial pattern known as a *fan*.

Photo courtesy of Ted Engler

Their elegance and economy has made them popular for middle-range spans. Following the extensive destruction of European bridges during World War II, rapid reconstruction of the highway infrastructure was critically urgent. Cable-stayed bridge design progressed rapidly in the years following the war since they were both economical and capable of being constructed more rapidly in comparison with similar spans using suspension designs.

A cable is not a self-supporting structure unless ways and means are found to absorb its tension, which in large spans may reach values of the order of thousands of tons. In suspension-bridge design, this result is achieved by channeling the tension of the main-span over the towers to the side spans, and by anchoring the cables in the ground. The heavy anchoring blocks of reinforced concrete are usually poured into rock, and resist tension both by the action of the block's substantial dead weight and by the reaction of the adjoining rock (Figure 6.13). Under these circumstances the optimal sag to minimize the cost of the entire structure is approximately 1/12 of the span, since a 3/10 sag-span

Figure 6.12   The Millau Viaduct

An example of one of the more common forms of the cable stayed bridge, a semi-fan, the cables are intermediate between the parallel lines of the harp shape and the more radial arrangement of the fan. Located in southern France, the Millau Viaduct is not only one of the longest multispan bridges in the world, it is also (at the time of this writing, or 2015) the tallest, the highest supporting pier exceeding the height of the Eiffel Tower.

Photo: Igor Plotnikov/123RF

Figure 6.13 Suspension bridge anchorage

This diagram illustrates the heavy concrete anchorage for the suspension cables of New York's Verazzano Narrows Bridge, which opened in 1964. The main suspension cables actually comprise numerous bundles of smaller cables, each of which in turn are made of multiple single wires. The cables are splayed within the anchorage to permit each wire bundle to have a connection within the anchorage structure. The enormous mass of the concrete anchorage counterbalances the tension from the main suspension cable.

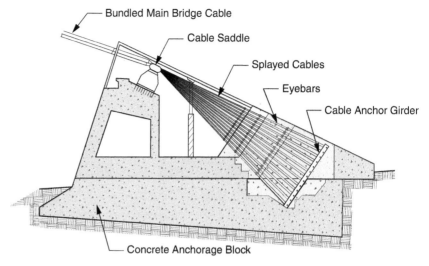

Figure 6.14 The New San Francisco-Oakland Bay Bridge East Span

Following the Loma Prieta earthquake of 1989, the San Francisco-Oakland Bay bridge (one of the world's busiest) suffered damage on its eastern span, which was constructed of rigid truss structures. Considered likely to catastrophically fail in the next major earthquake, a replacement span was called for. The final design employs the same self-anchoring principle of the Burgo Paper Mill (Figure 6.16) but on a massively larger scale. The bridge is the largest in the world to employ a self-anchored suspension cable, and is further unique in that the cable is one continuous piece. Starting at one anchorage at the east end, it passes over the tower, then loops around the western end of the span and returns back to the tower and then down to the anchorage. The two end anchorages are completely contained within the eastern end bridge structure, with the road deck acting as a massive compression structure.

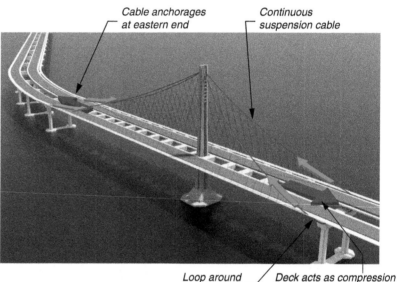

ratio leads to very tall, expensive towers. Compression in the towers, bending in the deck (often comprised of trusses), and shear in the anchorages are essential to the stability and strength of the tensile cables of suspension bridges. In *self-anchored* bridges the cables are anchored to the ends of the stiffening trusses or beams, thus compressing them, while the ends of the trusses are anchored to the piers to develop downward reactions (see Figure 6.14).

Cable bridges are essentially unidirectional spanning structures, which is appropriate for bridges. But this does not mean that the use of cables is constrained only to one direction. On the contrary, complex structures involving cables at multiple angles that cover a surface as well and other tensile structures are frequently used in building structures, and are considered in the next section.

6.2 CABLE ROOFS

The exceptional efficiency of steel cables suggests their use in the construction of large roofs such as sports stadia and transit terminals. This relatively recent twentieth century

development has brought about a number of new solutions in which tensile cables are the basic element in what may otherwise be a complex structural system. As noted above, the relative flexibility of cables is a primary consideration. There are numerous design approaches, each with different aesthetic qualities and fit with the architectural requirements, but each design solution is essentially aimed at stabilization of the cable system.

The most simple tensile roof design consists of a series of cables hanging from the tops of columns or buttresses, as though taking a number of suspension bridges and placing them side-by-side. Unlike the massive anchorages of a suspension bridge, however, the columns must be capable of significant bending resistance if the cable terminates at the column, due to the cable's large horizontal force component. An alternative design solution that prevents this column bending is to pass the suspension cable over the tops of compressive struts (i.e., columns) and anchor them to the ground, instead of terminating the cable at the column. In either design strategy, straight beams or plates then connect the parallel suspension cables, thus creating a polygonal or

Figure 6.15   The David Lawrence Convention Center in Pittsburgh, PA

This roof structure exemplifies the use of a suspension cable in a building structure. The cables are supported by anchorages similar to a suspension bridge at the low side of the building, and by stayed columns at the high side. The suspension cables carry the weight of the roof structure in a similar manner to a suspension bridge.

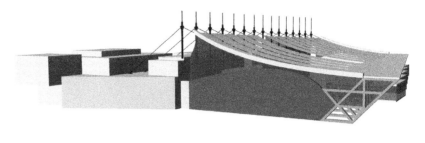

Figure 6.16   The BurgoPaper Mill in Mantua Italy

The requirements for a column-free interior space of this manufacturing plant necessitated the creation of an external structural support system. The ingenious solution by famed Italian engineer Pier Luigi Nervi employs the same structure as a suspension bridge, with two towers carrying main cables (which are actually solid rectangular bars that support suspender cables, which in turn support a trussed roof). The primary difference between this roof structure and a suspension bridge is the lack of a heavy anchorage at the ends. The horizontal force components are instead resolved by compression into the trussed roof deck, making it a self-anchoring structure.

Figure 6.17   Dulles Airport Terminal, Chantilly, Virginia

A masterpiece of modern architecture designed by famed Finnish-American architect Eero Sarrinen, the roof slab is directly supported on parallel cables anchored to the exterior angled columns. The massive weight of the columns, plus their outward tilt, acts to counterbalance the cable tension forces.

Photo courtesy of Deborah Oakley

inverted barrel roof surface (Figure 6.15). The simplicity and low cost of this suspension-bridge scheme would make it popular, but for the fact that the straight elements connecting the cables are usually light and tend to oscillate or "flutter" under the action of wind. To avoid flutter, the roofing material must be relatively heavy, or the cables must be stabilized by guy wires or stiffening trusses.

The suspension-bridge principle was directly adopted in a structure designed in1961 by Nervi and Cover for an Italian paper manufacturing plant (Figure 6.16). This structural scheme was an ideal fit with the architectural program because of the linear nature of the paper-making process. Unlike most suspension bridge designs, this is a self-anchoring structure (similar to the much later bridge of Figure 6.14), which balances the cable tension with internal compression in the roof structure rather than directing the tension to large anchorages in the ground. The reinforced concrete towers of each roof section (100 feet (30.5 m) wide and 830 feet (253 m) long) are

inclined posts supported by shorter compression struts that provide overall lateral stability to the building. The horizontal roof consists of reinforced concrete beams and slabs prefabricated on the ground and is suspended from the cables by means of wire hangers. The dead weight of this relatively heavy roof acts as a stiffening truss. It provides the necessary stabilizing force for the suspension cables, plus resists their tensile anchorage force though in-plane compression. This brilliantly simple structure permitted the economic roofing of a10,000 square feet (930 m^2) area without a single intermediate support or interior column. The peripheral walls are independent of the roof structure, and wind pressure against these exterior walls is resisted by vertical columns. Although vertical, these act more like beams than columns, since most of their structural action is in bending.

Regarded as a twentieth century architectural masterpiece, Eero Saarinen's 1962 Dulles Airport Terminal near Washington, D.C. (Figure 6.17), is similar in structural

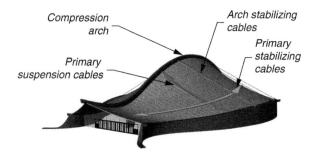

Figure 6.18   The David S. Ingalls Hockey Rink at Yale University
Designed by architect Eero Sarrinen and engineered by Fred Severud (who in engineering circles was nearly as famous as Sararinen in architecture), the central concrete rib supports suspension cables that are anchored by a curved perimeter concrete wall. Since the roof is lightweight wood, in order to stabilize the cables from uplift, *upward arcing* stabilizing cables run perpendicular to and above the suspension cables along the length of the structure. For aesthetic reasons, the central concrete rib was made very thin; however, this created problems of instability for the rib. Also visible are thus six straight cables that connect the rib to the perimeter. This was a minor consolation to architecture that deviated slightly from a more pure structural aesthetic, where the rib would be designed such that these cables would not be required.

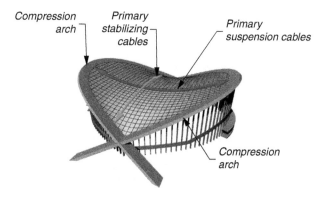

Figure 6.19   The J.S. Dornton Arena in Raleigh, North Carolina (Originally the North Carolina State Fair Livestock Judging Pavilion)
The arena uses the principle of a doubly curved structure with upward arcing suspension cables and downward arcing stabilizing cables (see Chapter 11). Engineered by Fred Severud (designer of the Ingalls Rink (Figure 6.18)), the two intersecting concrete compression rings function in a manner analogous to the principle of a director's chair: Tension in the cables prevents the rings from falling down, while this very tendency to collapse generates the tension in the roof structure, which acts in a manner similar to the seat of the analogous chair. The structure was designated an Historic Civil Engineering Landmark by the American Society of Civil Engineers in 2002.

principle to Nervi's paper mill, but consists of multiple parallel cables. The roof structure is a heavy concrete slab supported directly by the suspension cables; however, it does not resolve the tensile force of the cables. Here instead, the massive concrete pylons supporting the roof are angled outward, their dead weight and tendency to outwardly overturn counterbalances the horizontal pull of the roof cables. Designed at the start for expansion, the terminal was more than doubled in length in the late 1990s by adding additional columns, cables, and roof structure matching the original design.

A related yet different principle was used by Saarinen in the Yale University Ingalls skating rink in 1958 (Figure 6.18). In this design, the cables are suspended perpendicularly from each side of a central concrete arch, which has an inverted curvature at the ends. The cables hang with a natural downward sag, and the outer ends are anchored to the rink's heavy peripheral walls that are curved outward in plan. The roofing material is wood: Unlike Nervi's paper mill structure or the Dulles Airport Terminal, its relatively lightweight does not stabilize the cables entirely. The stabilizing is therefore accomplished by *upward curving* tension cables that pull *downward against* the primary suspension cables. The central concrete spine was too slender to be laterally stabilized by the roof structure alone, and so several additional straight cables tie it directly to the perimeter walls—a slight concession to the architectural desire of making the spine as slender as possible.

The 1952 solution by Mathew Nowicki for the roof of the Raleigh, North Carolina, arena illustrates an early implementation of the concept of interlocking countercurved cables. The building profile is dominated by two large inclined, intersecting concrete arches. In a manner similar to the Dulles Airport Terminal pylons, these massive parabolic arches simultaneously support a series of main suspension cables and provide for cable tensioning by their deadweight (Figure 6.19). Stabilization of the main downward-sagging cables is obtained by upward-curving cables at right angles to, and on top of, the primary cables—this then creates a mesh for support of the roof structure. The roofing consists of lightweight corrugated metal plates permanently anchored to the cable mesh. The concrete arches are themselves stabilized by vertical columns. By means of compression along their curvature, the arches resist the inward pull of the main cables. The surface thus defined by the cables resembles a saddle and is more stable under wind loads than a barrel-shaped roof. Notwithstanding the shape of the roof, in order to avoid flutter of the lightweight corrugated plates, it was still necessary to stabilize the roof panels by guy wires connecting a number of internal mesh points to the outer vertical columns. The interior of the structure is completely column free, and the essentially free-floating roof allows for abundant daylight through the perimeter walls.

In designing the roof for the 1956 Cilindro Municipal stadium in Montevideo, Uruguay, Leonel Viera invented an inexpensive and stable system that was well suited to cover large circular areas (Figure 6.20). In this roof, a series of radial cables connects a lower, central tension ring of steel to an outer compression ring of concrete. The outer ring resists all the tensile force of the cables through circumferential compression, and is supported by a cylindrical exterior concrete wall. With all tensile forces resolved within the roof structure itself, the exterior walls were required to only carry vertical loads, and thus were relatively thin. The roof decking consisted of a large number of prefabricated, wedge-shaped concrete slabs, which were supported on the radial cables by the hooked ends of their own reinforcing bars.

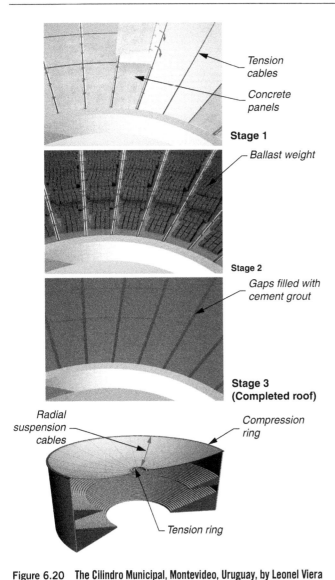

Stage 1

Stage 2

Stage 3
(Completed roof)

Figure 6.20 The Cilindro Municipal, Montevideo, Uruguay, by Leonel Viera

In order to provide permanent rigidity in the essentially simple cable-supported roof structure, Viera's construction process innovation was to hang wedge-shaped tiles on cables with gaps in between. The tiles were then covered with additional ballast weight, which caused the cables to stretch. Cement grout was filled within the gaps and the ballast was removed after curing. The lessoning weight caused the cables to shrink, but with this movement prevented by the now cemented roof, a permanent compressive prestress in the roof surface was created. The roof structure functioned for 54 years until a fire inside the stadium caused its collapse in 2010. The remaining structure was subsequently demolished.

Figure 6.21 The Oakland-Alameda County Coliseum Arena (Presently the Oracle Arena)

Designed by Skidmore, Owings and Merrill, the arena employed radially-arranged suspension cables similar to the Cilindro Municipal of Figure 6.20, the perimeter functions as a compression ring, which is supported on a diagonal gridwork of perimeter columns that provide lateral and torsional stability. Constructed in 1966, the structure has gone through substantial renovations to increase its capacity, which unfortunately detract from the clarity of its structural concept.

Photo courtesy of Deborah Oakley

The genius of this roof structure was due as much to its construction technique as to its structural concept. Here, to reduce cable instability, during construction the slabs were loaded with a ballast of bricks or sandbags after placement, which temporarily over-tensioned the cables. The resulting stretch of the cables created gaps between the wedge-shaped slabs. These radial and circumferential gaps were subsequently filled with cement mortar. Once the mortar had set, the entire roof became a monolithic concrete "dish." When the temporary ballast was taken off the dish, the cables tended to shorten, but were prevented from doing so by the monolithic concrete roof in which they were embedded (see also Section 12.7). The inverted roof was thus prestressed by the cables, and showed little tendency to flutter. Similar roofs, like that of Madison Square Garden in New York City and the Oakland-Alameda County Coliseum Arena (Figure 6.21), have been successfully and economically built in the United States and elsewhere on this principle (though with different construction technique), which Viera also applied to suspension bridge design.

The dish shape of downward sagging cable-supported roofs presents an architectural challenge of how to address the natural tendency for such shapes to collect rainwater. In the case of the roof of the Dulles Airport Terminal, three large sculptural elements dominate the interior roof surface of the terminal, which conceal vertical drainpipes. In the case of arenas, such centrally located drainpipes would be completely counter to the need for having an open clear span in the first place. The drainage of Viera-type roofs is obtained by the somewhat inelegant solution of pumping the rainwater to drain pipes located on the outer rim of the roof. In the Madison Square Garden and Oakland arenas, structures with very low-slope roofs housing offices and mechanical rooms were constructed atop the downward curving main roofs. Rainwater on these slightly sloped roofs drains naturally by gravity to the perimeter of the structures, thus avoiding the need for mechanical pumping.

Unlike the stadium or terminal roofs described above, in the design of airplane hangars it is essential to provide large front doors that slide open smoothly. These sliding doors cannot be supported by columns and are usually hung from a top beam, longer than the hangar opening. The structural problem to be solved for this type of building consists in covering a rectangular area having columns on

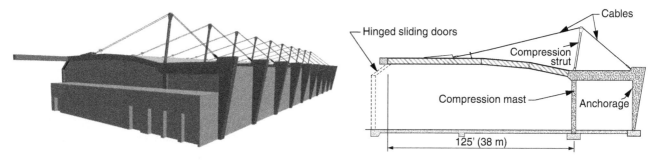

Figure 6.22 The TWA Maintenance Hangar at Philadelphia International Airport
One of the key requirements of an aircraft hangar is a large column-free space. Using the cable-stayed principle, this structure achieved a tremendous span with minimal materials use owing to the efficiency of tension cables. The rear of the structure consists of heavy concrete L-shaped moment frames (see chapter 8), which form a counterbalancing anchorage off which the front of the structure cantilevers 125 feet (37.8 m).

not more than three of its sides. A solution, realized by Ammann and Whitney in 1955 for the TWA Maintenance Hangar at the Philadelphia Airport, used a scheme in which a cantilevered span is equivalent to one half of a stayed bridge (Figure 6.22). The space between "tower" and "anchorage" was used for offices or shops, while the cable-supported cantilevered roof consisted of "folded-plate" reinforced concrete construction (see Section 10.8). The limited weight and high resistance to bending and compression of folded plates, together with the tensile strength of cables, make such structures capable of economically cantilevering as much as 125 feet (38 m).

A related solution was adopted by Tippets-Abbett-McCarthy and Stratton in roofing the 1960 Pan-American Worldport terminal at John F. Kennedy International Airport in New York. If one could imagine taking the cross section of the Philadelphia hangar, and rotating that section about a central point, this essentially describes the Pan-American terminal concept, though the form was elliptical versus circular in plan. Considered revolutionary in its day, the roof consisted of reinforced concrete plates supported by radial beams, an outer elliptical ring and an inner elliptical ring (Figure 6.23). The beams are cantilevered from the outer ring, and supported by cable stays running over compression struts at the outer ring,

Figure 6.23 The Pan Am Worldport Terminal at JFK Airport

Applying a similar principle to the cantilevered roof structure of the hangar of Figure 6.22, the roof of this building was constructed as a radial cantilever from a central ring. Considered revolutionary when it opened in 1960, the terminal's capacity was soon exceeded and subsequent additions detracted from the purity of its structural concept. The most recent owner, Delta Airlines, moved out of the terminal into a new facility, and, despite opposition by preservationists, the structure was demolished in 2013.

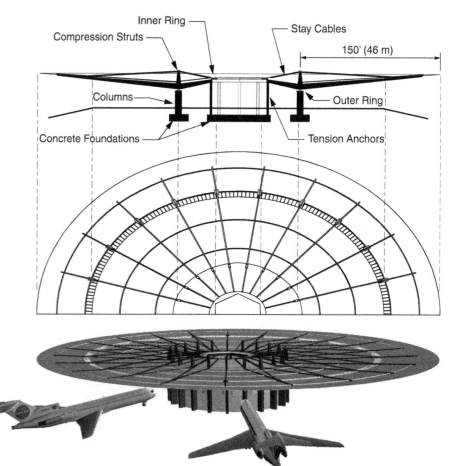

Figure 6.24   Tensegrity Sculpture by Kenneth Snelson at the Hirshhorn Museum, Washington, D.C.

Unlike traditional systems, tensegrity structures have compression members that do not contact one another. Though initially popularized by famed designer Buckminster Fuller, it was his student, Kenneth Snelson who originated the concept in 1948. In his more than 50-year-long career, Snelson's works appear in communities and galleries around the globe.

Photo courtesy of James Dymond

Figure 6.25   The Kurilpa Bridge in Brisbane, Australia

This pedestrian bridge with a main span of 394 ft (120m) is the world's longest bridge to use the principle of tensegrity as its primary structural system. The design is by the famous engineering firm Arup.

Photo courtesy of Terri Meyer Boake

and anchored at the inner ring. The outer ring is supported by compression columns; the inner ring is anchored to tension struts attached to heavy concrete blocks in the ground. The 150-feet-long cantilevers sheltered the planes during the embarkation and debarkation of the passengers.

In 1948, the sculptor Kenneth Snelson exhibited his first *Tensegrity* sculpture, consisting of a space truss with tensile members of prestressed steel cable and compressive members of steel pipe. In the principle of tensegrity, compression members separate tensile members connected at both ends, but no compression member is connected directly to any other compression member. In the following years, his sculptures assumed the dimensions of large structures (e.g., Figure 6.24). The Tensegrity principle has been successfully applied to large tensile roofs since 1955, but the structural theory of Tensegrity structures was only given in 1972 by G. Minke. More recent examples of the application of tensegrity to utilitarian structures include the 2009 Kurilpa pedestrian bridge in Melbourne Australia, designed by Ove Arup & Partners (Figure 6.25).

A noteworthy example of a tensile structure, built by means of radial tensile elements connecting inner and outer rings, is the bicycle wheel. The two sets of spokes are tensed between the tensile circular hub and the compressed circular rim, forming a structure with high "locked-in" stresses, which is stable against both in-plane and transverse loads. Since the circle is the funicular curve for a compression arch acted upon by radial forces (see Section 6.4), the entire (unloaded) wheel is a funicular prestressed structure. A groundbreaking project applying this principle is the roof of the auditorium in Utica, New York, designed by Lev Zetlin in 1955. It is based on the bicycle-wheel principle (Figure 6.26), but a wheel turned on its side. Two series of cables (with different cross sections because of different force magnitudes) connect the outer compression ring to the upper and lower rims of a central hub, consisting of two separate tension rings connected by a truss system. The cables are kept separate and posttensioned by compression struts of adjustable length. Each pair of cables can be correctly tensed by turning the turnbuckles of the struts, and the roof, covered by a series of prefabricated metal plates, is practically free of flutter because the high tension in the cables makes them more rigid than if they were freely hanging. The reader may note the similarity to the Viera-designed stadium roof; however, in the Utica design the upward-bowed stabilizing cables are stressed against the lower suspension cables, providing the stabilization that the Viera roof achieves by deadweight. This design also eliminates the issue of roof drainage because of the upward bow of the roof, providing for natural drainage to the outer rim.

David Geiger has designed tensegrity *domes* spanning up to 800 feet (244 m), the first of which for St. Petersburg, Florida, covers a circular stadium with a seating capacity of 43,000 people and a diameter of 680 feet (207 m). Its structure consists of radial trusses with diagonals and chords of steel cables, prestressed against verticals of steel piping, which span between an inner tensile ring of steel and an

Figure 6.26 The Utica Memorial Auditorium

Using the circular suspension cable concept similar to the roofs of figures 6.20 and 6.21, engineer Lev Zetlin added *upward* stabilizing cables separated from the suspension cables by compression struts. Similar to spokes on a bicycle wheel, both the top and bottom surfaces thus comprise tensile cables. An advantage of this system is that the upper surface has a natural outward bow that sheds rainwater to the perimeter. The structure was the first of its kind when it opened in 1960, and was designated an Historic Civil Engineering Landmark by the American Society of Civil Engineers in 2011.

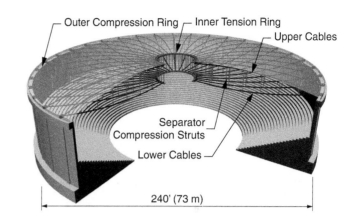

Figure 6.27 Tropicana Field, St. Petersburg, Florida (Originally the Florida Suncoast Dome)

The roof structure of this indoor baseball stadium was designed by engineer David Geiger and employed a variant of the tensegrity principle known as a "cable dome" with some similarities to the Utica Auditorium of Figure 6.26. In this structure, concentric tension rings increase in elevation toward the center, and each carry compression masts that support tension cables along concentric radial ridges. Diagonal cables connecting to each ring stabilize the masts, and these attachment points create circumferential radial valleys. The roof is clad in translucent fabric. A larger version of the dome structure was created for the 1988 Olympic Games in Seoul, South Korea.

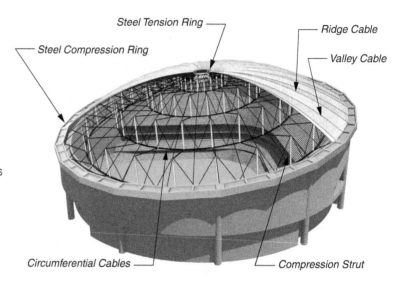

outer compressive ring of reinforced concrete (Figure 6.27). The vertical pipes, sheathing the continuous cables, are supported by the tensed diagonals, and kept vertical by concentric steel rings connected to their foot and inclined stays connected to their top of concentric steel rings. Radial-tensed cables, running between the compression and tension rings along the vertical pipes' bottoms, constitute the lower chords of the trusses. Similar radial cables (the "ridge cables") running along their tops constitute the upper chords of the trusses. The prestress in the radial ridge cables increases the tension in the lower chords due to the roof loads, and is high enough to entirely cancel the compression in the upper chords due to these loads. A third set of prestressed radial cables (the "valley cables") runs from the compression to the tension ring along the "valleys" formed by the intersections of the inclined stays at the concentric steel rings. All the components of the structure can be bought "off the shelf." The undulating roof surface is a membrane of silicone fiberglass fabric wrapped over and attached to the ridge and valley cables. Since the tension ring is at a higher level than the compression ring, these roofs are shallow tensile domes similar in principle to "bicycle-wheel" roofs. The 1992 Georgia Dome by Weidlinger's Matthys Levy uses a similar principle, and at a span of just

under 800 feet (244 m) it is presently (2015) the largest cable-supported domed stadium in the world. The 2012 La Plata dome (Figure 6.28) is the latest iteration of the Levy-type "Tenstar Dome."

Pneumatic roofs present one of the most interesting applications of cables to the reinforcing and stiffening of membranes. They consist of air-supported or air-inflated plastic fabrics stretched over a network of cables and can span hundreds of feet. Geiger and Levy-style tensegrity domes can also be considered as membrane roofs supported by trusses rather than air pressure: Elimination of the fan system and revolving doors (needed to maintain air pressure) reduces the cost of these roofs in comparison with that of pure pneumatic roofs. Chapter 11 covers these membrane structural types in detail.

The tensile solutions mentioned above are just a few of the most recently adopted to roof large areas. Whereas the largest span covered by a compressive roof to date (2015) is the 715 feet (218 m) C.N.I.T. double concrete shell in Paris (see Section 12.7), tensile roofs can easily span much larger distances. In view of the exceptional potentialities of cable roofs, it is to be expected that their use will increase with the increasing dimensions of future areas to be covered.

Figure 6.28   La Plata Dome

The La Plata Dome in Argentina is an extension of the Geiger principle by engineer Matthys Levy of Weidlinger Associates. Though similar in principle to the Geiger dome, the *Tenstar Dome* uses triangulated regions between cables that enable it to be formed around irregular nonround arena shapes.

Photo courtesy of Federico García Zúñiga

6.3 TRUSSES

At the beginning of this chapter, a flexible cable supporting a load at midspan was found to be a pure tensile structure (figs. 6.3 and 6.29a). Structural stability of this basic cable system requires that the ends be fixed to supports that prevent movement of the cables. If one support, for example, were on a roller, the tensile pull on the cable would cause an immediate collapse by horizontal movement of the roller (Figure 6.29b). If instead a wooden strut

is placed between the two supports, this horizontal movement will be prevented, and the tensile force will be balanced by compression within the strut (fig 6.29c). This is the simplest form of the structural type known as a *truss*. The triangular shape described by the cables and compression strut is referred to as a *panel*—larger trusses described below are multipanel trusses. The outer structural elements of the panels are referred to as *chords* of a truss. In these initial examples, the chords are horizontal, but sloping chords are common as well.

Consider now the structure that is created by *inverting* the original cable structure and strengthening its inclined sides to make them capable of resisting compression. The "negative sag," or *rise* changes the nature of the stresses, and the inverted cable becomes a pure compressive structure (Figure 6.30).

The load at the top of the truss is channeled by the compressed struts to the supports, which are acted upon by downward forces equal to one-half the load and by *outward* horizontal *thrusts*. By inverting the structure, therefore, all of the forces are exactly equal in magnitude, but opposite in direction, to the original tensile truss. The horizontal thrust of the inverted truss can be absorbed either by compression in buttresses of a material such as stone, masonry, or concrete,

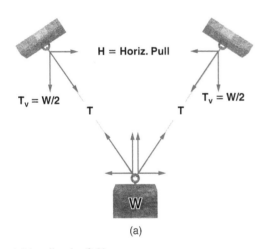

(a)

Figure 6.29a   Hanging Cable

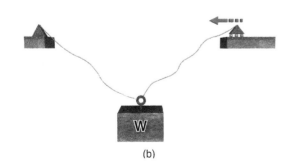

(b)

Figure 6.29b   Weight on cable with roller support

If the rigid anchorage of the hanging cable is replaced on one side by a roller connection, a collapse mechanism results due to the tension of the cable. It is an inherently unstable structure.

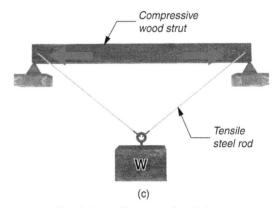

(c)

Figure 6.29c   Tensile truss with compressive strut

The horizontal pull in the hanging cable is resisted by compression developed in the strut, instead of transferring this force to the support. This is the most simple form of a truss, consisting of a single panel.

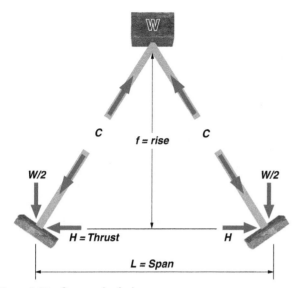

Figure 6.30   Compressive Arch
Inverting the hanging cable of figure 6.28a results in a mirror image of the forces, where previously tensile members become compressive. The inward pull from the cable likewise now becomes an outward thrust against the support. If the rise of the two structures is the same, the forces will be exactly equal in magnitude, but opposite in direction, for the same applied load.

or by tension in an element such as a wood or a steel *tie-rod* (see Figure 3.15). Such elementary trusses, constructed of wood compression members with iron tie-rods, were built in the Middle Ages to support the roofs of churches, and there is reason to believe that the Greek temples were covered by wooden structures of similar design. The introduction of iron and later steel allowed the use of very slender tensile members. As truss spans became larger, it was found practical to hang the tie-rod from the top of the truss to eliminate the large sag of this relatively flexible element (see Figure 3.15).

Figure 6.31a   Two Disconnected Truss Panels
Two truss panels connected at a central joint, but lacking a bottom bar member will create a collapse mechanism and cannot support a load.

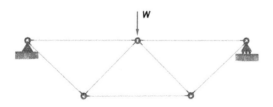

Figure 6.31b   Three Panel Truss
The same truss above with the addition of a bottom member creates a complete and stable truss system. The bottom member becomes a tensile tie preventing the outward movement of the two lower joints.

Trusses capable of spanning large distances by means of members that experience *only* tension and compression forces (with no shear or bending behavior like beams—see Chapter 7) are obtained by combining multiple elementary single-panel triangular trusses. For example, if two of the most basic triangular trusses are joined at an upper *joint* (Figure 6.31a), they cannot support a load unless a tensile bar prevents the vertices at the bottom from moving apart. The addition of a tie rod at this location thus creates a larger truss, now in the form of a three-panel truss, capable of spanning a greater distance (Figure 6.31b). By adding more single-panel trusses and connecting them with tension ties, ever larger trusses can be constructed.

The conceptual development of a more complex multi-panel truss can be understood through the sequence of Figures 6.32a–6.32c. Starting in Figure 6.32a with a single-panel truss supporting one concentrated load at the center, the one panel truss is supported at the two upper end joints. Note that the center vertical member is drawn as a dotted line, indicating that it is not carrying any force *under this loading*. This member is connected into the horizontal top chord of the single panel, and since these chord members are completely horizontal they are incapable of resisting a vertical force—only another vertical or sloping member connected to this joint would be capable of this. Such elements that carry no load are referred to as *zero*

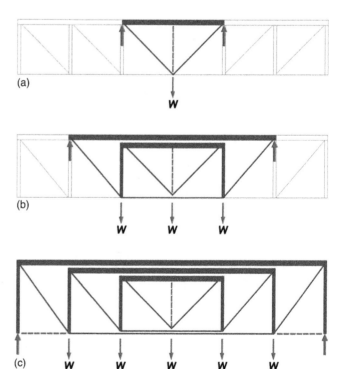

Figure 6.32   Conceptual Development of a Multipanel Truss
By adding successive cables and struts to support smaller truss units, increasingly longer span trusses are created. The additional elements provided to support the lower-ordered elements reflect the magnitude of forces at the varying truss locations, with the highest forces in the outer chords at the center of the span, and the highest forces in the diagonal cables in the outer panels.

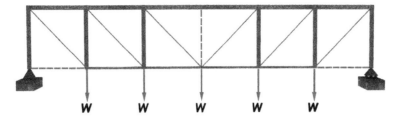

Figure 6.33   Truss with Tension Diagonals

Developed in 1844 by father and son designers Caleb and Thomas Pratt, the diagonal members of the truss are in tension while the vertical members are in compression. This arrangement is well suited for iron and steel structures to take advantage of their high tensile capacity, and all-steel Pratt trusses became a standard design by the early 20th century. The dotted lines indicate theoretical zero-stress members when the bottom chord is loaded equally at each panel point. In practice, only the center vertical member can safely be eliminated, because the two horizontal members are required for overall stability of the truss.

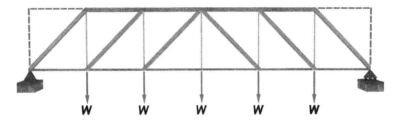

Figure 6.34   Truss with Compression Diagonals

Inverting the direction of the diagonals of the Pratt truss in Figure 6.33 will also cause the internal forces to invert. Providing the same equal load at each panel point as the Pratt truss, the diagonal members switch from tension to compression members, and the shorter vertical members switch from compression to tension. Created by the designer William Howe in 1840, the construction typically combined heavy wood diagonal members with light iron vertical members, which simplified construction compared to the all-wooden trusses of the era. The dotted lines represent theoretical zero-force members. If all loads are applied at the lower panel points, the horizontal and vertical end members may be eliminated from the structure altogether since they carry no force. This is the form of many railway and highway bridges of the late 19th and early 20th centuries.

force members. This vertical member would be necessary, however, to support a load placed at the upper joint, making the formerly zero-force member a compression member.

If the smaller truss of Figure 6.32a is itself supported by a larger cable and strut structure (Figure 6.32b), then a longer-span truss of six panels is created. Note that the "doubling-up" of the compression member of the top chord is reflective of the increase in force in this member. If the six-panel truss is in turn supported yet by another cable and strut structure (Figure 6.32c), then a still longer truss consisting of 12 panels is created. Again, the additional horizontal top strut and bottom cable elements in the larger structure are reflective of the increase in force in the members toward the center of the span. A complete *Pratt* truss is created in this conceptual process as illustrated in Figure 6.33.

The bars of trusses are considered *hinged* at their connection or panel points. Because they are hinged, they are free to rotate at their ends and do not develop bending if properly loaded. If, as is desirable, the loads on the truss are applied only at the panel points, and the dead load of the truss members is negligible, all the truss members are only tensed or compressed. When loads are applied *between* the panel points and the dead load is not negligible, the truss bars develop some bending also (see Section 7.4). Therefore, for maximum efficiency, the heaviest loads should be located only at panel points. Numerous structural failures have occurred through loading that

created bending forces on members that were designed for only tension or compression.

In Figure 6.33, we can see that merely reversing the inclination of the interior diagonal members causes these bars to work in compression, and the vertical members in tension, much as the single panel compressive arch of Figure 6.30 is a mirror of a hanging cable. The reader may wish to diagram the conceptual assembly of this truss type in the same manner as the diagrams of Figure 6.32a-c to follow its load-channeling mechanism. In this truss—*under this specific loading*—the end verticals and last upper chord bars are unnecessary and are thus theoretical zero-force members. Similarly, the end verticals and the last lower chord bars would have been unstressed (and thus redundant) if the truss of Figure 6.33 had been supported at the upper left and right joints. The scheme of Figure 6.34 is commonly used in highway bridges, and that of Figure 6.33 in railroad bridges, where the truss is often below the roadway. Additional insight on truss behavior may be gained by analogy with beam behavior, as shown in Section 7.2.

Upper chord members, as well as interior verticals and diagonals, may buckle under compression unless properly designed. One of the first studies on the buckling of trussed bridge structures was prompted by the failure in buckling of a Russian railroad bridge at the end of the nineteenth century. Since, moreover, moving loads may produce *either* tension *or* compression in the *same member*, depending on the

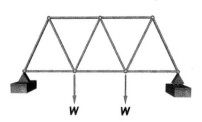

Figure 6.35   A Warren Truss

Named for its inventor, James Warren, who patented it in 1848. Well suited for prefabricated construction, there are no vertical members and all diagonals are of the same length. The diagonals must be designed for the maximum compression force, however, since a heavy moving load may cause a tension member to become compressive, or vice-versa.

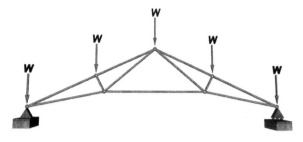

Figure 6.36   A Gable Truss

Commonly used in residential construction, this truss form provides for a vaulted ceiling. The angle between the upper and lower chords at the supports is limited by large forces developing as the angle becomes smaller. At small angles, very high shear forces develop at the connection that must be transferred between each wood member. Typically, bolting and/or a variety of metal connector pieces are used in wood construction to augment the wood itself.

Figure 6.37   Truss gusset plate

A common type of connector in a truss is known as a *gusset plate*. A gusset plate is a means of physically connecting members while transferring large forces between them. They may be used in a decorative manner, such as this heavy timber truss shown being fabricated, or may be hidden within the member thickness in so-called *knife plate connections*. Only the bolts are visible on the outside of the members in the latter case, and sometimes even these are countersunk and the holes plugged for a completely smooth appearance.

Photo courtesy of Deborah Oakley

load location, some trusses are built with both tension and compression diagonals, so that loads may be always carried by tension diagonals with purposely slender compression diagonals that are simply considered inactive when buckled.

The combinations of tension and compression bars capable of producing practical trusses are extremely varied. Figures 6.35 and 6.36 illustrate two of the many forms of trusses possible.

The bars of a truss are joined by being bolted or welded to a "gusset plate" at their panel points (Figure 6.37). In either case, the restraint against relative rotation produced by the gusset plates transforms the truss bars from pure tension or compression members into elements developing a minor amount of additional bending and shear stresses. These so-called *secondary stresses* are considered in Section 7.4.

Trusses are used in bridge design to span hundreds of feet between supports. They may be cantilevered from piers and in turn carry other simply supported trusses (Figure 6.38). Bridge trusses with curved top chords behave very much like suspension bridges (Figure 6.39). Parallel trusses are commonly used in bridges and in steel design to cover large halls (Figure 6.40). *Open web joists* are "off the shelf" light steel

Figure 6.38   The Forth Railway Bridge, at the Firth of Forth, Scotland

This bridge, consisting of two main spans of 1,710 feet (520 m) each, was inaugurated in 1890. Designed by engineers Sir Benjamin Baker and Sir John Fowler, it is constructed as three double-cantilever spans, each supported at their centers by enormous stone and wood caissons sunken into the riverbed. The two outermost cantilever arms are separated from the central structure by much shorter and smaller trusses, simply supported by the cantilever arms of the center and outer structures. The span is thus increased without adding even more depth to the already enormous main cantilever trusses, allowing ships to pass freely. It remains the second longest cantilever bridge in the world even after more than 125 years of continuous service.

Photo: Ramonespelt/Fotolia

Figure 6.39   The Macombs Dam Bridge, Spanning the Harlem River in New York

Opened in 1895, the Macombs bridge is a double cantilever supported from a central pivot point, referred to as a *swing bridge*, since the entire bridge can pivot about the central support to allow ships to pass on either side. The curving shape of the top chord reflects the nature of the increasing forces toward the central support, and in fact bears resemblance to a cable suspension bridge design.

Photo courtesy of Davie Fried

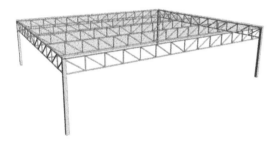

Figure 6.40   A Parallel Chord Trussed Roof

Rows of parallel chord trusses are a common and practical method of creating roofs or floors over long spans such as auditoriums, concert halls, stadiums, and so on. Smaller spaces may utilized preengineered/prefabricated members selected from a catalog, based on span and load requirements. Larger spaces (and thus longer clear spans) commonly feature custom fabricated trusses designed for a single specific project.

Figure 6.41   An Open-Web Steel Joist

Basically a premanufactured truss, the open-web steel joist (OWSJ) is a mainstay of efficient modern construction. Fabricated in a range of standard depths, lengths, and capacities, designers can select the required member type from a catalog to suit the span and load capacity requirements of a specific design.

trusses used to span small distances, either as roof or as floor structures (Figure 6.41). Vertical trusses are used in high-rise steel buildings to stiffen their frames against wind and earthquake forces (see Section 8.3).

Although not as popular today as they were in the 1800s, at the beginning of the structural steel era, trusses are still to this day one of the most essential components of large structures.

6.4 FUNICULAR ARCHES

Returning once more to the pure funicular forms introduced at the start of this chapter to consider the inverse of the tensile form: the funicular arch. The parabolic shape assumed by a tensed cable carrying loads uniformly distributed horizontally may be inverted to give the ideal shape for an arch developing only compression under this type of load (Figure 6.42). The arch is an essentially

compressive structure. It was developed in the shape of a half-circle by the Romans (the greatest road builders of antiquity) to span large distances. An aqueduct supported by a series of semicircular arches survives today in Segovia, Spain (Figure 6.43). Arches are also used in a variety of shapes to span smaller distances. The arch is one of the basic structural elements of all types of architecture.

The ideal shape of an arch capable of supporting a given set of loads by means of simple compression may always be found as the overturned shape of the funicular polygon for the corresponding tension structure: It is by this method that the Spanish architect Antonio Gaudi determined the form of the arches in the Church of the Sacred Family in Barcelona and many of his other structures (see Figure 14.15).

The funicular polygon for a set of equal and equally spaced loads *converging toward* or *diverging from* a common point is a regular polygon, centered about this point (Figure 6.44a). At the limit, when the loads become infinite

Figure 6.42   The Roosevelt Lake Bridge

Located northeast of Phoenix, Arizona, the graceful arch of the bridge represents a nearly perfect funicular compression form to carry the roadway suspended below it.

Photo courtesy of Deborah Oakley

Figure 6.43   Roman Aqueduct

The Roman Aqueduct in Segovia, Spain, is still standing after two thousand years, owing to the efficiency and strength of carrying loads compressively through stone arches.

Photo: Mrallen/Fotolia

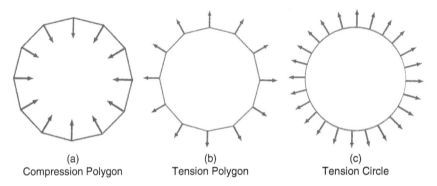

| (a) | (b) | (c) |
| Compression Polygon | Tension Polygon | Tension Circle |

Figure 6.44   Funicular polygons for equal radial load

The ideal form for a radial load, whether directed inward in compression (6.44a) or outward in tension (6.44b), begins to approach a circle as the number of loads increases (6.44c). Such forms are the basis for tension and compression rings employed in structures such as those illustrated in Figures 6.20, 6.21,and 6.27. Perimeter rings will be in compression for hanging cable structures to resist their inward pull, while the center rings will be in tension for such structures. In rigid domelike structures, the forces will be reversed, with the outer ring acting in tension and the inner ring in compression.

in number and infinitesimally small, the funicular polygon becomes the funicular curve for a *radial inward pressure* or outward *suction,* and is a circle (Figure 6.44c). Since arches can be funicular for only one specific set of loads, the circle cannot be the funicular curve for a vertical load uniformly distributed either along the horizontal or along the circular arch. Hence, the Roman arch develops some bending and, although a compression structure, is not *uniformly* compressed across its depth.

The shape of a masonry arch is usually chosen to be the funicular of the dead load. But whenever an arch is used in a bridge, a variety of moving loads must be assumed to travel over the arch, and a state of stress other than simple compression is bound to develop: The arch also develops some bending stresses. A cable can carry any set of loads in tension by changing shape; a rigid arch cannot change shape and, hence, cannot be funicular for all the loads it is supposed to carry. The stability of the arch implies lack of adaptability. The relative importance of bending stresses in arches is considered in Section 8.4.

The channeling of vertical loads along a curve by means of compression is today such an elementary concept that it is hard to realize how slow its evolution has been. Practically unknown to the Greeks, it produced the daring and lovely lines of the Gothic cathedrals as well as those of modern bridges (see Section 8.4). The ultimate consequences of the use of curved structural elements are even more recent and interesting: They are considered in Chapters 11 and 12.

KEY IDEAS DEVELOPED IN THIS CHAPTER

- Cables are flexible structures that support loads in pure tension.
- Because of their flexibility, cables change their shape under moving loads.
- A cable hung from two supports and loaded by a concentrated force in the middle will form an inverted triangular shape; the sides of which will be straight lines.

- The inclined reactions at the support will match the cable angle, and can be described by two components: one vertical and one horizontal. The vertical components balance the vertical load by carrying a portion of the load on each end, while the horizontal components prevent the cable from collapsing inward.
- The horizontal components are always equal and opposite to each other, and the magnitude is inversely proportional to the sag of the cable. The smaller the sag of the cable, the greater will be the horizontal force components at the supports.
- When additional loads are applied, the cable readjusts with straight lines between loads and supports. The vertical reactions add up to the total load; the horizontal components will increase but remain equal and opposite to each other.
- Under an evenly distributed vertical load, the cable forms a smooth parabolic shape.
- Suspension bridge cables are parabolic.
- The cable ends of suspension bridges are typically embedded into heavy buttresses (*anchorages*) in order to supply the necessary horizontal reaction. Because of the inherent instability of cables, stiffening trusses are used to stabilize suspension bridge cables.
- Cables supporting roofs are laid parallel to each other or are arranged radially. In either case, the cable ends are anchored or are held by a compression ring.
- When a cable that supports a load at midspan is turned upside down, it becomes a compressive structure. Its sloping sides are in pure compression, and horizontal buttress reactions prevent the ends from moving.
- If a third horizontal, tension member is added to replace the horizontal support reactions, a basic three-bar truss panel is created.
- Several panels connected to each other may form larger and more complicated trusses. In ideal trusses, the members are in either pure tension or in pure compression.
- Cables loaded at several points when turned upside down become arches, with their members in compression.
- Stone and concrete arches that can support compression have been built since Roman times as bridges, aqueducts, and cathedrals.
- The longest-spanning arch bridges of today are made of structural steel elements
- Circular arches such as bicycle wheels and roofs of similar shape are in circumferential compression and are loaded by radial tension members or even membranes.

QUESTIONS AND EXERCISES

1. Using a piece of string approximately 12" (30 cm) long, hold the string between the thumb and forefinger of each hand while suspending a small weight of about one pound (4.5 N) from the string (a sturdy coffee mug will work for this). Observe the tension felt in the string through the fingers. Notice how the force decreases as the hands are brought together and the string gets deeper, and how the force increases as the hands move apart. Is it possible to make the string completely flat with enough tension? Why or why not?

2. Tie the weight (e.g., a coffee mug) at a point not in the middle at the string (it needs to be tied to prevent sliding on the string). Lift the string again with two hands and feel the different forces on your hands. Which one is greater? Why?

3. Take a flexible ruler, bend it into a shallow arch end place the ends between two heavy books. Press down on top of the arch. Why do the books move?

4. Observe spanning systems in buildings and bridges. This is ideally done in person, but many structural images can now be found on the Internet. Which structures use suspension cables, which use stay cables, which are trusses, and which use arches? Is one type more common than the others? How does it appear that the spanning distance or usage of the structure impacts the choice of the structural system?

FURTHER READING

Allen, Edward and Zalewski, Waclaw. *Form and Forces: Designing Efficient, Expressive Structures*. John Wiley & Sons, Inc. 2009. (Chapters 1 and 2)

Sandaker, Bjorn N., Eggen, Arne P., and Cruvellier, Mark R. *The Structural Basis of Architecture*, 2nd Edition. Routledge. 2011. (Chapter 10)

Schodek, Daniel and Bechthold, Martin. *Structures*, 6th Edition. Pearson Prentice Hall. 2008. (Chapter 5)

BEAMS

INTRODUCTION

As discussed in Chapter six, pure tension members are the most efficient structural type for carrying loads, since the entire cross section is fully utilized and there is no possibility of a buckling failure. For this reason, they are frequently employed in long-span bridges and roofs where efficiency is a critical design factor. This efficiency comes with a necessary trade-off, however: In order for a cable to transfer a vertical load across a horizontal distance, a sag is required that necessitates (for bridges) tall towers, or (for roofs) an upward or downward curvature. For those structural types, these are acceptable trade-offs. In the case of floors, though, these limitations are typically restrictive enough that cable structures do not suit this most common architectural function.

Most loads in building structures are vertical, and the majority of usable surfaces must be horizontal; furthermore, sloping members will tend to interfere with the interior clear space, which is typically undesirable. The structural requirements of cables are thus at odds with proper functioning for the floors of most buildings, and so another structural element more appropriate to the task is required: one that can transfer vertical loads across a horizontal distance within an essentially flat plane. Beams are such elements. As will be demonstrated, they are far less efficient than either funicular tension or compression members, yet because they commonly feature two flat surfaces, they are well suited for the construction of floors and ceilings. Their internal structural mechanism involves a *combination* of bending and shear. Simple cantilevered beams effectively demonstrate these principles most clearly and so will be considered first.

7.1 CANTILEVERED BEAMS

7.1.1 Basic Observations

The reader is encouraged to follow along in this simple experiment, which requires a thin steel or plastic ruler. Clasping it in its middle with one hand, while pressing down with a finger on the free end, the following observations can be made:

a. For the same applied load, the deflection of the loaded tip of the cantilevered ruler increases rapidly with an increase of the cantilevered length; in fact, doubling the cantilever length, by holding the ruler at the very end does not merely increase the deflection two times, or even four times, but increases it by a factor of *eight*—which is the cube of the ratio between the original and increased length (Figure 7.1). It may seem incredible, but increasing the length by three times would thus increase the deflection by *twenty-seven times*.

b. The deflection at the tip is much larger when the width of the ruler is kept horizontal than when it is kept vertical (Figure 7.2). Accurate measurements show that the

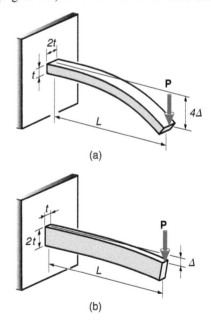

(a)

(b)

Figure 7.2 Influence of depth and width on cantilever deflection
The orientation of a member influences its deflection. The amount of deflection is inversely proportional to both the width and the cube of the depth.

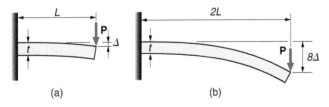

(a)   (b)

Figure 7.1 Influence of length on cantilever deflection
The length of a cantilever has a profound impact on deflection. Because the amount of deflection varies in direct proportion to the cube of the length, doubling the length will increase the deflection by eight times.

deflection is *inversely* proportional to the horizontal side of the ruler's cross section, and to the *cube* of its vertical side. If the ruler has a width of two units and a thickness of one unit, in the first case the deflection is inversely proportional to 2 times 1 cubed (i.e., 2); in the second, to 1 times 2 cubed (i.e., 8): the deflections are four times larger in the first case than in the second.

c. The deflections under the same load of two identical rulers made of different materials, such as steel and aluminum, are inversely proportional to the elastic moduli of the materials. The aluminum ruler deflects three times as much as the steel ruler (Figure 7.3) because its elastic modulus is one-third that of steel (see Chapter 3).

d. Finally, the tip deflection of the cantilevered ruler increases as the load moves from the support toward the tip of the cantilever (Figure 7.4).

7.1.2 The Effect of Beam Shape on Bending

Another geometrical factor influences the magnitude of the beam deflections and merits detailed consideration. This is an influence unique to bending that does not apply to funicular tension elements: namely the *shape* of the beam's cross section. In the case of a true funicular tension element, the cross section is uniformly stressed in tension. For a given cross-sectional area, whether it is round, square, or even an irregular shape has no real bearing on the level of stress (ignoring secondary effects such as localized deformation) (Figure 7.5). In bending, however, the shape of a cross section has a significant influence on the capacity of a member and, as has been noted, simultaneous compression and tension stresses exist in different portions of the member's depth. First, a beam of rectangular cross section will be considered.

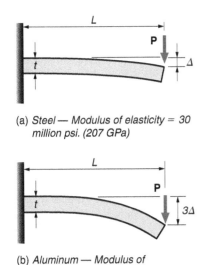

(a) *Steel — Modulus of elasticity = 30 million psi. (207 GPa)*

(b) *Aluminum — Modulus of elasticity = 10 million psi (69 GPa)*

Figure 7.3   Influence of material on cantilever deflection
Deflection will vary inversely with material stiffness. Aluminum has only one-third the stiffness of steel; consequentially, it will deflect three times more than steel.

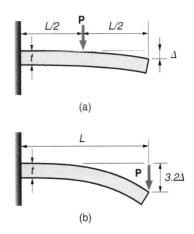

(a)

(b)

Figure 7.4   Influence of load location on cantilever deflection
The moment arm of a load increases with distance from the support, and so the magnitude of deflection will increase.

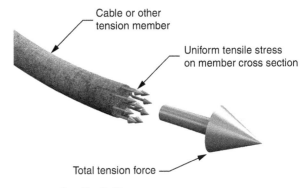

Figure 7.5   Pure Tensile Stress
A tensile member will experience uniform stress under a load, which is to say that each "fiber" of its structure is being pulled upon equally. "Fiber" in this case may in fact be actual plant fiber or animal hair, or be a metaphor when completely homogeneous/nonfibrous materials such as steel are used. These do not have individual fiber-like materials such as wood, although in the case of cables many individual strands may be bundled together to make larger and stronger cables.

The deflection of the loaded tip of the cantilever is due to the deformation of the originally straight beam into a slightly curved element. Such deformation requires the elongation of the upper fibers and the shortening of the lower fibers of the beam, and produces the state of stress defined in Section 5.5 as *simple bending*. Within the elastic range of a given material, the bending stresses in a cantilever beam vary linearly from a maximum tensile value at the top fibers to a maximum compressive value at the bottom fibers, and vanish at the middle fibers, the so-called *neutral surface* of the beam (Figures 7.6a and 7.6b). The neutral surface intersects the beam's cross section at a line referred to as the neutral axis, which is positioned on the cross-sectional shape at its balance point, known as the centroid.

The internal forces in the cantilever can be observed to be in a state of rotational equilibrium. The load, through its lever arm about any point along the length of the cantilever, tends to rotate the beam clockwise as pictured. This clockwise rotation is equilibrated by the internal bending stresses, through the lever arm of their resultants *T* and *C* (which equals two-thirds

Figure 7.6a   Simple bending of a cantilever beam

The external load, P, on the cantilever end induces a rotational moment (M_{Acting}) about the supported end on the left, which provides a countering moment ($M_{Resisting}$) that is required for rotational equilibrium. The net result is tension on the top surface of the beam that transitions to compression on the bottom surface. The internal stresses are maximum on the outer surfaces with a so-called *neutral surface* at the mid-depth where the stress changes from tension to compression. The neutral surface coincides with the centroid of the beam cross section, and on the cross section is referred to as the *neutral axis*.

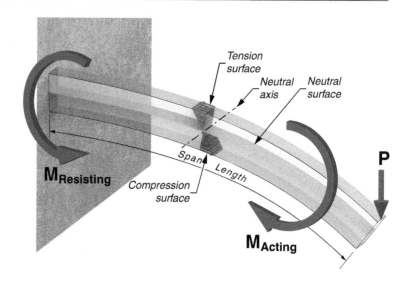

Figure 7.6b   Bending stress resultants

Stressed within the elastic range of a material, the shape of the stress block is closely approximated by a triangle, and the centroid of the total force on the stress block is therefore one-third the distance from the large end. This means that the effective moment arm distance is two-third the depth of the beam, and is thus the internal moment arm available to resist bending forces on the beam. The internal tension and compression forces form a couple that is equal and opposite in magnitude to the moment induced by the applied force, P.

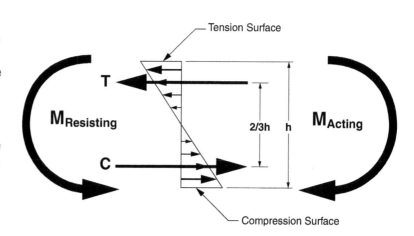

of the beam depth *h*) (Figure 7.6b), which develop a counter-clockwise rotation. The deeper the beam, the larger is the lever arm of the bending stresses. Thus, for a given value of the allowable stress at the top and bottom fibers, the deeper beam is capable of equilibrating a larger load. Since stresses are constant across the width of the beam, an increase in depth is much more profitable in resisting bending than an increase in width: how profitable is shown by the decrease in deflection with the *cube* of the depth (see Figure 7.2).

The behavior described above is typical of beams with a length at least *twice* their depth. *Deep beams*, that is, beams with a length less than twice their depth, support the load mostly by shear and do not exhibit a linear variation of stress across their depth.

From the viewpoint of use of material, a beam of rectangular cross section is very inefficient in bending, since most of its fibers are not fully stressed to the allowable stress; usually, only its top and bottom fibers at one point along its span reach maximum values, while all other fibers are unavoidably understressed as illustrated in Figure 7.6. This cross-sectional inefficiency can be remedied by relocating most of the beam material near the top and bottom of the beam. The cross section of an I-beam (or S section) or of

a "wide flange" (or "W section" as the most commonly used rolled steel sections in the United States are called), with most of the material in flanges connected by a thin web, has this characteristic (Figure 7.7). The large areas of the flanges develop maximum tensile and compressive stresses up to the allowable limit with a lever arm almost equal to the full depth of the beam, and contribute substantially to the bending action. The smaller web area is nearer the neutral axis and contributes negligibly to bending. If *all* the material of a beam of rectangular cross section could be shifted toward its top and bottom, creating an "ideal I-section" without a web, its rigidity would increase by a factor of three. This means that deflections and stresses would decrease by a factor of three or, to put it the other way around, that the beam could support three times as much load with the same allowable stress and deflection (Figure 7.8).

Splitting the web of an I-beam in two and shifting the material laterally produces another cross section just as efficient in bending but far more efficient in torsion, the "box section" (Fig 7.9). Box sections may be constructed of multiple separate plates in a *built-up* section, from face-to-face channel sections welded together, or made from a single long plate folded into a square or rectangular shape and

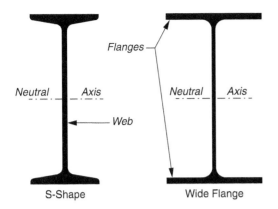

Figure 7.7   Rolled steel sections

These two structural forms are approximately the same weight per unit length; however, the wide flange section has been optimized for greater bending resistance through its geometry. The S-Section beam is a much older form and is more closely resembling what is referred to as an I-Beam in the vernacular.

welded closed along its length (known as *Hollow Structural Sections (HSS)*). Built-up sections are normally custom designed for a specific use, whereas a wide variety of standard off-the-shelf shapes of prefabricated tubular sections are available (and can be found in references such as manufacturer's catalogs and in the American Institute of Steel Construction's (AISC) *Manual of Steel Construction*).

The symmetrical I-section and the box section are typical forms when constructed of materials with equal strength in tension and compression; different—*asymmetric*—shapes become more logical when these strengths differ. For example, the compressive strength of concrete is approximately one-tenth of the tensile strength of structural steel; hence, in a reinforced concrete beam the area of the compressed concrete must be much larger than the area of the tensed steel. This leads to a T-section, in which the wide bar of the horizontal compression flange is made of concrete, and the narrow area of steel is concentrated in the web. In the case of cantilevered beams, the T-beam is necessarily inverted: The flange is at the bottom and the web at the top of the beam, since the compressed fibers are at the bottom of the beam (Figure 7.10). The reinforced-concrete T-beam acts very much like an I-beam in which the area of one of the flanges is shrunken because its material is stronger, but T-beams of steel, obtained by connecting two L-shaped sections or cutting an I-section in two, are also used, mainly to facilitate connections with other beams (Figure 7.11).

The bending stiffness of a cross-sectional shape is gauged by a property called its "*moment of inertia*." Though it can initially be unclear for those new to the subject, this property must not be confused with the *moment of a force*, described in Chapter 4. While the area of a cross section and the balance point of the area (the *centroid*) are easy to visualize, moment of inertia is a more abstract concept that

Figure 7.8   Variation of strength with cross-section shape

An ideal I-Shaped section distributes the cross-sectional area as far away from the neutral axis as possible, resulting in a significant improvement in strength by increasing the internal moment couple. In reality, a web is required to transfer shear between the top and bottom flanges (see Figure 7.20).

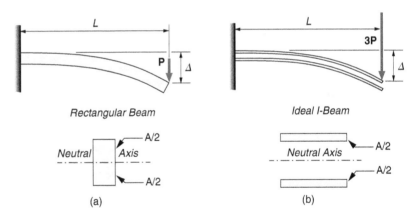

Figure 7.9   Box Sections

Box sections are commonly made from (a) built-up sheets of steel plate, or (b) from channels welded face-to-face, or (c) as tubular sections that are made from a single sheet of steel that is folded into the box shape and welded at one seam to form a hollow structural section (HSS). The built-up box section offers the greatest flexibility and is used for larger structures, whereas the hollow structural sections are widely available from a catalog of standard sizes.

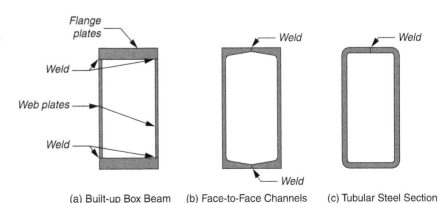

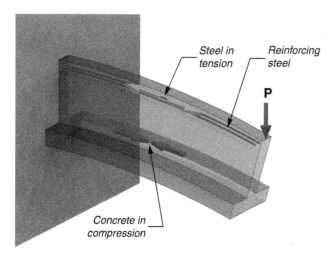

Figure 7.10   Reinforced concrete T-Section

As a cantilever, the T-Beam concrete section is inverted with the greatest area of concrete in compression, and the top surface employing steel reinforcing to withstand tension.

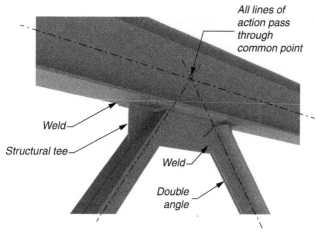

Figure 7.11   Steel T-Section connection

A T-section is sometimes used to facilitate connections between members. In this case, a diagonal brace is created from angle sections placed back-to-back on either side of the tee and welded to it. Ideally, the centerline of force action of the angles and the primary beam all intersect at a point to eliminate moment on the joint due to load eccentricity.

(a) *Wide stance = high moment of inertia = slower spin*

(b) *Narrow stance = low moment of inertia = faster spin*

Figure 7.12   Moment of inertia

Moment of inertia of an area is a geometric property of a cross section that cannot be "seen" in the same way that the cross sectional area of a shape can be seen. A related concept is *mass moment of inertia*, which is associated with volume and material density. An ice skater executing a spin with arms extended has a lower rotational speed due to the greater resistance of mass further from her center (7.12a). As she pulls her arms inward, the rate of spin increases with more mass near the center, which conserves angular momentum (7.12b). Similarly, with more material distributed further away from the centroid, a geometric shape develops a higher area moment of inertia, and thus a greater resistance to bending. It is one of the most fundamental concepts of bending resistance in a beam.

cannot be "seen" in the same way. Nevertheless, understanding this principle is very important to truly understand how beams behave.

A related concept may help to clarify the meaning of moment of inertia: The reader may have watched an ice skater perform a grand finale by starting in a slow spin with arms extended, and then rapidly increasing the spin rate as his or her arms are pulled inward (Figure 7.12). This is an example of how changing the *mass* moment of inertia influences rotation. A distribution of material nearer to the center creates a lower moment of inertia, and thus less resistance to rotation (which in this example leads to a faster spin rate); conversely, moving material away from the center increases the moment of inertia and increases the resistance to

rotation. Moment of inertia for a beam is *area* moment of inertia; however, it is directly analogous to mass moment of inertia. Separating the material further from the neutral axis of a beam will increase its resistance to bending in the same way that one's arms held outward (especially with weights (see Exercise 1) will increase resistance to rotation.

Moment of inertia is a property of geometry in the same way that area is a property of geometry. Mathematically, moment of inertia is proportional to the area of the cross section, (i.e., to the amount of material used) and also—and more significantly—to the square of a property called its *radius of gyration*. If it were possible to lump all of the material of the cross section into one location above and below the neutral axis, yet maintain the same resistance to bending as the actual beam section, the radius of gyration is the theoretical distance to these points (see Figure 7.13). As alluded to in the preceding paragraph, the deflections and bending stresses in a beam are *inversely* proportional to the moment of inertia of its cross section. Since rolled beam sections are made in standard shapes (e.g., wide flange beams, channels, and angles) this property (along with others including dimensional and area information) can be found in standard section tables. A designer can thus gauge the strength and rigidity of a given beam without lengthy calculations, simply by referring to these tabulated values.

The bending stresses and deflections of a cantilever depend on the location of the loads. The greater the lever

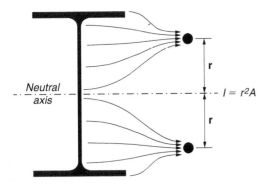

Figure 7.13 Radius of gyration and moment of inertia
If all of the material could be lumped in one location that would have the same resistance to bending (i.e., the moment of inertia would remain the same), this distance "r" (measured from the neutral axis) is known as the *radius of gyration*. Moment of inertia is proportional to the square of this distance multiplied by the cross sectional area, or $I = r^2A$. Because these are properties of geometry, the values of the radius of gyration and of moment of inertia are listed in tables for standard shapes for materials such as timber and steel.

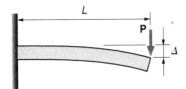

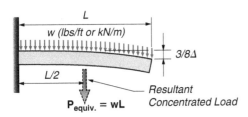

Figure 7.14 Deflections due to concentrated and distributed loads
Note that while the resultant concentrated load from the uniformly distributed load (b) is a useful means of understanding the reacting forces on the beam, it is *not* the same as (a) true concentrated load of the same magnitude at this location. The bending stresses in the beam would be much higher than if the same total load amount is uniformly distributed.

arm of the load, the greater the bending stresses it produces in tension and compression. Equilibrium in rotation requires that the rotational moment of the tension and compression forces be equal to the rotational moment of the load (see Section 4.2 and Figure 7.6). Consequently, the largest stresses occur at the support of the cantilever, the point farthest from the load, where the load develops the largest moment. For example, if the same total load is evenly distributed over the beam instead of being applied at the tip of the cantilever, the bending stresses are reduced in the ratio of two to one, and the deflections in the ratio of eight to three (Figure 7.14).

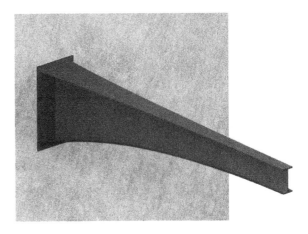

Figure 7.15 Tapered cantilever beam
A more optimal form to a cantilever increases the distance between the top and bottom flanges, as well as their cross-sectional area, toward the supported end where the moment forces are highest.

Because the bending stresses vary from largest values at the support to minimums at the free end, a beam with a constant cross section along its length is an inefficient use of material. By varying the shape, efficiency can be improved. The change in section may be obtained in several ways. One is by increasing the area of the flanges from the beam tip to the support while maintaining a constant beam depth. Another is by maintaining a constant area of the flanges while increasing the beam depth. A third combines the first two by increasing both the flange area and the beam depth (Figure 7.15). Thus, as the lever arm of the load increases, either the tensile and compressive forces in the flanges increase or else their lever arm also increases, keeping the beam in equilibrium without a substantial increase in stress.

The sloping bottom of tapered cantilever beams is the visual evidence of the increased lever arm of the flange forces necessary to counterbalance the increasing lever arm of the loads. With the increased ability to create cost-efficient custom sections through computer-aided design and manufacturing techniques, it is likely that in the future there will be more structures elegantly shaped to better match their loading with a reduction of material. This shaping of a cantilever is but one example of how knowledge of structural principles can benefit the designer to create a richer architectural design while potentially saving cost (though significant savings are realized only through repetition of a detail).

7.1.3 Lateral-Torsional Buckling

Whenever depth limitations permit, it is most economical to select an I-beam with flanges as far apart as possible (i.e., a deeper section), thereby reducing the flange cross-sectional area to a minimum by increasing the internal moment arm and decreasing the force due to bending. It must also be remembered, however, that *any* thin compressed element has a tendency to buckle (see Section 5.3); consequently, if the compression flange or the web of an

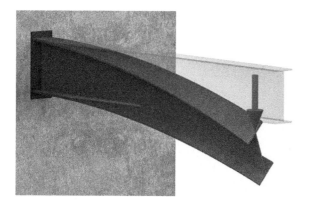

Figure 7.16 Lateral-torsional buckling
An unsupported cantilever end is potentially subject to an out-of-plane twisting under a load. Proper bracing of the beam is required to prevent this, or otherwise loads must be limited to low values. The ghosted shadow in the image represents the original nondeformed shape.

I-beam is too thin, it may buckle. The compression flange buckles by bending out in its own horizontal plane, since the web prevents its otherwise easier buckling in a vertical plane. When the web buckles, its compressed fibers bend out of the vertical plane (the compression zone being located below the neutral axis for a cantilever with gravity loading). In either case, the cross section of the beam rotates and the beam twists (Figure 7.16).

This condition is known as *lateral-torsional buckling*, due to the tendency of the beam to twist about its length in addition to buckling out of plane. Such buckling of an entire cross section, due to the buckling in compression of some of its fibers, may take place during construction when very long I-beams are gripped at midspan and lifted, so that each half of the beam acts as a cantilever. In order to prevent lateral-torsional buckling, the beam cross section must be sufficiently strong in horizontal bending and in torsion (see Section 5). It is less critical in completed construction since the floors and roofs of a building will tend to prevent it from occurring. In lighter-weight construction, such as wood joists or a flexible roof, it is necessary to provide elements that will resist this type of structural action.

7.1.4 Shear

So far only the *bending* effects of a load applied at the free end of a cantilever have been considered. As pointed out in Section 5.3, however, this load also tends to *shear* the beam, for example, at the point where it is secured to the supporting wall. Actually the tendency to shear occurs across the entire length in a cantilever beam; it is also independent of the beam length for a concentrated load at the end. If the beam were shorter with the same load, the load would have the same tendency to shear it (Figure 7.17). When the load is evenly distributed on the cantilever instead of being concentrated at its tip, the amount of load on the beam, and hence the tendency to shear, increases from the tip to the support of the beam (Figure 7.18).

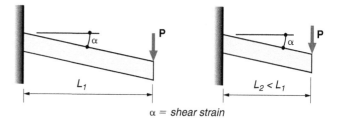

α = shear strain

Figure 7.17 Shear distribution along a cantilever beam with a concentrated load
For a concentrated load at the tip of a cantilever, the magnitude of shear is constant along the length of a beam. The angular deformation (shear strain) of the beam is thus independent of its length in this loading.

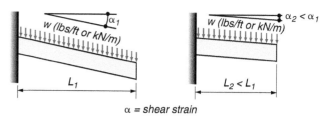

α = shear strain

Figure 7.18 Shear distribution along a cantilever beam with a uniform load
For a uniformly distributed load, the magnitude of shear progressively increases from zero at the free end to the full load of the beam at the supported end.

The presence of vertical shear in the beam becomes evident if one notices that a vertically sliced beam is unable to carry vertical loads, but that its carrying capacity is restored if the vertical slices are bound together by horizontal wires, so as to prevent their relative sliding. The reader may perform an experiment of the same nature by trying to lift a row of books: This will not be possible unless one presses the books together, thus creating between them a large amount of friction, that is, shear, which prevents their sliding (Figure 7.19).

Figure 7.19 Vertical shear in a row of books
A row of books can be picked up if they are pressed together, causing friction between the individual books to prevent them from falling apart. These frictional forces are vertical shear on the surfaces that keep the books in place. The heavier the books are, the more force is required to generate adequate friction between the books.

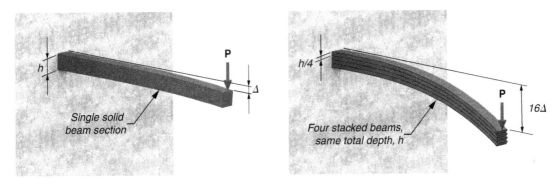

Figure 7.20 Superimposed shallow beams

A solid beam section acts as a single unit, whereas a beam of the same depth but consisting of multiple unattached layers will simply act as separate beams that slide passed one another in horizontal shear. The magnitude of deflections for a stacking of four beams, each ¼ the thickness of an equivalent solid beam, will be 16 times larger than the solid section.

It was shown in Section 5.4 that vertical shear is always accompanied by horizontal shear (see Figure 5.13). The influence of horizontal shear on the rigidity of a rectangular beam may be visualized by considering the beam "sliced" horizontally into a number of rectangular beams of limited depth. Under the action of a tip load, this assemblage of superimposed beams presents large deflections, because each slice, free to slide with respect to the others, acts as a shallow, flexible beam (Figure 7.20b). The sliding of the various slices shows the tendency of the beam fibers to shear along horizontal planes. This tendency is prevented by shear stresses along the horizontal planes, obtained either by gluing the various slices together or by inserting vertical pins through the entire depth of the parallel slices.

When shear resistance is developed, the slices act as a single beam, and the deflections under the same load are considerably reduced. The moment of inertia of a rectangular beam is proportional to the cube of its depth, and so a comparison between a single monolithic member and one comprised of slices can therefore be directly made. If one assigns a unit depth of one to a monolithic beam (Figure 7.20a) and compares this to a beam of four slices each one-fourth this depth (Figure 7.20b), the moment of inertia of the first beam is proportional to the cube of one (1^3 — which is one), and the moment of inertia of the second beam is proportional to four times the cube of one fourth (4 $(\frac{1}{4})^3$) = 4(1/64) = 1/16, or one-sixteenth). *The monolithic beam is thus sixteen times more rigid than the beam comprised of four separate slices.*

The essential function of the web in the I-section is to develop the shear stresses necessary to have the two flanges work together. Without the web, each flange of an "ideal I-beam" would act as a *separate* shallow beam, and the moment of inertia of the cross section would be small. The flanges connected by the web act monolithically so that the I-section acquires a large moment of inertia. It is thus seen that *the transmission of shear through the web from flange to flange is essential to the bending action of the flanges*, and that *shear and bending are **interdependent** beam actions.*

Bending stresses are distributed linearly across the depth of the beam section, whatever its cross-sectional shape, as illustrated in Figure 7.6. Shear stresses, however, may be shown to be distributed parabolically across the depth of a rectangular beam, with a maximum value at the neutral axis (Figure 7.21) This maximum is one and a half times the average shear stress, that is, one and a half times the shear force divided by the area of the cross section. In I-beams, bending resistance is essentially provided by the flanges, and shear resistance by the web. The maximum shear stress is approximately equal to the load divided by the area of the web. The shear stresses increase from the tip to the support in proportion to the accumulated load, and hence are usually maximum at the support.

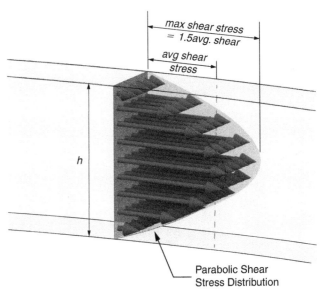

Figure 7.21 Shear stress distribution across the depth of a rectangular beam

Unlike flexural stress, which is greatest on the outer surfaces of a beam, shear stress reaches its maximum value at the mid-depth of the beam. The distribution is also not linear, but is instead parabolic for a rectangular section.

7.2 SIMPLY SUPPORTED BEAMS

7.2.1 Cantilever Beam Analogy

An understanding of the behavior of cantilever beams provides us with a background to understand the most commonly used beam type, the *simply supported* beam. The term "simply supported" means that, unlike a cantilever, *both* ends are supported and, further, the ends are connected in such a manner that each is free to rotate, *and* the beam is free to expand or contract longitudinally (Figure 7.22).

For a concentrated load applied at the midspan of a simply supported beam, it is evident that one-half of this load is carried by one support and one-half by the other. As with a cantilever, the load causes it to bend and deflect but, because of symmetry, the midspan section moves straight downward and remains vertical. The behavior of the simple beam can be visualized effectively as two upside-down cantilevers, with their moment-resisting supports at the middle of the beam attached to one another. The end reactions of the actual beam are envisioned as upward-acting concentrated loads at the free ends of the effective cantilevers, each of which is one-half the length of the actual simple beam (Figure 7.23). Comparing the effective cantilever in this analogy to an actual cantilever beam supporting the same load magnitude

as, and of length equal to, the actual simple beam, some important observations can be made:

Recall from Section 7.1 that cantilever deflections are directly proportional to the load magnitude and to the cube of the length. As illustrated in figure 7.24, for a simple beam of length "L" and supporting a load "P" at midspan, when compared with a cantilever of the same length and carrying the same load at its tip, the deflections are seen to be only one-sixteenth as large. By the same token, the half load reaction on the effective cantilever (acting through a lever arm half the length of the actual beam) produces maximum bending stresses one-fourth of those found in a cantilever the same length as the simple beam.

A simply supported beam is thus substantially stronger and stiffer than a cantilever of the same length. For the same cross-sectional dimensions, it can carry at midspan a load four times as large, and will deflect only one-fourth

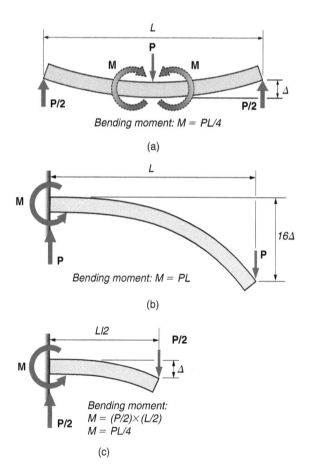

Bending moment: M = PL/4

(a)

Bending moment: M = PL

(b)

Bending moment:
M = (P/2)×(L/2)
M = PL/4

(c)

Figure 7.24   Relationship between stress and deflection of simple span vs. cantilever beams

For a span length of "L," compared with a cantilever beam of the same length and carrying the same load, a simply supported beam (7.24a) will experience an internal moment only one-fourth as large (and thus a stress only one-fourth as high) as the cantilever beam (7.24b). The deflections of this same simply supported beam will only be one-sixteenth as large as the cantilever beam of the same length. The benefit of providing support at both ends of the beam is thus apparent. It can also be seen that the half-length cantilever beam with half the applied load (7.24c) produces the same moment and deflection as the simple span beam. The reaction moment, M, on the cantilever is thus equivalent to the internal bending moment of the simple span beam.

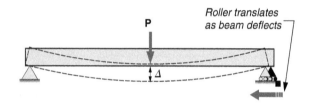

Figure 7.22   Simply supported beam

An ideal simply supported beam has one pinned support and one roller support. As a load is applied, the beam deflects (Δ) and with the increasing curvature the length becomes shorter, sliding inward slightly on the roller support. Were the roller support made into a pin instead, additional stresses would develop in the beam that would restrain it from this movement.

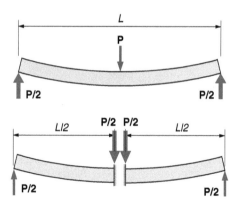

Figure 7.23   Concentrated load on a simply supported beam

The bending stress on a simply supported beam with a concentrated load at midspan (7.23a) can be visualized as two cantilever beams back to back, each ½ the total span length and supporting ½ of the total load, which are applied as upward acting forces at the free ends of the cantilevers (7.23b).

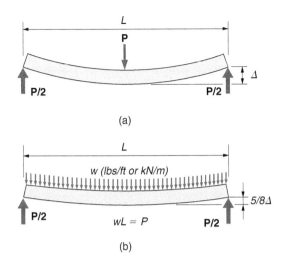

(a)

(b)

Figure 7.25 **Concentrated vs. distributed load on a simply supported beam**
An equivalent total load distributed uniformly along the length of a beam (7.24b) produces half the internal stress, and five-eighth the total deflection of the same total load applied as a concentrated force at the midspan (7.24a).

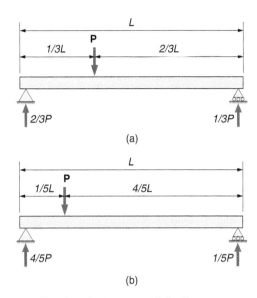

(a)

(b)

Figure 7.26 **Reactions due to asymmetric loading**
Rotational equilibrium ensures that all clockwise and counterclockwise rotational tendencies are in balance. In order for this to be true for asymmetric loadings, the magnitudes of the reactions must vary in direct proportion to the placement of the loads.

as much under this larger load, than the end-loaded cantilever (Figure 7.24).

7.2.2 Effects of Loading

The distribution of load is another important consideration in beam behavior. If the same total load is evenly distributed over the length of a beam, the stresses in the beam become one-half, and the deflections five-eighths, as large as when the load is concentrated at midspan (Figure 7.25).

In a uniformly loaded beam the bending stresses are also *maximum at midspan*, the point corresponding to the supports of the two half beams considered as upside-down cantilevers (Figure 7.23). The shearing action is maximum in the neighborhood of the supports, where the support reactions tend to move the beam up while the *total* load tends to move it down. The shear *vanishes at midspan*, where there is no tendency for adjoining sections to slide either up or down with respect to one another. The vanishing of the shearing action at the point where bending stresses are greatest is a characteristic of beam behavior for any and all beams other than cantilevers.

Unsymmetrical loads produce similar bending and shearing actions in a simply supported beam. Concentrated loads near the supports have a reduced bending influence, owing to the reduction of their lever arms, but an increased shearing effect. In fact, as a load approaches one of the supports, a larger fraction of the load is transferred to this support, while the other is gradually unloaded (Figure 7.26) (see also Exercise 1 of Chapter 4). When the load is in the immediate neighborhood of a support, the shearing force equals the total load, whereas it was equal to only half the load when the load was at midspan.

It was noted in Section 5.3 that shear may be considered as a combination of tension and compression at right angles and at 45 degrees to the shear plane (Figure 7.27). The presence of maximum shearing effects near the supports of a

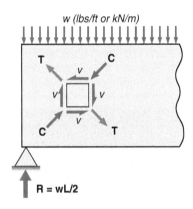

Figure 7.27 **Beam end shear**
In a simply supported beam with a uniform load, shear forces will be highest at the end supports. As described in Section 5.4, the horizontal and vertical components of shearing forces can be resolved into resultant tension and compression forces. Depending on the type of material, either the tension or compression resultant will be an influencing factor.

beam is clearly shown by 45-degree cracks appearing near the supports of reinforced-concrete beams that have inadequate shear reinforcement. Isolating a beam element in this region and applying to it the corresponding shear stresses or the equivalent tensile and compressive stresses at 45 degrees, it is seen that the cracks are perpendicular to the direction of the equivalent tensile stress (Figure 7.28).

In order to avoid shear cracks, reinforcing bars must be added toward the supports, ideally running at a 45-degree angle and capable of absorbing the tensile stresses due to shear (Figure 7.28a). More commonly, due to the challenges posed by constructing beams with long bent bars, the horizontal shear stress is resisted by "sewing together" a beam by means of vertical steel hoops called *stirrups* (Figures 7.28b

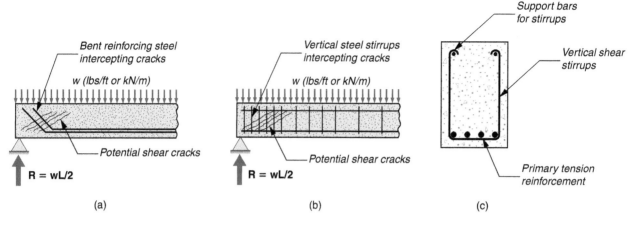

Figure 7.28   Reinforced concrete simply supported beam

Potential tension shear cracks at the end of a concrete beam are at a 45° angle, and thus the ideal position for steel reinforcement to carry the tension forces is perpendicular to that direction (7.28a). Owing to the difficulty of working with long bent reinforcing bars, however, the more common and practical arrangement of shear reinforcement is to use vertical "stirrups" that intercept the cracks at an angle (7.28b). The stirrups are often more closely spaced at the end of the beam where shear forces are highest. These bars are frequently bent into a "U" shape that wraps around the primary tensile reinforcement as can be seen in cross section (7.28c).

and 7.28c). Similarly, a portion of the thin web of a steel beam near its support may buckle, as shown by a wavy deformation in a direction perpendicular to that of the equivalent compressive stresses (Figure 7.29). To prevent such buckling (referred to as "*web crippling*"), the web may be strengthened by *vertical stiffeners* (Figure 7.30). The stiffened web sections then act as mini-cruciform-shaped columns that resist these compression forces.

The influence of the cross section's shape on stresses or deformations is the same for simply supported beams as for cantilevers or any other type of beam. Small beams may be built with a uniform cross section; large beams usually have a variable cross section, obtained by increasing the area of the flanges or increasing the depth of the beam toward midspan where the bending stresses are usually larger. *Built-up girders* or *plate girders* are beams with variable flange areas, built of plates, and used to carry heavy loads over spans of fifty or more feet (15.3 m) (Figure 7.30). To simplify fabrication, it is common to increase the flange area rather than making the beam depth higher near the center of the span, even though the latter technique is more effective in strengthening the beam.

In contrast to downward-loaded cantilevers, simply supported beams under downward loads develop tension in the bottom fibers and compression in the top fibers. The reinforcing bars in a simply supported concrete beam therefore run along its bottom and, if necessary, are increased in number or in area toward the center of the beam (see Figure 7.28).

In steel beams, the horizontal compressive stresses due to bending (which are usually highest at midspan) may produce lateral buckling of the upper flange and of the upper part of the web, thus causing a twist of the entire cross section (see Lateral-Torsional Buckling in Section 7.1). For example, one of the larger rolled W-sections used as beams in the United States is 36 inches deep (0.9 m) and weighs

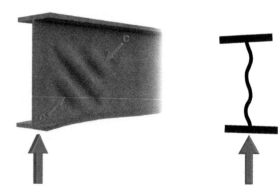

Figure 7.29   Web crippling of steel beam under high shear force

Shear failure in a steel member with a thin web results from the resultant compression forces acting at 45°.

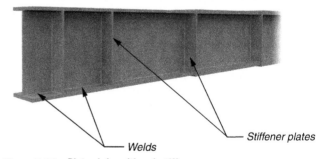

Figure 7.30   Plate girder with web stiffeners

A beam built up of many plates can be created for very large sections. These frequently have vertical plates that stiffen the web section against buckling. These effectively behave like small cross-shaped columns in plan that carry shear forces. As with reinforced concrete stirrups, the spacing of the stiffener plates may be decreased near the supports where shear forces are highest. The flange plates may be increased in thickness toward the center of the span where the moment is highest.

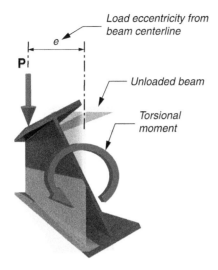

Figure 7.31 Twist-producing loads
Loads applied eccentric to the centerline of a beam will cause twisting action on the beam, with a similar potential result as lateral-torsional buckling of Figure 7.16 when a critical load magnitude is reached. Eccentrically applied loads will cause this phenomenon at a much lower force value, however.

Figure 7.32 I-Beam with openings
Removing material that is not required for shear in the web of a beam, a trusslike structure emerges where the remaining web material acts like the interior diagonal members of a truss.

Figure 7.33 Castellated beam
By cutting the web of a flanged beam in the pattern illustrated, and then reattaching the top cut to the bottom with an offset equal to half of the linear cut distance, a castellated beam is created. The beam has the same weight as the original, but more than three times the moment of inertia, and becomes somewhat trusslike as in Figure 7.32. The spaces created allow room for pipes, electrical conduit, and small ducts to pass through as well. Other patterns (such as circular) are also used.

300 pounds per foot (4.38 KN/m). This section, when made out of high-strength structural steel, may carry a load of 6000 pounds per foot (87.6 KN/m), besides its own dead load, over a span of 50 feet (15.2 m).

The largest distance such a W-section could span is one for which the beam can safely carry its own dead load alone, and no additional live load. For an allowable bending stress of 30,000 pounds per square inch (206.8 MPa) this maximum span is 271 feet (82.6 m). But this distance can be spanned *only* if the twisting of the beam in buckling is prevented by the restraint of adjoining structural elements, such as transverse beams or slabs. When the beam is free to twist, the danger of lateral buckling reduces the maximum span to nearly *one-half* the previous value, or 150 feet (45.7m). Twisting action is also induced by loads displaced sideways over the upper flange of a beam, due to their lever arm with respect to the web (Figure 7.31).

The large amount of material inefficiently used in the neighborhood of the neutral axis of a beam suggests the elimination of some of it by cutouts, which also makes the beam lighter (Figure 7.32). As the cutouts become larger, the beam gradually is transformed into a truss. Thus, in a truss considered as a beam with cutouts, the upper and lower chords act as the upper and lower flanges of an I-beam, while the diagonals and the verticals absorb the shearing forces and act as a discontinuous web. Through cutting the web in a diagonal pattern, then separating the two halves of the beam and reattaching them, a castellated beam is created which has the same weight but twice the depth and more than three times the moment of inertia (Figure 7.33). The openings displace material not required for shear and also allow space for building utilities like piping, electrical conduit and small ducts to pass through them. By varying the vertical distance between the chords, or by properly distributing the area of the chord

members according to the variation of the bending action, and by locating the verticals according to the variation of the shear, trusses could be theoretically designed for maximum efficiency (Figure 7.34). In practice, however, the verticals of a parallel chord truss are usually spaced equally for economy of fabrication.

Beams must be designed for *stiffness* as much as for strength. It is customary to limit beam deflections to about one 360th of their span in buildings, and one 800th in bridges. Thus, a plate-girder bridge 200 feet long (61 m), should not deflect more than three inches (76 mm) under maximum loads, while a beam 30 feet long (9.1 m) is unacceptable if it deflects more than 1.2 inches (30 mm). When the dead load represents the largest share of loads, beams may be built with a *camber*, that is, an upward deflection equal to the downward deflection due to the dead load; the live load alone then induces smaller deflections, easily contained within acceptable limits (Figure 7.35).

Combinations of beams with different support conditions are often used to span long distances. A pioneering reinforced concrete bridge in Venezuela, the General Rafael Urdaneta Bridge over Lake Maracaibo, designed by the Italian engineer Riccardo Morandi, has main spans consisting of beams continuous over a support with cantilevered stayed sections, and of intermediate beams simply supported at the tips of opposing cantilevers (Figure 7.36). Thus, an efficient combination of tensile and bending structures allows the spanning of 771 feet (235 m) between towers, while the effects of thermal

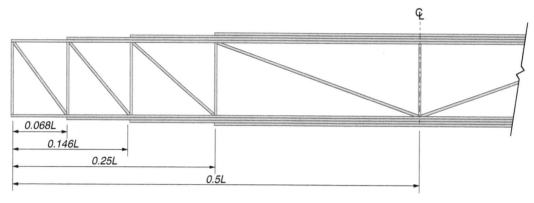

Figure 7.34   Maximum-efficiency truss

Shaping a truss to carry loads most effectively requires more material on the top and bottom chords near the center of the span (where the moment is highest), and more material in the web verticals and diagonals at the end of the span (where shear is highest). The resulting optimally shaped truss may be compared with the conceptual construction of a truss from successive panel additions of Figure 6.31.

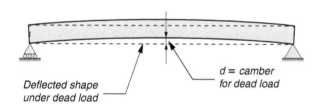

Figure 7.35   Beam camber

Larger beams are often fabricated with an upward bow (camber) to compensate for deflections under dead loading, and thus minimize the total deflection under dead plus live loads.

Figure 7.36   Combination of simply supported and stayed cantilevered beams

The General Rafael Urdaneta bridge at Lake Maracaibo in Venezuela is a stunning example of cantilever bridge construction. The majority of the span is supported by the diagonal stays, but between the two towers a shorter and much lighter simple span is supported on the ends of the stayed cantilevers. It was designed by the Italian engineer Riccardo Morandi in 1958.

Photo courtesy of Wilfredo R. Rodríguez H.

expansion are eliminated by the use of a rocker connection between the stayed and the cantilevered spans.

7.3 FIXED BEAMS AND CONTINUOUS BEAMS

It was noted in the preceding section that a uniformly loaded, simply supported beam is an inefficient structure. This is because the bending stresses in it reach their allowable value only in the extreme fibers at midspan, where the beam fibers (that have downward curvature all along the beam) have their greatest curvature (Figure 7.37).

An effective way of improving the beam efficiency is by shifting the supports toward the center of the span whenever it is feasible to do so. The beam thus acquires two cantilevered portions, and the load on the cantilevers then partly balances the load between the supports. When this is done, the deflection at midspan is reduced and the curvature near or over the supports reversed; the beam may or may not curve up toward the middle of the span, but always curves down toward its ends (Figures 7.38(b–d)).

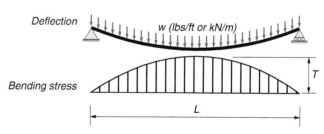

Figure 7.37   Deflected beam axis

The bottom fiber stress in a simply supported beam with gravity loads will always be in tension (T). The diagram below the deflected beam form is a graph of relative stress magnitude along the length of a beam. For a uniform load, it will be parabolic in shape, with a maximum at midspan and zero at the supports.

Figure 7.38 Bottom fiber stress and deflections in beams with cantilevered ends

The bending stress graph shows the stress magnitude and sense (tension or compression) on the vertical axis, and distance on the horizontal axis (i.e., span of the beam). The various configurations illustrate changes to the stress on the *bottom* fibers of the beam. When supported on a single point, the same beam with the same load of figure 7.37 will experience compression (C) on the bottom surface equal in magnitude but opposite in sense to the tension force (T) of the simple beam. If overhangs are provided on either side, the bottom fiber stress will always be compressive over the supports, but gradually become tensile at the midspan. As the cantilever ends become longer, the relative magnitude of compressive stresses over the supports decreases and tensile stress at the center span increases until they become equal (7.38d).

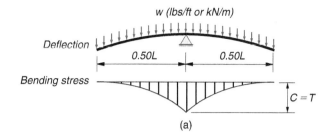

(a)

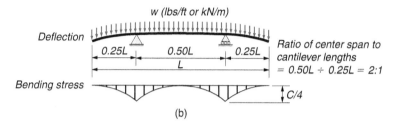

(b)

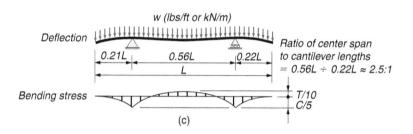

(c)

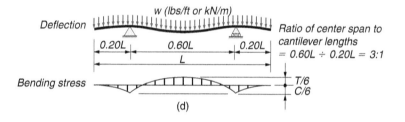

(d)

It is interesting to examine the beam behavior as the length of the cantilevers decreases. When both supports become a single support at the midspan section, the two halves of the beam are cantilevered, and the curvature is down everywhere. The largest stresses occur at the support, and are equal and opposite in value to those developed at midspan in the simply supported beam (Figure 7.38a). (This is, of course, an unstable position, and so forms a theoretical basis for the current discussion only.)

When the ratio of cantilevered length to supported span is 1 to 2, the center deflection is small but upward and the entire beam still has a downward curvature. The largest stresses occur over the supports, and are about one-fourth as large as those in the beam supported at the ends (Figure 7.38b).

When the ratio of cantilever to span lengths is approximately 1 to 2.5, the sections of the beam over the supports do not rotate; the stresses over the supports are about twice the stresses at midspan but of opposite sign, and about one-fifth as large as the stresses in the same beam supported at the ends (Figure 7.38c).

When the ratio of cantilever to span lengths is 1 to 3, the stresses over the supports are *equal and opposite* to those at midspan and about one-sixth as large as in the beam supported at the ends (Figure 7.38d). This is the *smallest value* of maximum stresses that may be obtained, as the cantilever spans have effectively balanced the center span bending action.

When the ends of a beam are built into a *rigid* element and are thus prevented from rotating (but not from moving horizontally one relative to the other), the beam behaves similarly to the supported span of a beam with cantilevers in the ratio of 1 to 2.5 (Figure 7.38c). Such a uniformly loaded, *fixed-end beam* has the largest bending stresses at the built-in ends. These stresses are twice as large as the stresses at midspan, and two-thirds as large as the stresses in the same span simply supported (Figure 7.39a). Provided the beam can develop tensile and compressive stresses both at top and bottom, it can carry a load *50 percent higher than an identical simply supported beam* (whenever shear does not govern). Moreover, the fixed beam is *five times stiffer* than the simply supported beam: Its midspan deflection is one-fifth that of the simply supported beam.

Figure 7.39 Bottom fiber stress and deflections in simply supported and fixed end beams

Relative to the same span and load, a beam with fixed ends (i.e., restrained against rotation) will have only one-fifth the total deflection of the simply supported beam. The deflection curvature will change from downward to upward at approximately 21 percent of the span length. The moment at this inflection point will be zero, changing from negative (producing downward curvature at the ends) to positive (producing upward curvature in the center). Correspondingly, the bottom fibers will be in compression on the ends and tension in the center. The built-in beam behaves the same as a central span supported by short cantilevers on either side.

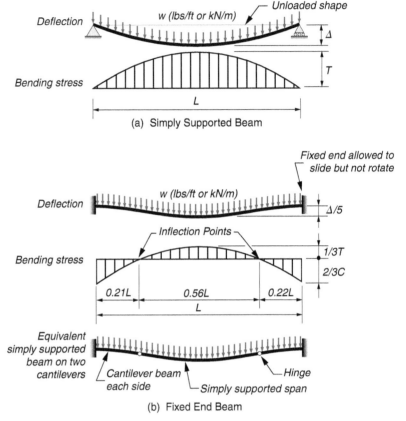

(a) Simply Supported Beam

(b) Fixed End Beam

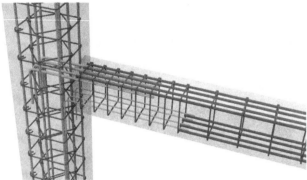

Figure 7.40 Fixed-end reinforced concrete beam

A reinforced concrete beam is made moment resisting (fixed) at the end by lapping reinforcing steel from the top bars into a column or wall. In this rendering, blue represents primary tension reinforcement in the beam, while green are the lapped bars connecting the blue to the column. Red indicates stirrups in the beam and ties in the column around the yellow vertical reinforcing bars. For a fixed-end beam with gravity loads only, the top bars need to extend approximately 21 percent of the span distance, while the bottom bars cover roughly 60 percent of the total length about the centerline of the span.

The deflected axis of a fixed beam changes from downward to upward curvature at two points, called *inflection points,* where bending stresses vanish, and a stress reversal occurs (Figure 7.39b). Accordingly, fixed beams of concrete must have the tensile reinforcement at the bottom near midspan and at the top near the supports (Figure 7.40).

The lack of bending stresses at inflection points indicates that the beam behaves as if it were simply supported

at such points, since bending stresses vanish at the hinged ends of a simply supported beam. Thus, a beam with fixed ends behaves as a simply supported beam of *shorter* span supported on two short cantilevers, and this explains its greater load capacity and its greater rigidity (bottom of Figure 7.39b). This is conceptually equivalent to a cantilever bridge supporting a shorter and lighter central simple span (Figure 6.37 and Figure 7.36).

Fixed beams provide the added advantage of a greater resistance to both longitudinal and lateral buckling from flexural compression. Whereas in a simply supported beam the top fibers are entirely compressed, in a fixed beam only about 58 percent of their length is in compression at the top and 42 percent at the bottom. Remembering that the buckling load is inversely proportional to the square of the compressed length, a compressed fixed beam and, hence, a fixed-end column are found to be four times as strong in longitudinal buckling as a simply supported beam or column (see Section 5.2). The resistance to *lateral* buckling is inversely proportional to the beam length between inflection points. A fixed beam is therefore about twice as strong against lateral buckling as an identical simply supported beam.

Simply supported and fixed-end beams present two extreme cases of support conditions. The first allows the unrestrained rotation of the beam ends; the second completely prevents such rotation. In practice, any intermediate condition may prevail. The wall into which the beam is built may not prevent the rotation entirely, or the beam may be continuous over more than two supports. In the last case, each span, if cut from the others, usually deflects down and curves up as a simply supported beam (Figure 7.41a).

Figure 7.41   Continuous beam

Under a uniformly distributed load, three simply separate supported beams will have no interaction with one another, and so their stresses and deformations are independent and always with tension on the bottom face (7.41a). When connected in a continuous span, however, compression on the bottom surface (and simultaneous tension on the top surface) develops over the supports (7.41b). Under a concentrated load, once again separate simply supported beams will behave independently (7.41c), whereas when connected as one continuous span, a heavy load placed on one span will affect the others. In the case of a single concentrated load placed at the center span, the center span will experience only bottom tension, whereas the two end spans will react by bending upward and developing bottom compression since the supports will prevent an upward displacement (7.41d).

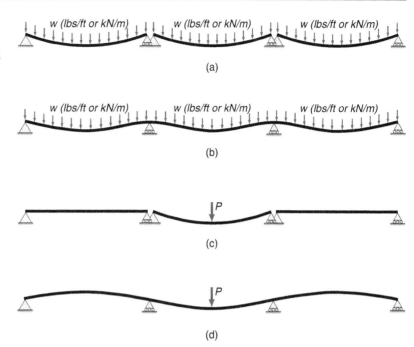

Figure 7.42   Change in curvature in continuous beam under moving loads

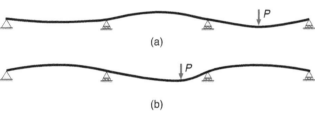

In the case of continuous spans that are subjected to a moving load (such as on a bridge), the effect of a heavy load along a span length will vary depending on placement, always with tensile stress below the load. In a three span condition, when positioned at the end span, the load causes a downward deflection below it, and upward deflection in the center span, then changing again to downward deflection in the opposite end span, though at a greatly reduced magnitude (7.42a). As the load moves into the center span, it begins to take on the appearance of figure 7.41d as it moves toward the center (7.42b).

Continuity with the other spans reverses the curvature near or over the supports and restrains the rotation of the ends common to two spans (Figure 7.41b). Depending on the relative lengths and rigidities of the beams in the various spans, the rotations at the supports are more or less prevented, and each beam develops stresses and deflections intermediate between those of simply supported and fixed-end spans.

The continuity of a beam over many supports introduces new characteristics in its behavior. If each span were simply supported and only one span were loaded, the load would be supported exclusively by the bending and shear stresses of the loaded span (Figure 7.41c). Continuity makes the loaded span stiffer by restraining its end rotations and introduces bending and shear in the unloaded spans. The entire beam participates in the load-carrying mechanism, and some of the load may be considered as transferred to the unloaded spans (Figure 7.41d)

The stresses due to continuity peter out as one moves away from the loaded span. The curvature in the continuous beam is greatest *under* the load and is "damped out" by the supports, so that stresses become practically negligible two or three spans away from the loaded span (Figure 7.42a). Continuity increases the resistance of a beam to concentrated loads, but its effect diminishes rapidly, becoming negligible a few supports away from the load.

It may finally be noticed that, depending on the location of the load, the curvature of a given span may change from downward to upward, or vice versa (Figure 7.42b). Hence, continuous beams of reinforced concrete may require tensile reinforcement at *both* top and bottom whenever loads may be located on different spans, as is the case, for example, with moving loads on bridges.

7.4 SECONDARY BENDING STRESSES

Compressive and tensile stresses of *constant value* over the entire cross section of a structural element are called *direct stresses* to distinguish them from bending stresses, which vary linearly through its depth from a maximum compression to a maximum tension. Direct stresses allow a more efficient use of the material than bending stresses because they develop the same stress in every fiber of the element. The hinged bars of a truss, for example, develop their maximum resistance at all points of their cross sections, since they are acted upon by simple tension or simple compression.

One of the essential characteristics of direct stress is to develop displacements that are extremely small in comparison with bending displacements. Presuming restraint against bucking, the shortening of a ruler by a compressive load cannot be gauged by the naked eye (Figure 7.43a), but the bending deflection of the tip of the cantilevered ruler under the

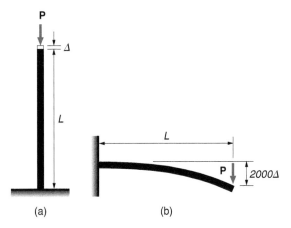

(a)   (b)

Figure 7.43   Compressive and bending displacements
Deformations due to purely axial forces are substantially smaller than those due to bending forces. In the case of a member of a given length 'L', the deformations due to cantilever bending (7.42b) may be two or three thousand times greater than the same length member carrying the same load in an axial manner (7.42a).

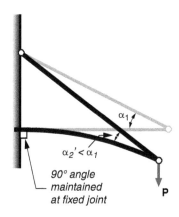

Figure 7.45   Bending action in bracket
If a joint at the support is fixed in place, or the hinge action becomes inhibited through rust or wear, then bending action will develop in the truss bracket. The top member will experience tension as in Figure 7.44; however, the bottom bracket will experience a combination of axial and bending forces.

same load can be noticeable (Figure 7.43b). *The bending deflection is often two or three thousand times larger than the compression shortening due to the same load*; in other words, a structural element is much stiffer in direct tension or compression than in bending. Unfortunately, pure compression or tension elements are not easily realizable in practice, as the example of a simple truss will clearly indicate.

The triangular bracket of Figure 7.44 is acted upon by a vertical load at its tip; it supports the load as an elementary truss by tension in the inclined bar and compression in the horizontal bar. Direct stresses develop in these bars only if their ends are free to rotate; otherwise, the lower bar, for example, would behave as a cantilevered beam and carry part of the load by bending action (Figure 7.45). Ideally, the bars of a

truss should be connected to the supports and to one another by joints allowing unrestrained rotations, or so-called *hinges*.

Some of the truss bridges of the nineteenth century were actually built with hinged bars. But because of weathering, rusting, and painting, the hinges often became locked, and the bars of modern trusses are all bolted or welded in order to avoid prohibitive maintenance costs. In this case, even if the small amount of bending due to the dead load of the bars is neglected, it may easily be seen that additional bending stresses are introduced by the welding or bolting (or in earlier eras, riveting) of the bar ends. In the bracket of Figure 7.44, under the action of the load the upper bar elongates and the lower bar shortens. If the bars were hinged, they would rotate so as to bring the tips of the lengthened and shortened bars together again, and these rotations would change the angle between the bars (Figure 7.44). Such angle changes are prevented by the rigid welded or bolted connection. In order to maintain the angle unchanged, the bars must

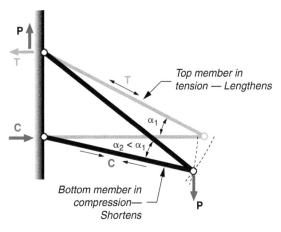

Figure 7.44   Truss bracket
In an ideal construction with pinned connections at all joints, the deformations of a cantilever truss bracket with a downward load at its tip will be purely axial in nature. The top member will undergo tension and therefore lengthen slightly. The bottom member will experience compression and thus shorten slightly. The supports will correspondingly experience tension on the top and compression on the bottom.

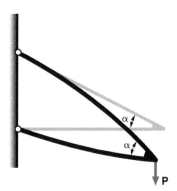

Figure 7.46   Bending deformations of bracket hinged to wall
If the joint at the tip is fixed, then both the top and bottom member will experience a combination of axial plus bending forces, with bending action particularly concentrated at the top.

Figure 7.47 Bending deformations of bracket fixed to wall
A fully fixed bracket subjects members to axial and bending forces throughout their lengths. The bottom member will experience a force reversal ("*contraflexure*") between the support and the tip, with a corresponding change from downward curvature at the support to upward at the tip.

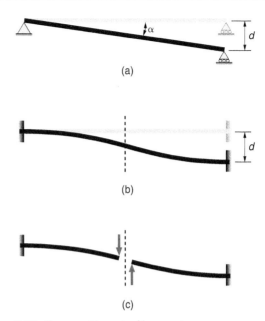

Figure 7.48 Uneven settlements of beam ends
A simply supported beam that experiences an uneven settlement (vertical displacement) from one end to the other will not be subjected to additional stresses because the ends are free to rotate (7.48a). For a member that is fixed into the supports (a "*built-in beam*"), this will not be the case. The downward movement creates secondary bending stresses with a reversal in curvature from one end to the other (7.48b). The result is effectively the same as two cantilever beams with equal and opposite loads applied to their tips (7.48c).

bend, and develop bending stresses (Figure 7.46). These so-called *secondary* bending stresses may reach values as high as 20 percent of the direct stresses.

The presence of bending stresses can always be detected by visualizing the deformation of a bar. In a continuous beam on four supports, loaded on the middle span only, the deflection shows that the center span develops bending stresses with tension at the bottom and compression at the top, toward midspan, while the outer spans and the portions of the middle span nearer the supports have inverted stresses (see Figure 7.41d). Similarly, in the bolted bracket the bending stresses produce tension in the outer fibers and compression in the inner fibers of both bars, if the ends connected to the wall are hinged (see Figure 7.46). When the connections to the wall are fixed, the horizontal bar behaves very much like a beam with fixed ends, and its stresses change from tension to compression at both the outer and the inner fibers (Figure 7.47).

The visualization of deflections is most useful in determining a state of locked-in stresses not caused by loads. Such is the state of stress due to the uneven settlements of the foundations of a building (see Section 2.5). If the right support of a simply supported beam settles more than the left, the beam adapts to this situation by rotating around the hinges and acquiring a slightly tilted position, but remains straight and hence *unstressed* (Figure 7.48a). If the beam has fixed ends, the end rotations are prevented, and the beam bends (Figure 7.48b). Its midspan section presents an inflection point and does not develop bending stresses. The two beam halves behave like two cantilevers carrying a load at their tips (Figure 7.48c).

The advantages of fixed-end beams are thus balanced by some disadvantages; fixed-end beams are more sensitive to settlements, which may not be easily assessed beforehand. Whenever the danger of foundation settlements is present, a structural system should be chosen capable of adapting to such conditions without straining and stressing.

Considerations of the same kind hold for *differential deformations due to temperature*. The outer columns of a

tall building elongate when the temperature of the outside air rises, while the interior columns and the core do not if the building is air conditioned. The ends of a beam connecting an outer to an inner column or to the core acquire different levels, and may develop additional bending stresses (Figure 7.49). Here again, a flexible structural system capable of adapting itself to deformations by developing minor stresses is superior to a rigid system.

Both rigidity and flexibility have their place in structural design, but must be given different weight depending on the loading conditions to be considered. Rigidity is well adapted to withstand loads; flexibility, to withstand deformations.

KEY IDEAS DEVELOPED IN THIS CHAPTER

- Beams are the structural components that can carry loads over horizontal distances.
- Contrary to cables and arches, which are pure tension and compression members, beams have tensile stresses at some depth of the cross section and compression at others.
- Bending stresses vary linearly from maximum tension at the top to maximum compression at the bottom, with zero stress at the neutral surface.
- The moment of the applied load is equilibrated by the internal moment of the tension and compression stresses.
- Beams with areas further from the neutral axis (i.e., flanged sections) are more efficient than beams with rectangular cross sections.

Figure 7.49   Thermal deformation of building frame

Significant temperature variations from the outside to the inside of a building with fixed connections leads to secondary bending stresses, particularly in the case of tall buildings. The cooler interior of the building remains relatively unchanged, whereas the exterior expands (or contracts in a cold winter) relative to the interior. If the joints are restrained against rotation, they will experience bending stresses.

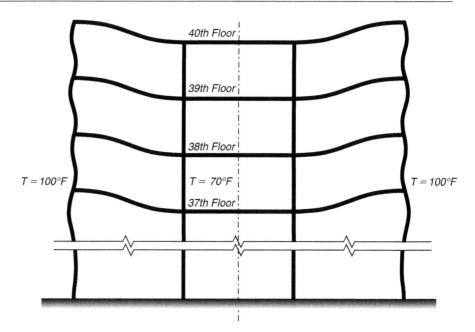

- Cantilever beams under gravity loads bend downward, and fibers in the top half area are in tension, while the bottom ones in compression.
- Stresses are lower in deep beams than in shallow ones.
- Stresses are inversely proportional to the moment of inertia.
- Deflections are proportional to the load and to the cube of the length.
- Both vertical and horizontal shear stresses are simultaneously present in beams.
- Shear stresses vary parabolically through the depth of the beam and are maximum on the neutral surface.
- Simply supported beams under typical downward gravity loads have an upward (bowl-shape) curvature, with tension in the bottom fibers and compression on top.
- The behavior of simple beams can be understood as two analogous inverted cantilever beams with supports back-to-back.
- Bending stresses are greatest at the center of the span where bending moments are largest. Here, shear stresses are zero. Shears are greatest near the supports.
- Because the webs of flanged beams carry less stress and are needed mainly for connecting the two flanges, their areas may be reduced by cutouts. Such beams resemble trusses.
- Deflections and stresses may be reduced in simple beams by moving supports toward each other, creating cantilevered ends and a shorter simple span between the supports.
- Fixed end beams can be considered as two short cantilevers supporting a short simple beam. Bending stresses are greatest at the supports, as are shear stresses.
- Deflections of fixed end beams are much smaller than those of simple or cantilever beams.
- The individual spans of beams that are continuous across several supports all share in carrying loads on any one span.

QUESTIONS AND EXERCISES

1. Moment of Inertia Exercise: One may develop a "feel" for moment of inertia by taking two equal-weight objects (such as light hand weights or two medium-sized hardcover books) and holding them at arms-length while standing up. Twisting back and forth around the waist (not spinning in a circle, but oscillating in place), the reader will literally *feel* the resistance the objects make to the twisting motion. Next, hold the objects directly overhead and twist once again. One will immediately notice how much easier it is to rotate. Can you sense how this relates to the idea of moment of inertia in a bending member?

2. Take a piece of flexible foam beam about 2 inches (51 mm) on a side in cross section, and mark horizontal and vertical lines on it. Notice how when the foam is bent, the vertical lines will pull apart in tension areas and squeeze together in compressive areas. The middle of the foam will stay unchanged, as this is the neutral surface.

3. Take a piece of brittle Styrofoam and bend it; notice how it easily snaps. Next, take a piece of tape and place it along the bottom surface; notice how it now becomes more flexible and does not break as easily. The tape has become a reinforcement for the Styrofoam, and the model works very much like a reinforced concrete beam.

4. Create cardboard I-Beams of varying depth and note the resistance to bending that each provides.

FURTHER READING

Moore, Fuller. *Understanding Structures*. McGraw-Hill. 1998. (Chapter 8)

Schodek, Daniel and Martin Bechthold. *Structures*, 7th Edition, Pearson. 2014. (Chapter 8)

Allen, Edward and Waclaw Zalewski, *Form and Forces: Designing Efficient, Expressive Structures*, John Wiley & Sons, Inc. 2009. (Chapters 16 & 17)

Millais Malcom. *Building Structures: From Concept to Design*, 2nd Edition. Spon Press. 2005. (Chapters 3 and 4)

Sandaker, Bjorn N, Arne P. Eggen, and Mark R. Cruvellier. *The Structural Basis of Architecture*, Second Edition. Routledge. 2011. (Chapter 6)

FRAMES AND ARCHES

8.1 POST AND LINTEL

From time immemorial the problem of sheltering human beings from the weather has been solved by an enclosure of walls topped by a roof. In prehistoric times, walls and roof were made of the same material, without any distinction between a supporting "structure" and the protecting "skin." A separation of the supporting and protecting functions leads to the simplest "framed" system: the post and lintel (Figure 8.1).

The lintel is a beam simply supported on two posts and carrying the roof load. The posts are vertical struts compressed by the lintel. The posts must also resist some horizontal loads, such as wind forces; this resistance stems either from a bending capacity in the case of wooden or steel posts or from their own weight in stone or masonry piers. Some connection between post and lintel must also be provided, lest the wind blow the roof away.

The foundations of the posts carry the roof and post loads to the ground by means of footings which spread the load and guarantee that soil settlements will be limited (Figure 8.1). In some cases, lateral loads are resisted by base fixity in the soil through a large spread footing or an embedded pole (Figure 8.2). In any event, the posts and the foundations are essentially under compression, and this is characteristic of the post-and-lintel system.

Post-and-lintel systems may be built one on top of another to frame multistory buildings. In this case, the lintels are supported by vertical columns, or bearing walls of stone or masonry as high as the entire building (Figure 8.3). Construction of this type, while capable of carrying vertical loads, is not well suited to resist horizontal loads, and is easily damaged by hurricane winds and earthquakes. This happens because masonry or stone elements have little bending resistance, and a strong connection between the horizontal and the vertical structural elements is not easily developed.

8.2 THE SIMPLE FRAME

The action of the post-and-lintel system changes substantially if a rigid connection is developed between the lintel and bending-resistant posts. This new structure, the simple or single-bay frame, behaves monolithically, and is stronger than the post-and-lintel against both vertical and horizontal loads (Figure 8.4). Referred to as a *rigid frame* or *moment frame*, it is used commonly in buildings for lateral support against wind and seismic forces. The rigid connection between the beams and columns leads to the following further behavior investigation.

Figure 8.1 Post and lintel

The most elemental structural frame is referred to the *post and lintel*, which consists of two simple columns and a horizontal beam spanning between them. As an elemental single bay structural unit, it is conceptually equivalent of the single panel truss of Figure 6.29c. In both cases, larger structures are created by adding modules. Its use dates back to the earliest records of human civilization, such as the Egyptian temple of Karnak in Luxor, Egypt, shown here.

Photo: Oleg Kozlov/Fotolia

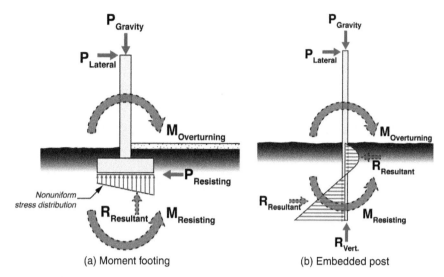

(a) Moment footing
(b) Embedded post

Figure 8.2 Fixed (i.e., moment-resistant) foundations

A post-and-lintel frame can often support gravity loads with nothing more than its basic components; however, it also requires the ability to resist rotation when lateral loads are applied. As later illustrated, one means is by providing fixity at the joint between the columns and beam (see Figure 8.4). The other means is to provide a moment connection at the base (Figure 8.2a). Columns are most frequently supported on footings that distribute the load over a larger area. When a lateral force is applied, an overturning moment is generated that must be resisted by an unequal pressure on the footing base. In some pole-type buildings, posts are embedded into the ground sufficiently deep to perform the function of rotational resistance under lateral loads (Figure 8.2b). Such foundations have a very different stress distribution, characterized by a pressure in direct opposition to the lateral force near the soil surface, but transitioning into a *reverse* pressure deeper in the soil, since the entire pole will tend to rotate within the ground itself.

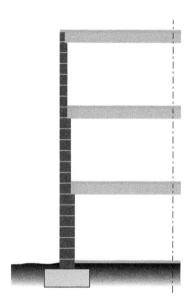

Figure 8.3 Multistory post-and-lintel system

A multistory frame is conceptually a stack of multiple single post-and-lintel frames. The walls typically thicken toward the base in order to carry the increasing load as the structure becomes taller.

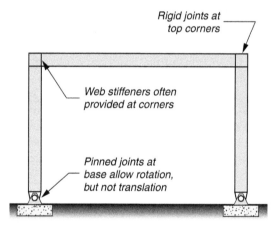

Figure 8.4 Simple moment frame

By making the corners of the single-bay frame capable of resisting rotation (moment), a simple frame that is stable against lateral forces is produced. The column base may itself be simply supported (i.e., permit rotation) or fixed as in Figures 8.2a and Figure 8.9.

Under a uniform load the lintel of a post-and-lintel system deflects, and its ends rotate freely with respect to the posts, which remain vertical (Figure 8.1). In order to grasp the action of the rigid frame under the same load, one may first consider the structure as supported by rollers allowing free horizontal movement of the columns at their bases. The horizontal beam is supported at its ends, with the columns rigidly connected to the rotated ends of the beam. The lack

of base fixity allows the columns to bow outward from the beam rotation so as to stick out in a straight, inclined position (Figure 8.5). In order to bring back the feet of the columns to their supports location, the columns must be forced inward by horizontal forces, and the ends of the beam must rotate partially backwards. In their final position, both the beam and the columns of the frame are curved and develop bending stresses (fig. 8.5b). The beam has partially restrained ends and behaves like the center span of a continuous beam on four supports (see Figure 7.41b).

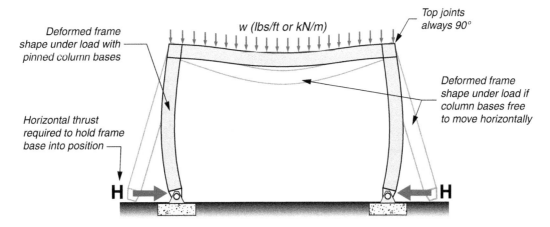

Figure 8.5 Hinged frame deflection

A characteristic of a moment frame acted on by gravity loads alone is the generation of a lateral force at the column bases. As the beam deflects under load, it will rotate at its ends. Since the columns and beams are rigidly connected, as the beam rotates, the columns will rotate as well. Because the columns are not free to slide laterally at the base, a horizontal thrust is generated, causing a bending action in the column.

Three consequences of *rigidly connecting the horizontal and vertical elements* to create a moment frame are immediately apparent: (a) the beam has partially restrained ends, it becomes more rigid, and is capable of supporting a heavier load in bending (see Section 7.3); (b) the columns are subjected not only to the compressive loads of the beam and of their own weight but also to bending stresses due to the continuity with the beam; (c) a new horizontal force is required to maintain the frame in equilibrium under vertical loads, the thrust, which brings back the columns to their bent vertical position, and, through the columns, introduces compression in the beam. The thrust is typical of frame action. It may be provided by the resistance of the foundation to lateral displacements, for example by the natural buttressing action of rock (Figure 8.6a); by buttresses of stone or masonry built for this purpose (Figure 8.6b); or by a tie-rod, which does not permit the frame to open up under the action of vertical loads (Figure 8.6c).

All three members of a simple frame under vertical loads are bent and compressed. Simple bending develops a linear stress distribution across the depth of the element, with maximum tensile and compressive stresses of equal value; compression adds a constant compressive stress. The combination of bending and compression together develops the resultant trapezoidal stress distributions shown in Figure 8.7. Usually compression prevails in the columns so that their stresses are entirely compressive (Figure 8.7a); bending prevails in the beam, so that tensile stresses are developed in some parts of the beam (Figure 8.7b).

The foot of the column may be either "hinged" or "fixed." In steel frames, the first type of connection may be provided by an actual hinge (Figure 8.8a) or by suitably located anchor bolts; in reinforced-concrete frames, it is provided by crossing the reinforcing bars so as to reduce the bending capacity of the section (Figure 8.8b). The deformation of the frame under vertical load shows that a hinged

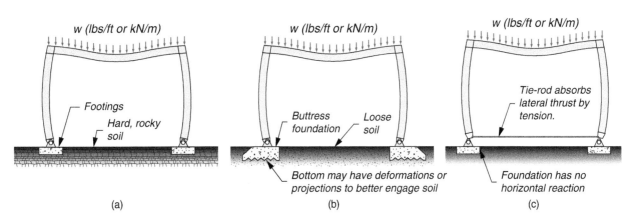

(a) (b) (c)

Figure 8.6 Thrust absorption

There are several approaches to resist the base thrust of a rigid frame under gravity loads. If the foundation is on a hard, rocky soil, the ground itself may be able to resist the lateral forces (8.6a). If the soil is softer, large and heavy footings that can develop sufficient friction with the ground may be used (8.6b). More economical than heavy footings, though, is to use a tie-rod between the two column ends to support the thrust in tension (8.6c). This is essentially similar in principle to a tie-rod employed in a timber truss (Figure 3.15).

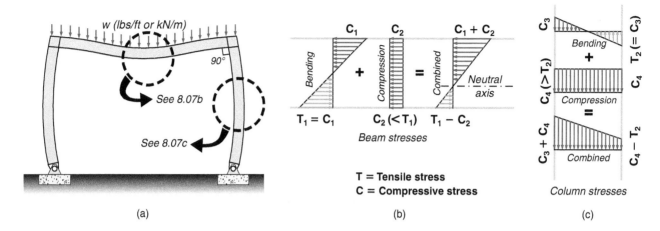

(a)   (b)   (c)

Figure 8.7   a) Stresses in a simple frame; b) Stresses due to compression and bending in beam, c) Stresses due to compression in right column
The stresses in a simple moment frame with only gravity load on the beam produces a combination of bending plus pure axial stresses in both the columns and beam. Figure 8.7a illustrates the internal stresses of the beam. In this member, bending stress predominates and the axial compression (as a result of the horizontal column thrust) subtracts partly from the tension side, while it adds to the compression side. In the column (Figure 8.7c), the axial forces are larger than bending. The tensile stresses of bending only partly offset the axial compression, and so the column remains in compression through its entire cross section. For both the beam and column, the resulting combined bending and axial stress is asymmetric over the member cross section. Furthermore, since bending stresses vary across the lengths of all members, the resulting combined bending plus axial stresses will likewise vary along their lengths.

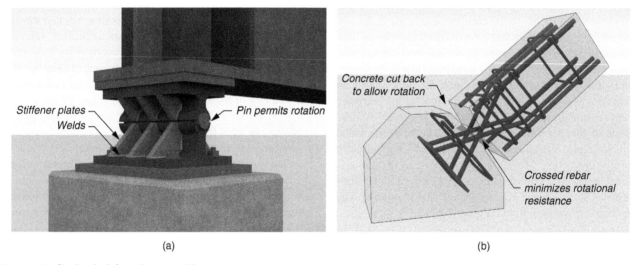

(a)   (b)

Figure 8.8   Steel and reinforced-concrete hinges
A hinge connection in structural steel will resemble a pivoting door hinge, but at a much larger scale (8.8a). There are numerous variations on this type of connection, and all allow rotation but resist translation(see also Figure 2.14). In concrete construction, a hinge must be made by intentionally locating the reinforcing bars close to the center where they have a much shorter internal moment arm to resist bending, thus allowing a degree of rotation to occur.

frame usually has smaller compressive stresses on the outer column fibers than on the inner fibers (Figure 8.7b).

The deformations of a fixed frame under vertical loads show that the columns develop inflection points (Figure 8.9). Since an inflection point is a point of no curvature, it is equivalent to a hinge where no bending stresses are developed. Hence, the fixed frame is roughly equivalent to a hinged frame with shorter columns and is stiffer than the hinged frame of Figure 8.6 (a,b,c). The thrust in the fixed frame is greater than in the hinged frame, since it takes a larger force to bring back its shorter, stiffer equivalent columns.

8.2.1 Lateral Frame Behavior

Frames are stronger against vertical loads than post-and-lintel systems, but their action is even more advantageous in resisting lateral loads. A post loaded by a horizontal load, such as wind pressure, acts as a single cantilevered beam without any collaboration from the lintel or the opposite post (Figure 8.10a). In the frame, instead, continuity with the beam transfers part of the wind load to the opposite column, as shown by the fact that both columns bend (Figure 8.10b). Even if the beam were hinged at the tops of the columns,

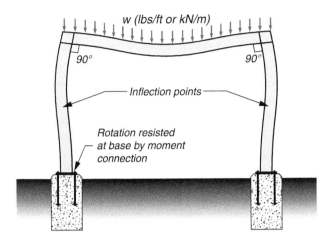

Figure 8.9   Simple fixed frame

A simple moment frame with its base *also* fixed becomes a simple fixed frame. Under gravity loads, this type of construction leads to a reversal of curvature in the columns known as *contraflexure*. The point where the direction of curvature changes is referred to as the *inflection point*. As in any moment frame, the joint between the beam and column maintains a right angle, and bending in the column is induced by beam rotation due to its deflection under load.

because of its rigidity in compression it would carry half the wind load to the leeward column, and cut the column bending stresses in half (Figure 8.11). Since, moreover, the rigidity of the frame connections compels the beam to bend together with the columns, an additional restraint is introduced in the columns, which become stiffer (Figure 8.10b). Their deflection is reduced, and so are their bending stresses.

The tendency of the frame to turn over due to the pressure of wind is balanced by additional reactions in the columns: tensile in the windward column and compressive in the leeward column (see Figure 8.10b); but these forces are usually small because their lever arm, equal to the width of the frame, is large compared to the lever arm of the wind pressure, which is half the frame height. The lateral displacement (or *sidesway*) produced in a frame by lateral loads is also present when unsymmetrical vertical loads are applied to the frame (Figure 8.12).

Whenever the beam of a fixed frame is much stiffer in bending than the columns, under lateral or unsymmetrical loads the inflection points in the columns appear approximately at mid-height; the shortened legs of the equivalent hinged frame show that the fixed frame is stiffer than the hinged frame against lateral loads also (Figure 8.13). In either type of frame, lateral loads must be considered acting from either side of the frame, so that the reinforcement of concrete frames must be located along *both* the inner and the outer fibers of the posts *and* the top and bottom fibers of the beam.

The bending-strength requirements of the columns are often so small as to make the entire frame flexible. If for functional reasons an open bay is required, a substantial increase in lateral rigidity of the frame can be achieved only by increasing the bending rigidity of its columns. On the other hand, lateral rigidity is inexpensively obtained by tensile or compressive elements; hence, tensile or compressive diagonals may be used to stiffen the frame with little increase in material (Figure 8.14) and with a large reduction of bending stresses in the columns and beam. In general, structural systems with rectangular meshes, such as frames, are more flexible than triangulated systems, but triangulated systems seldom meet the functional requirements of a modern building.

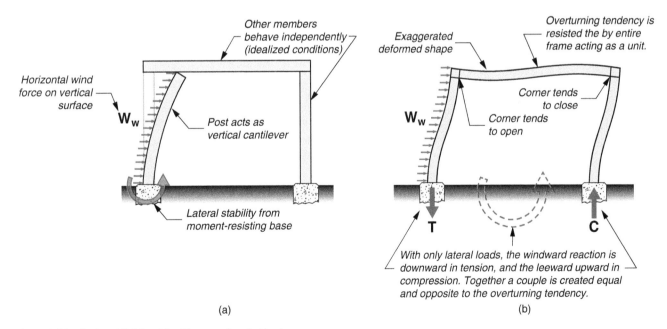

(a)   (b)

Figure 8.10   Post-and-lintel and fixed frame under wind load

Under lateral loads, a simple post-and-lintel frame will behave as independent components when not designed to transfer these forces. Walls will support loads individually without contribution from one another or the horizontal beam. In contrast, a fixed frame will share load among all members, thus reducing stresses and resulting deflections.

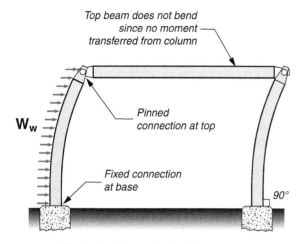

Figure 8.11 Wind load on frame with hinged bar

Even if a simple frame has pinned connections at the beam, lateral forces will be transferred through the beam to the opposite column, and both columns will share in carrying the load.

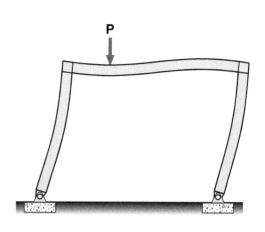

Figure 8.12 Sidesway under vertical load

One aspect of a rigid frame structure under asymmetric gravity loading is the propensity to shift laterally, due to asymmetric internal members stresses. Thus, even without an actual lateral force, lateral deformation is generated by the load placement alone.

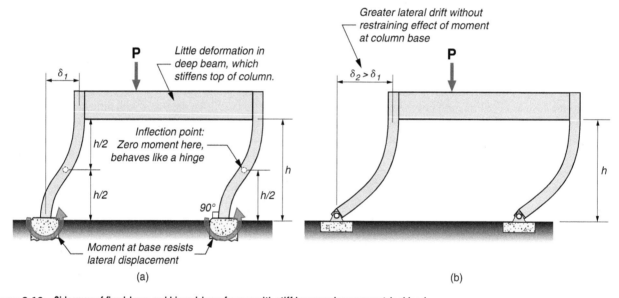

(a)   (b)

Figure 8.13 Sidesway of fixed-base and hinged-base frames with stiff beam under asymmetrical load

Comparing a fixed base frame with an equivalent pinned base frame, when constructed using a stiff beam that will exhibit little deflection, it can be seen that the fixed base frame will have much less tendency for sidesway. The moment at the foundation tends to keep the columns vertical, leading to a reversal of curvature of the columns at approximately mid-height, and a stiffer structural system overall.

Figure 8.14 Diagonally stiffened frame

By far the most efficient means of stabilizing a structure laterally, a diagonally stiffened frame can use all pinned connections, resulting in a reduction of material and fabrication costs. Such a frame is essentially a truss structure cantilevering vertically from the ground. While diagonals in tension are the most efficient, compression members may also be employed.

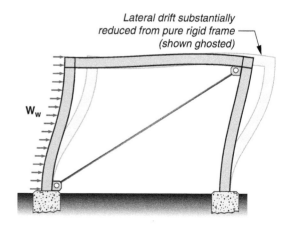

8.3 MULTIPLE FRAMES

The advantages of continuity can be compounded by the use of multiple frames, in which a horizontal beam is supported by, and rigidly connected to, three or more columns.

In view of the rigidity of compressed elements (see Section 5.2), the vertical deflections of the columns of a multiple frame are small, and the behavior of its beam, when loaded vertically, is similar to, although not identical with, that of a continuous beam on rigid supports (see Section 7.3). Under the action of loads concentrated on one span, the multiple frame develops curvature not only in all other spans, as a continuous beam, but in all columns as well. The frame as a whole also exhibits a lateral displacement, referred to as sidesway, which is absent in the continuous beam (Figure 8.15).

If the spans and the vertical loads of the multiple frame do not differ substantially from bay to bay, the thrusts in adjoining frames act in opposite directions and tend to cancel each other. Theoretically, only the two outer bays need to be tied or buttressed. In practice, some means must always be provided to take care of uneven loadings on the various spans of the multiple frame: The excess thrust on each column is often resisted by the action of the soil on the foundations.

Multiple frames are efficient in resisting lateral loads. The rigidity of the beam against compressive loads makes the lateral deflections of the tops of all the columns practically identical. Hence, if the columns are identical, the lateral load is carried equally by all the columns of the frame, each acting as a cantilever and additionally stiffened by the bending rigidity of the beam. The overturning tendency due to wind action is equilibrated by tensile reactions on half the frame columns on the windward side, and by compressive reactions on the half on the leeward side (Figure 8.16). In view of the greater number of column reactions and their large lever arms, these forces are relatively small. Multiple frames in the external walls or in the core walls of a building are often used to carry the wind load on the faces perpendicular to these walls (Figure 8.17).

A single-bay frame with a beam rigidly connecting the base of the columns is a closed structural element capable of spanning a distance while supporting both vertical and horizontal loads (Figure 8.18a). A multiple-bay frame with continuous beams at the top and the bottom of the columns is also a closed system, and may be used as a type of truss to span large distances (Figure 8.18b). In this case, the multiple frames may be thought of as a deep beam with compression and tension flanges, with shear forces being resisted by the columns. The "secondary" bending stresses in the horizontal and vertical elements of such truss-frames (see Section 7.4) are much greater than in the members of triangulated trusses, but the simplicity of their connections and their unencumbered bays have made these truss-frames popular.

Such Vierendeel frames, named after their Belgian inventor, are commonly used in bridge design. Often referred to as Vierendeel *trusses*, they are not true trusses because they lack triangulation. Vierendeels are also used in buildings when the structure is supported on wide spans and unencumbered bays are essential. In this case, the frame consists actually of the columns and floor beams of the building (Figure 8.19).

Multistory frames are commonly used in tall structures. Their action against vertical loads is similar to that of

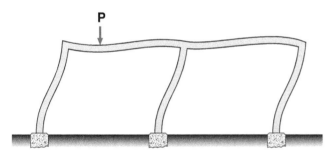

Figure 8.15 Sidesway of multiple frames
Similar to a single frame with an asymmetric load, sidesway will be produced in the entire structure, and all members will be affected. Because all columns are connected at their tops, the lateral displacement will be equal for each. Bending effects, however, dissipate with increasing distance from the load.

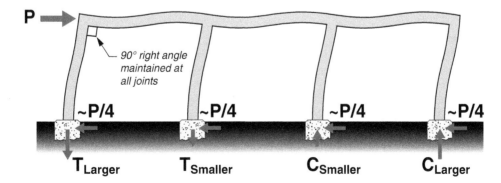

Figure 8.16 Lateral resistance in multibay frame
Lateral resistance in a multibay frame is shared by all members. Lateral loads (collected at the roof and floor levels) develop an overturning tendency for the frame. The overturning is resisted most by the exterior columns in a moment couple and secondarily by interior columns with a smaller moment couple. All columns share approximately equally in linear equilibrium to the lateral force, P.

Figure 8.17 Typical lateral frame locations in building plans

Lateral framing structures are commonly located at the perimeter of the structure (8.17a) or within one or more central cores (8.17b) (see also Figure 8.28). The core location will typically coincide with the location of elevators and stairways, mechanical shafts, bathrooms, mechanical and electrical equipment closets, and so on. Countless variations between a pure perimeter or core structure are possible.

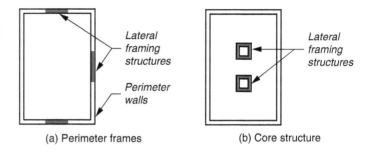

(a) Perimeter frames   (b) Core structure

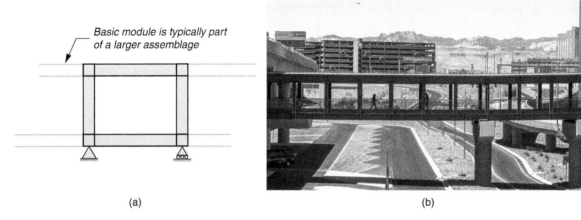

(a)   (b)

Figure 8.18 Vierendeel frame and bridge

A Vierendeel is essentially a complete rigid frame with moment connections at all four corners and members on all sides, forming a basic unit capable of spanning a distance (8.18a). The lack of diagonals makes it an attractive alternative to spans using a truss, even at the expense of material efficiency (they will be heavier than trusses). They can be used for bridge structures such as this pedestrian bridge in Las Vegas (8.18b).

(b) Photo courtesy of Deborah Oakley

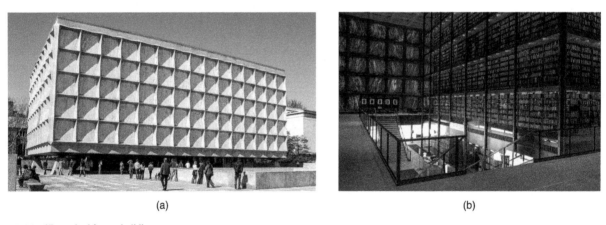

(a)   (b)

Figure 8.19 Vierendeel frame building

Vierendeels can be stacked both vertically and horizontally to form deep beam structures. One of the most famous such structures is the Beinike Rare Book Library at Yale University. The building is constructed of four deep perimeter Vierendeel frames, each supported on two corner columns, and so the entire building façade is supported by only four columns in total (8.19a). One of the most elegant features of the building is found in how it takes advantage of the openings in each frame section to fill the space with thinly sliced marble panels. Filtered sunlight passes through the thin marble, illuminating the interior with a soft glow that highlights the stone's natural beauty while protecting the delicate books stored inside (8.19b).

Photo courtesy of Michael McGlynn

single-bay frames, with the added advantage that the horizontal beams act both as load-carrying elements for a given floor and as tie-rods for the frame above it.

The action of a single-bay, multistory frame under lateral loads is similar to the action of a cantilevered Vierendeel frame: The leeward columns are in compression and bending, the windward columns are in tension and bending, while the floor beams or slabs transmit the shear from the tension to the compression elements (Figure 8.20). One may also think of a multibay, multistory frame as a gigantic I-beam,

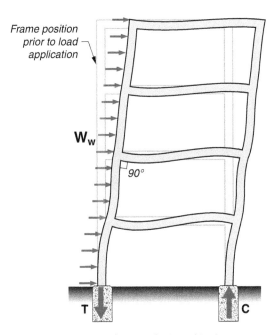

Frame position prior to load application

W$_w$

90°

T   C

Figure 8.20 Multistory rigid frame under lateral loads

A multistory frame can be conceived of as a Vierendeel cantilevering from the ground. All members participate in carrying the forces. They may also be envisioned as building-sized I-Beams, where the outer columns form the flanges and carry moment, while the beams transmit shear.

bending in both the horizontal and the vertical elements. A concentrated vertical load is felt in the entire structure, each part of which collaborates in carrying it (Figure 8.21), although the effects rapidly peter out within a short distance of the loaded member.

As the height and the width of the building increase, it becomes practical to increase the number of bays so as to reduce the beam spans and to absorb horizontal loads more economically. The resisting structure of the building thus becomes a frame with a number of rectangular meshes, allowing free circulation inside the building, and capable of resisting both vertical and horizontal loads. A number of such frames, parallel to each other and connected by horizontal beams, constitute the cage structure encountered in most steel or concrete buildings today. These three-dimensional frames act integrally against horizontal loads coming from any direction, since their columns may be considered as part of either system of frames at right angles to each other (Figure 8.22).

The high-rise building, or skyscraper, is one of the great conquests of modern structural design, made possible by the multi-story frame and the high strength of steel and concrete. The Empire State Building in New York, built in 1930, has 102 floors and is 1050 feet high (320 m), not including the 200-foot (61 m) upper tower originally designed to anchor dirigibles, and the 222-foot (68 m) TV antenna.

The eminent structural engineer Fazlur Khan of Skidmore, Owings and Merrill developed new and innovative concepts to address increasing forces of wind in super tall buildings, while maintaining the greatest economy of the structural system. The 1,127 ft (344 m) John Hancock

cantilevered from the ground, in which the depth equals the width of the frame and the floor beams act as a discontinuous web. The continuity between the floor beams and the columns makes the entire structure monolithic and introduces

Figure 8.21 Multistory frame under concentrated load

Due to their moment-resistant joints, large-scale moment frame structures transmit forces throughout the assemblage. A large load placed on one beam will affect a large portion of the structure, the effects dissipating with distance from the load. Frames are subject to sidesway due to loading asymmetry, so must be designed to minimize potential lateral movement.

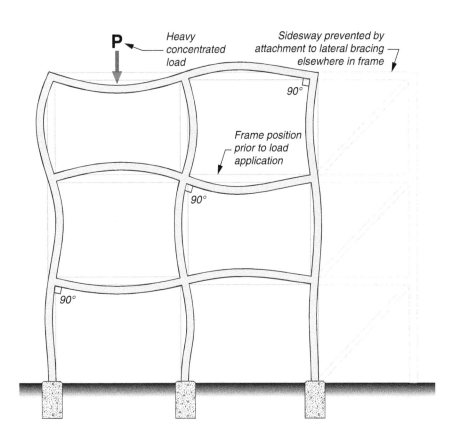

P

Heavy concentrated load

Sidesway prevented by attachment to lateral bracing elsewhere in frame

90°

Frame position prior to load application

90°

90°

Figure 8.22   Three-dimensional frame

The basic planar unit of a moment frame can be extended into the third dimension to create a spatial structure. The concept for the frame developed rapidly in the city of Chicago, Illinois, in the decades following the Great Chicago Fire of 1871 that leveled the city. It became recognized that first iron and later steel frames could carry the weight of exterior cladding rather than increasingly thick walls as buildings grew taller. The evolution of that idea has led to the proliferation of tall buildings throughout the world today.

Photo courtesy of Deborah Oakley

Figure 8.23   Hancock Tower

The Hancock Tower uses a framed "tubular" design. The diagonal trussing around the perimeter forms a closed vertical tube, behaving very much like a cantilevered truss under lateral loads. Each diagonal "X" spans up to 18 stories. Condominiums and offices with window views disrupted by the diagonals were, counter to initial expectations, proven to be popular and in demand.

Photo: MaxyM/Shutterstock

Figure 8.24   Willis Tower

The tubular construction of the Willis Tower (formerly the Sears Tower) is evident in this photograph. Nine three-dimensional moment frames (each 75 ft (23 m) on a side), work in unison to achieve a strong, yet highly economical, structure. The black banding most evident toward the lower third of the building are multistory high trusses that enclose mechanical spaces. The trusses serve to tie the tubes together, as well as redistributing loads from the varying height tubes.

Photo: Keith Levit/123rf

Insurance Company building in Chicago (Figure 8.23) uses a trussed tube configuration. Under lateral loads, the tower behaves very much like a cantilevered tubular truss, with each diagonal brace spanning as many as 18 stories. The Willis Tower, also in Chicago (Figure 8.24), at 1450 ft (442 m) was the tallest building in the world for nearly 25 years. Kahn's structural scheme for this building introduced the concept of *bundled tubes* consisting of a series of nine square moment frames, each topping out at different heights. Similar to shear forces in a beam (see Figure 7.20), combining several smaller building frames together makes them function as a unit for enhanced strength and resistance to bending deflections for the building as a whole.

Tubular high-rise steel buildings have also been built by bundling together a number of nonrectangular tubular frame sections. The bundled trussed triangular tubes in the case of the Bank of China building in Hong Kong (Figure 8.25), engineered by Leslie Robertson, are prominent on the façade. Like the Willis Tower, the four triangular tubes terminate at varying heights.

An evolution of the previous structural concepts is the *diagrid*. These structures utilize the braced frame tube concept of an earlier generation, but the in this case eliminate perimeter vertical columns. The diagonals of a diagrid thus

Figure 8.25   Bank of China Building

The 72-story Bank of China Building uses the bundled tube concept, here in the form of four triangular tubes constructed with diagonally braced multistory frames. Like the Willis Tower, each of the four triangular-shaped tubes terminates at a different height.

Photo courtesy of Terri Meyer Boake

Figure 8.26   New York: Hearst Tower

The Hearst Tower, global headquarters of the Hearst Corporation, employs one of the more recently developed variants of framed structures, known as a *diagrid*. At first glance, the 46-story tower appears similar to the taller Hancock Tower of Figure 8.23, but with smaller diagonals. On closer inspection, though, it can be seen that the diagrid structure actually eliminates *vertical* perimeter columns. Here, the diagonal members thus function to carry gravity loads as well as lateral forces.

Photo courtesy of Terri Meyer Boake

carry both lateral loads *and* gravity loads. Diagrids have gained popularity in recent decades in such famous projects as Foster and Partner's 30 St. Mary Axe Building (Swiss Re) in London, engineered by Arup, and their Hearst Magazine Tower in New York City (Figure 8.26), engineered by WSP Cantor Seinuk. The expressed diagonal grid on the façade gives these buildings a distinctive appearance.

Innovations continue in the area of super tall structures. Engineer William Baker of Skidmore, Owings and Merrill developed a "buttressed core" design for the current tallest building in the world (2015), the 2716 feet (828 m) Burj Khalifa in Dubai, UAE (Figure 8.27). This will be surpassed upon completion of the Kingdom Tower in Saudi Arabia engineered by Thornton Tomasetti, anticipated at 3,304 ft (1007 m) when completed in 2018. If executed as planned, the Kingdom Tower will be the first to break the symbolic height of one kilometer.

The foregoing are all steel structures consisting of frames with bays of narrow width on the outside of the building. Such structures behave essentially as cantilevered steel tubes of rectangular cross section. The early years of the twenty-first century have seen a veritable explosion of super tall buildings, all built with variations of multiple frames, and this trend will only continue into the foreseeable future.

Reinforced concrete skyscrapers typically cannot reach the height of steel skyscrapers, but have been built to heights of over 1300 feet: the tallest to date (2015) is the 1,389 foot (423 m) Trump International Hotel and Tower in Chicago. Such towers usually consist of outer frames and of an inner core built by means of concrete walls. The lateral stiffness of the boxed core is usually so large relative to that of the outer frame that the lateral forces due to wind or earthquake are predominantly resisted by the so-called shear walls of the core. (These walls resist the horizontal loads as cantilevered

Figure 8.27   Burj Khalifa

Currently (2015) the world's tallest building, at 2,717 ft (828) and 163 floors. Using a *buttressed tube* concept designed by engineer William Baker of Skidmore, Owings and Merrill, the framed structure is here taken to its greatest extreme to date.

Photo courtesy of Jocelyn Hidalgo

Figure 8.28   Steel frame with concrete core

As illustrated in Figure 8.17b, core structures are frequently used for lateral support. While the core may be constructed as a steel frame, it is also very frequently made of concrete. Heavy core structures have become important design concepts in all recent high-rise buildings to serve as the vital lifeline to safely evacuate buildings in the event of an emergency, as well as the strongest element to maintain overall structural integrity.

Photo courtesy of Terri Meyer Boake

beams in bending; their misleading name indicates that they resist the wind shear.)

Because of their lateral stiffness, concrete core walls are also used in high-rise buildings with outer steel frames (Figure 8.28), thus combining the advantages of smaller steel columns to resist vertical loads only and of concrete walls to resist lateral loads. The steel frame of the core of a steel building may also be stiffened by steel diagonals or by prefabricated concrete panels inserted in its bays.

A skyscraper is nothing but a slender, vertical, cantilevered beam resisting lateral and vertical loads. The vertical floor loads accumulate from the top to the bottom of the building, requiring larger columns as one moves down. Were it not for the high compressive strength of modern structural materials, the area of the columns would encroach on the usable floor area to the point where this might become nil.

The efficiency of frame and shear-wall action is required by skyscrapers to resist large horizontal wind loads with minor deflections. A constant wind pressure of 30 pounds per square foot (1.4 kPa) due to a 90 mile per hour wind (145 km per hour)) represents a total load of more than nine million pounds (40 MN) on the 1127 foot (344 m) cantilever

of Chicago's Hancock Tower. For buildings located on coastal areas, the wind forces are even higher. Under the action of this wind load the top of the building sways less than ($\pm$)12 inches (0.3 m) each side of vertical owing to the rigidity of the trussed tube. Many more flexible structures of similar or greater height can have lateral sway of several feet (1 m). In order to avoid discomfort to the occupants of the top floors of a high-rise building, the top deflection or drift must be limited to between one thousandth and one five-hundredth of its height. Such stiffness is required because human discomfort is severe when the period of the swaying is in resonance with the period of human internal organs.

The columns, beams, and walls of a framed structure are its resisting "skeleton." In order to enclose the space defined by the skeleton and to make it usable, the exterior of the building is covered with a "skin" and the floor areas are spanned by horizontal floor systems. The skin of modern buildings is often made of metal or concrete, and glass. It is called a *curtain wall* due to it being suspended from the floors like a fabric curtain is from a rod. A curtain wall of prefabricated concrete or brick panels may also be used to enclose buildings. Such systems must be connected to the structure so as to allow expansion and contraction of the panels under differences in temperature between the exterior and the interior of the buildings. The floor structure consists of long-span beams connecting the outer frames to the core, of secondary beams or joists spanning the distance between the main beams, and of slabs of concrete or steel decks spanning the distance between the secondary beams. Floor systems are analyzed in Chapter 10.

8.4 GABLED FRAMES AND ARCHES

All three members of a single-bay frame under vertical loads are subjected to compressive and bending stresses (Figure 8.7). The columns are compressed by the loads on the beam and bent by the rotation of the joint connecting them rigidly to the beam. The beam is bent by the loads on it and compressed by the thrust from the base of the columns. With the usual proportions of beam and columns, compression prevails in the columns and bending in the beams, since the columns are relatively slender and the beam is relatively deep.

Whenever for functional reasons the top member of the frame must be horizontal, this type of design is efficient and economical; but it may be improved when the upper member does not have to be horizontal. In gabled frames, the top member consists of two inclined beams. If these were hinged at top and bottom, they would act as the compressed struts of a triangular truss, and the columns would be bent by the thrust from the upper members, besides being compressed by their vertical reactions (provided the column bases are fixed at the ground to create overall stability) (Figure 8.29). Continuity between the gable and the columns introduces bending in the gabled member, so that the frame transfers loads by a combination of compression and bending in all its members (Figure 8.30). The higher the rise of the gable, the smaller is the thrust and hence the bending in it and in the columns. The load-carrying capacity of the gable may be split between the two mechanisms of compression and bending, gaining in efficiency, and saving material in comparison with a beam in bending.

The principle used in gabled frames may be extended. The columns may be shortened and the upper member folded so as to have more than two straight sides, with an increase in compression and a reduction in bending (Figure 8.31). In the limit, the polygonal frame with an infinite number of infinitesimally short sides becomes an arch.

It was noted in Section 6.4 that for each set of loads there is a particular arch shape (the so-called funicular shape) for which the entire arch is under simple compression. Such shape can be determined by hanging the arch loads from a cable and by turning the resulting curve upside-down (Figure 8.32).

Funicular arches are at one end of the structural scale, where there is uniform compression and no bending; beams

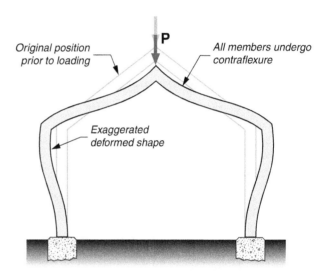

Figure 8.30 Continuous gabled frame
Adding rigidity at the joints of the gabled frame increases structural efficiency. The sloping members are more effective at carrying forces in combined bending and axial stress than a simple beam in bending.

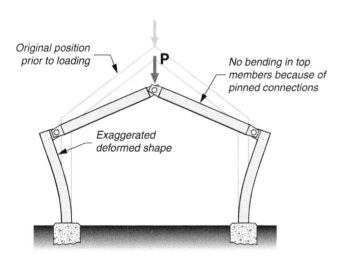

Figure 8.29 Hinged gabled frame
A gabled frame takes the horizontal beam of a simple frame and splits it into two sloping members. Lateral thrust from the sloping members acting at the tops of the columns, however, must be accounted for in the design of such frames. Tie-rods, such as Figure 3.15, are commonly used for this function, or otherwise the columns must function as vertical cantilevers supporting the horizontal thrust.

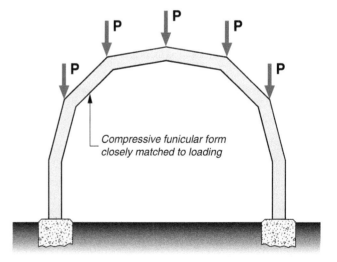

Figure 8.31 Polygonal frame
Adding folds to a gable frame results in a polygonal frame, with the eventual shape being a semi-circular arch as the number of sides is increased. The shape that is most closely matched to its funicular form will maximize efficiency by minimizing bending action.

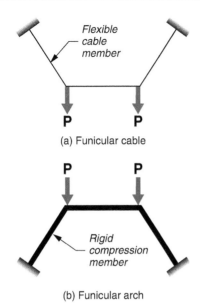

(a) Funicular cable

(b) Funicular arch

Figure 8.32   Funicular cable inverted to arch shape

For any specific loading, the funicular form is that which a cable will take under this loading (8.32a). The cable will naturally reshape into *one and only one* shape based on the load position and direction, distance between the loads, and the position of the supports. Inverting the form of this structure and applying exactly the same loads results in the funicular arch, a purely compressive structure with no bending action. It must be stressed, however, that the form is funicular for *one and only one* loading. All other loads will cause varying degrees of bending.

are at the other, where there is bending and no uniform compression. Any other downward curved structural member will carry loads by a combination of compression and bending. Even if an arch is funicular for one set of loads, it cannot be funicular for all sets of loads it may have to carry: A combination of compression and bending is always present in an arch.

In masonry design, the arch is heavy and loaded by the weight of walls; its shape is usually the funicular of the dead load, and some bending is introduced in it by the live loads. In large steel arches, the live load represents a greater share of the total load than in masonry arches and introduces a larger amount of bending; but bending is seldom critical, in view of the tensile strength of steel.

The arch thrust is absorbed by a tie-rod whenever the foundation material is not suitable to resist it. When the arch must allow the free passage of traffic under it, whether it is a bridge or the arched entrance to a hall, its thrust is absorbed either by buttresses or by tie-rods buried under ground.

The stationary or moving loads carried by the arch of a bridge are usually supported on a horizontal surface. This surface may be above or below the arch, connected to it by compression struts (Figure 8.33a) or tension hangers (Figure 8.33b). The compression struts are often restrained laterally against buckling by diagonals or transverse beams, particularly when the arch is deep.

The shape of an arch is not necessarily chosen for purely structural reasons. The half circle, exclusively used by the Romans, has convenient construction properties that justify its use. Similarly, the pointed Gothic arch has both visual and structural advantages (it develops lesser thrusts), while the Arabic arch, typical of the mosques and of some Venetian architecture, is "incorrect" from a purely structural viewpoint, but often only visually so because it conceals a Gothic arch within the wall itself. (Figure 8.34).

The shape of the arch may be chosen to be as close as possible to the funicular of the heaviest loads, so as to minimize bending. On the other hand, it is often more important to minimize the arch thrust so as to reduce the dimensions of the tie-rod, or to guarantee that the soil will not move under the pressure of the abutments. The thrust may be shown to be proportional to the total load and to the span and inversely proportional to the rise of the arch. To minimize the thrust for a given distance to be spanned, the arch must be as light as possible, and have a rise as high as economically feasible (Figure 8.35).

The arch may have hinged or fixed supports. The hinged supports allow the rotation of the arch at the abutments under load or temperature changes. Hinged arches are relatively flexible and do not develop high bending stresses under temperature variations or soil settlements.

(a)

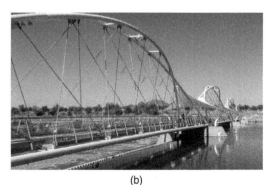

(b)

Figure 8.33   Arch Bridge structures

Arches have been used for centuries as bridge structures. In the past, they would be massive stone structures. New developments in materials and in structural understanding have enabled increasingly efficient structures to emerge based on this ancient structural form. Today, arches are used below road decks with compressive struts to support the road deck from below as in the Infante D. Henrique bridge in Porto, Portugal (8.33a), or suspended from above as in the pedestrian bridge crossing the Town Lake reservoir in Tempe, Arizona (8.33b).

Photo courtesy of Deborah Oakley

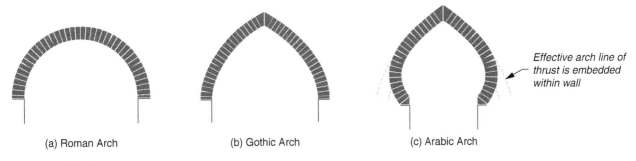

(a) Roman Arch   (b) Gothic Arch   (c) Arabic Arch

Effective arch line of thrust is embedded within wall

Figure 8.34   Arch shapes

Arches come in many forms, not always based on structural efficiency. The Roman arch, for example, is simply constructed by employing reusable semi-circular formwork to temporarily support the arch under construction. The pointed Gothic arch is more efficient as the form is closer to the funicular shape. The Arabic arch can be envisioned as a form of the Gothic arch where the lower portion is actually contained within the wall itself.

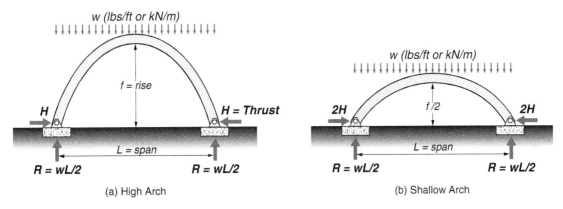

(a) High Arch   (b) Shallow Arch

Figure 8.35   Variation of arch thrust with rise

As with funicular cables, there is an inverse correlation between arch thrust and rise (see also Figure 6.4). The higher the rise, the lower the arch thrust will be.

Steel hinges are used at the abutments of most steel arches (see Figure 8.8a). Reinforced-concrete hinges were developed by the French engineer Mesnagier, who first crossed the reinforcing bars to create a section where the bending resistance is substantially weakened by the lack of lever arm between the steel bars (see Figure 8.8b).

Fixed arches are built both in steel and in concrete. They are stiffer than hinged arches and consequently are more sensitive to temperature stresses and settlements of the supports.

If temperature or settlement stresses are to be entirely avoided, a third hinge may be introduced, usually at the crown of the arch: The three-hinge arch is entirely free to "breathe" or to settle unevenly without bending deformation of its two halves (Figure 8.36). Such arches, commonly used at the end of the nineteenth century, have lost their popularity because of the improved knowledge of arch behavior. Eliminating the crown joint also minimizes maintenance; many structural problems and failures can be traced to worn or deteriorated joints that were not properly maintained over the life of the structure (see Chapter 13).

When bending materials were costly and easily deteriorated by adverse weather conditions, the arch was the structural member most commonly used to span even modest distances. Roman and Romanesque architecture are immediately recognized by the circular arch motive. Romans used arches extensively in the extensive system of aqueducts throughout the Empire (Figure 6.41). The Gothic, high-rise arch and the buttresses required to absorb its thrust are typical of one of the greatest achievements in architectural design, the Gothic cathedral (Figure 8.37). In these magnificent structures, built of stone capable of resisting compression only, a combination of arches and vaults channels to the ground the loads of roofs spanning up to one hundred feet (30.5 m). The high rise of the arch reduces the thrust, and lightened flying buttresses, in the shape of half-arches, channel it from the inner to the outer pillars whose heavy weight, combined with that of the pinnacles, deflects it to an almost vertical direction. A Gothic cathedral is essentially a framed building in which the curved frame elements develop almost exclusively compressive stresses.

It was only with the development of modern bending-resistant materials that greater structural achievements were obtained. Roman circular arches spanned about 100 feet (30.5 m) and medieval stone bridges up to 180 feet (54.9 m), but the New River Gorge Bridge in West Virginia spans 1700 feet (518.2 m) (Figure 8.38). The world's longest steel-arch span to date (2015) is the Chaotianmen Bridge in Chongqing, China, spanning 1811 feet (552 m). The largest single-arch span in reinforced concrete built to date is the 1380 foot (420 m) span Wanxian Bridge in China, which in 1997

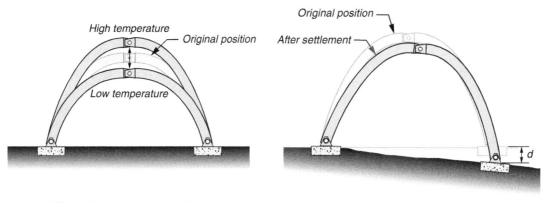

(a) Three-Hinged Arch Temp. Change   (b) Three-Hinged Arch Settlement

(c) Three-Hinged Arch Bridge

Figure 8.36   Three-hinge arch

A three-hinged arch consists of two half-segments connected by pinned base supports and a pin connecting the two halves at the arch apex. Owing to their ability to reshape without inducing additional stresses, such structures are relatively insensitive to temperature induced thermal deformations (8.36a), as well as movements due to foundation settlement (8.36b). Less commonly utilized than in previous eras, it is still a popular choice for certain material systems, such as arches and frames constructed of glued laminated timber. The Keystone Wye Bridge in South Dakota is an example of a three-hinged arch bridge (8.36c).

Photo: Photosbyjam/Shutterstock

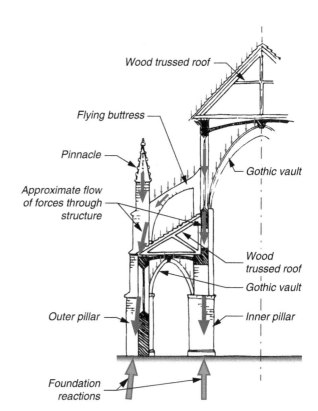

Figure 8.37   Section of Gothic cathedral

The Gothic Cathedral is a masterpiece of medieval design, utilizing stone in compression only through a series of vaults. The lateral thrust of the central nave vault is channeled to the ground through flying buttresses—half-arches that receive the upper wall thrust and transfer it to the outer pillars. Additional dead loads provided by masonry pinnacles create greater stability by directing the force nearly vertically.

Figure 8.38 New River Gorge Bridge

Spanning 1700' (518m), the New River Gorge Bridge has remained the longest steel span arch bridge in the Western Hemisphere since it opened in 1977.

Photo courtesy of West Virginia Division of Culture and History

surpassed the Krk Bridge in Croatia, the previous longest at concrete arch at 1365 feet (416m) (Figure 8.39). Combinations of trussed arches with cantilevered half-arches connected by trusses were built to span as much as 1800 feet (548.6 m) in the Quebec Bridge in 1917 (Figure 8.40). To this day, no other structural element is as commonly used to span large distances as the arch.

8.5 ARCHED ROOFS

It was seen in the previous section that the roof of a Gothic cathedral is a combination of arches and vaults, capable of channeling the roof load to the ground. Arches are used in a variety of combinations to support curved roofs. One of the simplest is a series of parallel arches connected by

Figure 8.39 Krk Bridge

The Krk bridge connects the mainland of Croatia with the island of Krk, and consists of two concrete arch structures, the longer of which at 1365 feet (416m) is second only to the 1380 foot (420 m) Wanxian Bridge in China.

Photo courtesy of Marin Lucic

Figure 8.40 The Quebec River Bridge, near Quebec, Canada

Constructed in 1917, the structure is still the longest cantilever bridge in the world. The short center span is supported by two massive cantilever arms, each of which is balanced by equally massive cantilever end spans. The clear span is 1800 feet (549m), and the 67 foot (20 m) width of the span is enough to accommodate two railroad tracks, two streetcar tracks, and two roadways. It was designated an Historic Civil Engineering Landmark by the American Society of Civil Engineers in 1987.

Photo courtesy of Julien Cochlin

transverse elements and roofed by plates, which constitute a barrel roof (Figure 8.41). The connecting elements transmit the roof load to the arches by bending or arch action, and these channel the load to the ground mostly by compression. This is the mechanism used in the medieval, Romanesque church roofs, and in the modern cylindrical concrete hangars, first built in Europe in the 1920s and in the United States in the 1940s. The shape of the arches is chosen to be the funicular of the dead load; any additional load due to wind or snow produces a certain amount of bending in the arches. The element connecting the arches is a brick vault in the medieval churches, and a concrete curved plate in the modern hangars. Barrel roofs supported by steel arches are also common in modern factory roof design.

Rectangular areas are often covered by structures in which the arches span the diagonals rather than the sides of the rectangle (Figure 8.42). Typical of this kind of structure are the vaults over the transept of medieval churches, and some reinforced-concrete structures used in large halls, as in the air terminal at St. Louis, Missouri, designed by Minoru Yamasaki and engineered by Anton Tedesko.

Areas with circular or otherwise curved boundaries may be covered by arches converging toward the center of the curved area, as in some Gothic chapels or modern wood roofs (Figure 8.43),the arches butt against a central ring or plate and may be restrained by a perimeter tension ring; their action is more integral than in the case of parallel arches. Arched domes of steel, wood, and reinforced concrete have

been used in modern times as roofs of exceptionally large halls for a variety of functional purposes (see Chapter 12).

In the lamella cylindrical roof structure, a series of parallel arches, skewed with respect to the sides of the rectangular covered area, is intersected by another series of skew arches so that an efficient interaction is obtained between them (Figure 8.44a). This system does away with the beams connecting parallel arches and constitutes a curved space frame. It is particularly efficient when the lengths of the sides of the rectangular area do not differ appreciably, since in this case the arches near the corners have small spans and are very rigid, thus reducing the effective span length of the arches butting into them (see also Section 10.3). Lamella roofs of wood, steel, and concrete are used to create barrel-shaped roofs. Among the largest concrete roofs of this kind are the hangar roofs designed and built by Nervi in the 1940s which spanned 1080 feet (329 m) by 427 feet (130 m) and consisted of prefabricated concrete elements, connected by welding the reinforcing bars at the mesh points of the roof and concreting the joints. None of these hangars exists today, since they were dynamited during World War II, but it is interesting to notice that after being dropped to the ground from a height of 40 feet (12 m), these roofs remained integral, except for a few among the many hundreds of joints.

In the lamella dome, the intersecting ribs are spirals (Figure 8.44b). Arched ribbed roofs behave structurally in a manner analogous to shells; their load-carrying mechanism is explained in Chapter 12.

Figure 8.41 Arched barrel roof
Austrian engineer Anton Tedesko is widely credited with introducing thin shell concrete construction in the United States in the early 1930s. Included in his many projects is this aircraft hanger at Rapid City, SD Ellsworth Air Force Base. In this type of construction, parallel concrete arches form the primary load-carrying structure that supports a roof spanning between the individual arches.

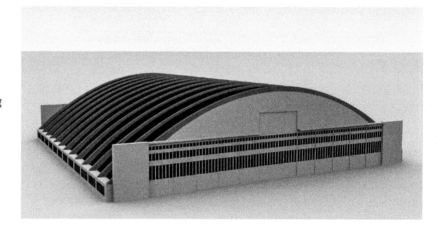

Figure 8.42 Diagonal-arch roof
Arched roofs can also span across diagonals, with secondary members supported by the arches, which span parallel to the base.

Figure 8.43 Radial-arch roof
Arches may also be arranged in a radial pattern around a central compression ring. Outward thrust is resisted by a perimeter tension ring.

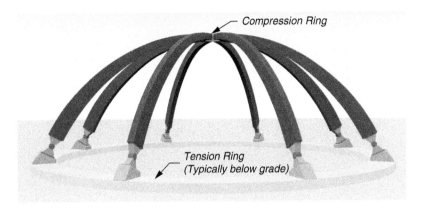

Compression Ring

Tension Ring
(Typically below grade)

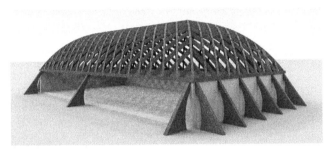

Figure 8.44a Lamella cylindrical roof
A lamella roof consists of primary members forming a cylindrical roof, which are arranged in a diagonal pattern that creates diamond-shaped supports for structural surface of the roof. They may be constructed of wood, steel, or concrete.

Figure 8.44b Lamella dome
Domes may also be designed with a lamella construction. The diagonally skewed arches form a spiral around a central compression ring perimeter tension ring similar to Figure 8.43.

KEY IDEAS DEVELOPED IN THIS CHAPTER

- Post-and-lintel systems can carry vertical loads, but are unstable under horizontal forces.
- Simple frames may consist of three bars connected with joints either allowing rotation or having the horizontal element rigidly connected to the two verticals. In the second case, all three members will experience bending, and the verticals will have additional compressive stresses when subjected to vertical loads.
- When beam to column connections are hinges, a horizontal force will only bend the columns. If the connections are fixed

against rotation and the frame is subjected to the same horizontal force, then all members will bend.
- A four-bar frame with all four corners fixed is called a Vierendeel frame, which is a closed system capable of carrying both horizontal and vertical loads.
- Vierendeel frames may be connected to each other and can be used as bridges or multistory structures.
- The top horizontal beam of a frame may be replaced by several short interconnected members in the shape of a segmented arch. All members are subject to bending and compressive stresses.
- If the bars are very short, a continuous curved arch is formed.
- Funicular arches with mostly compressive stresses may span large distances and serve as the primary support for a bridge.
- Placed parallel to each other, arches can support barrel-shaped roofs; arranged radially they may resemble cupolas.

QUESTIONS AND EXERCISES

1. Place a flexible ruler on books standing upright at its ends. Press down on the ruler and observe its deflection. Apply more force until the books topple over. Why did the books fall?
2. Get an open-top cardboard box from a grocery store. Cut out two opposing sides. You now have a frame. Place each leg between two books lying flat (not too tightly). Apply about a pound weight (a can of soup will do) in the middle of the top of this simple frame and observe the deformations of the beam and the legs. Move the weight around and observe again. Apply a horizontal force to a top corner. How did the deformations change?
3. Repeat the above experiment, but this time sandwich the legs tightly between the books for a fixed end frame.
4. Cut your box in half so that you will have two identical frames. Glue the frames to each other into a double-bay frame. Repeat the above experiments by loading one frame at a time. Observe the deformations of the unloaded parts.
5. Repeat the above, this time with a strip of cardboard bent into an arch.

FURTHER READING

Boake, Terri M. *Diagrid Structures*. Birkhäuser. 2014.

Ching, Francis D. K., Onouye, Barry S., Zuberbuhler Douglas. *Building Structures Illustrated: Patterns, Systems, and Design*, 2nd Edition. Wiley. 2013. (Chapter 5)

Sandaker, Bjorn N., Eggen, Arne P. and Cruvellier, Mark R. *The Structural Basis of Architecture*, 2nd Edition. Routledge. 2011. (Chapter 9)

Schodek, Daniel and Bechthold, Martin. *Structures*, 7th Edition, Pearson. 2014. (Chapter 9)

SOME FINE POINTS OF STRUCTURAL BEHAVIOR

9.1 HOW SIMPLE IS SIMPLE STRESS?

The structural systems considered in the previous chapters react to external loads with states of stress called "simple." A cable develops "simple tension," a funicular arch "simple compression," a beam "simple bending," or "simple bending" combined with "simple shear." A more careful look at the ways in which loads are actually carried by even the simplest element will prove, instead, that the stresses developed in a structure are most of the time anything but simple.

If a weight is to be suspended from a cable, a connection must be developed between the weight and the cable. One of the devices commonly used to achieve this purpose is called a "conical connector." This ingeniously simple device consists of a short cylinder with a slightly tapered inner surface, within which are located two or three conical segments, each engraved with serrated teeth (Figure 9.1). The cable is gripped by the teeth of the segments, which are pulled into the cylinder by the friction developed between the cable and the segments; the larger the pull of the cable, the greater is the friction developed. Because of the conical shape, the greater that friction pulls the segments deeper into the conical cylinder, the

tighter is the grip on the cable. Thus, the conical connector is a self-locking device, and is widely used to secure high strength steel tendons in posttensioned concrete construction after hydraulic jacks apply the requisite tension.

The section of cable gripped by the connector is acted upon by frictional shears parallel to its surface and by compressive stresses perpendicular to its surface (Figure 9.2). The compressive stresses squeeze the cable, and the frictional shears actually transfer the load from the cable in tension to the connector. In the neighborhood of the connector, the stresses in the cable present a complicated pattern of compression and surface shear, but they *rapidly* merge into a simple-tension pattern away from the connector (Figure 9.2).

As shown by the French engineer de St. Venant, the tendency of complicated stress patterns to *become* simple occurs often in correctly designed structural elements and justifies the elementary approach to stress used in the preceding sections. The stresses in the portion of cable gripped

Figure 9.1 Conical connector
Conical connectors are used in a variety of tension cable applications, most notably for securing high-strength posttensioning strands for concrete beams. They are ingenious in their simplicity. Consisting of two or three serrated wedge segments sized to the cable diameter, the wedges clamp onto the cable and form a self-locking mechanism. Owing to their conical shape being placed in a conical hole, the greater the tension on the cable, the more tightly the conical connector grips the cable.

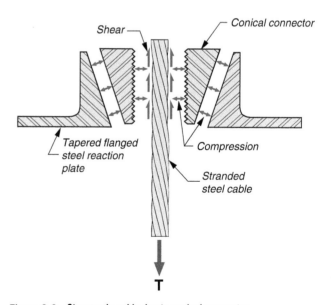

Figure 9.2 Stresses in cable due to conical connector
This diagram illustrates the forces present at the conical connector itself, which involves a complex interaction of simultaneous compression and shear forces. At a distance of approximately two to three times the cable diameter, though, the complex pattern of stress will have resolved itself into a condition of simple stress.

by the connector are complicated, but at a short distance from the connector equal to two or three times the diameter of the cable, the stress pattern, to all practical purposes, is one of simple tension. The cable is designed for simple tension because only a small portion of it is under a more complicated state of stress. On the other hand, the stresses in the portion of cable gripped by the connector may be much higher, and of a more dangerous kind than simple tension, and cannot be ignored.

A condition similar to that existing in the cable arises in a truss bar under tension. The connection between the bar and the gusset plate is often bolted. The bolts develop compressive stresses, called bearing stresses on part of the inner surface of the bar holes, and shears due to friction between the gusset and bar surfaces; these stresses are transformed into simple tension away from the connection (Figure 9.3). Truss bars are designed as if they were under simple tension or compression, but the complicated stress pattern in the neighborhood of the gusset plate must be taken into careful account.

When a concentrated load is applied to a beam the stress pattern under the load is again complicated, but a short distance from the load the beam is in simple bending.

9.2 THE LARGEST STRESS

It was shown in Section 5.3 that pure shear is equivalent to compression and tension at right angles to each other and at 45 degrees to the direction of the shear. Some essential implications of these equivalent states of stress must now be considered.

Since shear stresses develop on the horizontal and vertical sides of a square element cut out of a beam, and tension and compression on the sides of a square element cut out of the same beam at 45 degrees to the horizontal, one may inquire as to what stresses are developed on the side AB of a square element oriented to any other angle (Figure 9.4). It seems natural to investigate *how the stress on one of its sides*, say AB, *varies as the cutout square element is rotated*, since for an angle equal to zero (horizontal position of the side AB) the stress is pure shear, for an angle of 45 degrees it becomes

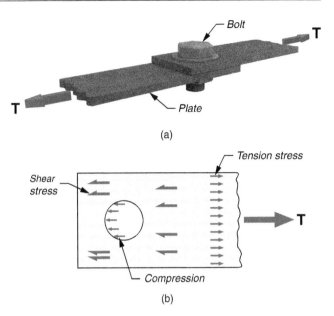

(a)

(b)

Figure 9.3 Stresses in bolted joint under tension

In a bolted joint that involves bearing of the bolt on the plate, stresses are concentrated at the contact point between the bolt and the edge of the hole. As with a clamped cable, at a short distance from the hole, these stress concentrations will be resolved into simple tension. Improved strength and performance can be gained by using so-called *slip critical* connections. These are bolted connections that grip the parts together with such a high camping force that an applied tensile load cannot overcome the friction forces. The bolt therefore never bears on the hole edge, and thus improved connection strength and performance is obtained with a more uniform stress patterns and no stress concentrations on the hole itself.

compression, for an angle of 90 degrees it is shear again, and for an angle of 135 degrees it becomes tension.

A general analysis of this question shows that, *whatever the stress pattern in the beam*, as the element is rotated: (a) the direct stress (tension or compression) on the sides varies from a maximum to a minimum value); (b) the maximum and minimum direct stresses occur on sides at right angles to each other; (c) shear is *never* developed on the sides of the

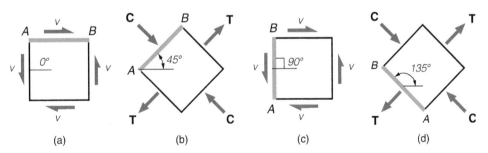

(a)   (b)   (c)   (d)

Figure 9.4 Stress variation with orientation of element

The stress on an element subjected to shear can be seen as equivalent to direct tension and compression stresses perpendicular to the surface of a block rotated by 45°. Looking at the element face labeled A-B, one can observe how the stress on that surface changes from shear (a) to direct compression and no shear at a 45° block rotation (b), back to shear but now in the vertical direction with another 45° rotation (c), and then to direct tension and no shear with another 45° turn, or a total turn of 135° (d). The stresses that are *direct tension and compression without shear* are referred to as *principal stresses*, and they are the maximum values for a given loading.

element for which *the direct stress is maximum or minimum*; (d) both shear *and* direct stress are developed for any other orientation of the sides. In the elementary case of "simple shear" of Figure 9.4, the maximum stress is tension (equal in magnitude to the shear stress) at an angle of 135°; the minimum stress is compression (considered as *negative* tension) at an angle of 45°; and there is no shear on the side of the elements in these two orientations.

The perpendicular directions for which the direct stresses become maximum or minimum are called *the principal stress directions*; the corresponding values of the stresses are called their *principal values*.

The principal directions of stress allow a visualization of complex stress patterns. In general, the two principal stress directions vary from point to point of a stress pattern and may be indicated by small crosses. Lines may, then, be drawn parallel to the principal directions at each point of the stress pattern. There are two families of such lines, and they intersect at right angles (Figure 9.5); they are called the *principal stress lines*, or *isostatic lines*. The stress-lines pattern indicates the flow of stress within the structural element and is useful in visualizing complex stress situations, like that due to a tension load applied to a plate through a bolt (Figure 9.5).

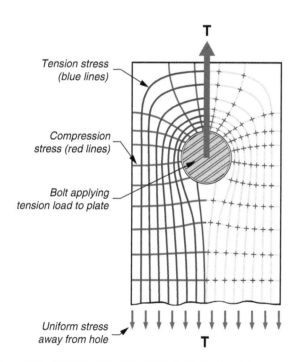

Figure 9.5   Principal stress lines (isostatics) in tension plate containing a hole

The angle at which a principal stress acts depends on the magnitude of shear at a given location of a member, but the principal tension and compression stresses always act 90° to one another. We can observe the directional changes of the principal stresses around an opening in a tension bar as force is applied. These lines of constant stress are referred to as *isostatic lines*. Another visual interpretation is to model the intersection of the tension and compression principal stresses as small "+" indicators. The angle that they make relative to the member at any given location is referred to as the principal stress direction.

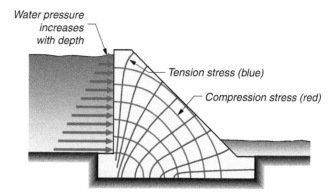

Figure 9.6   Isostatics of dam

A gravity dam uses its deadweight to resist the pressure of water behind it. Water pressure increases with depth, as does the pressure due to the self-weight. The isostatic lines indicate that the compressive forces from the water are being redirected to the foundation. The blue lines of tensile stress intersect the compression isostatic lines at 90° for each line.

Figure 9.6 illustrates the stress lines in a dam loaded by the pressure of water on its vertical side. Since water exerts only normal pressure on the vertical wall, the dam elements at the wall do not develop shear. Hence, the principal stress directions at the wall are vertical and horizontal. The family of compression isostatics starts horizontally at this wall. The inclined dam wall is altogether free of stress and, in particular, of shear stress. Hence, the direction of the inclined wall is one of the principal directions there, and the other is perpendicular to it. The heavy red stress lines show how the pressure of the water is channeled down the dam to its foundation. The light blue stress lines of the tension family cross the compression lines at right angles and show that in these directions the dam acts as a series of cables in tension. Thus, the action of the dam is seen to be a combination of tension and compression at right angles, in varying directions.

The stress lines for a flat bar in simple tension constitute a mesh of vertical tension lines, and of horizontal compressive lines due to Poisson's phenomenon (see Section 5.1) (Figure 9.7a). If a hole is introduced in the bar, the straight tensile lines cannot go through the hole and are diverted around it. In so doing, the tensile stress lines crowd in the neighborhood of the hole, and the compression lines bend so as to remain perpendicular to the tension lines (Figure 9.7b). The flow of the tensile stresses is similar to the flow of water in a river: The flow pattern is made up of parallel lines when the river banks are parallel and there is no obstruction in the river, but the lines curve when a circular pile is embedded in the current, compelling the water to move around the pile. The similarity between these two flow patterns is quite significant: Just as the water of the river increases its speed to move through the reduced section created by the obstruction, the tensile stress increases its value because of the obstruction created by the hole. Thus, the tensile stress across the two sections at the sides of the hole is not uniform, but higher near the hole than at the boundary of the bar (Figure 9.8).

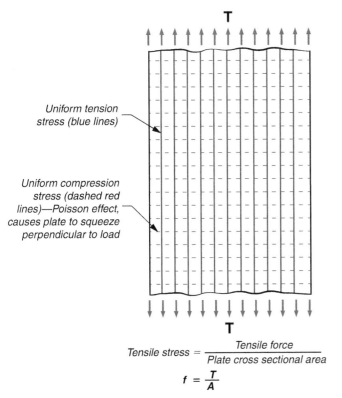

Uniform tension stress (blue lines)

Uniform compression stress (dashed red lines)—Poisson effect, causes plate to squeeze perpendicular to load

$$\text{Tensile stress} = \frac{\text{Tensile force}}{\text{Plate cross sectional area}}$$

$$f = \frac{T}{A}$$

Figure 9.7a   Isostatics of tension bar

For a purely tensile member, the lines of principal stress run directly parallel to the load in tension and perpendicular across the bar in compression. It is the latter that is responsible for the Poisson effect of a member growing thinner as the load is increased.

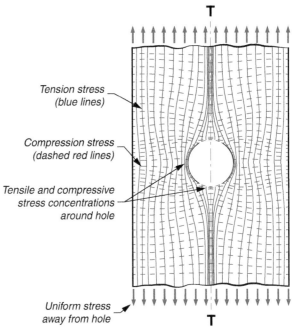

Tension stress (blue lines)

Compression stress (dashed red lines)

Tensile and compressive stress concentrations around hole

Uniform stress away from hole

Spacing of stress lines indicates relative magnitude of stress levels. Closer spacing = higher stress.

Figure 9.7b   Isostatics of tension bar with hole

When a hole is introduced into a tension member, the lines of principal stress must "flow" around it, similar to how water will flow around rocks or other objects in a stream. The flow lines become compressed around the hole, indicating a higher level of stress, just as the flow of water will increase in speed in the narrowed area around obstructions.

Figure 9.8   Stress concentration in tension bar

Upon close inspection, it can be seen that the tensile stresses increase considerably near an opening. Such concentrations are significant, as they can locally exceed the permissible stress values of the material and initiate a failure.

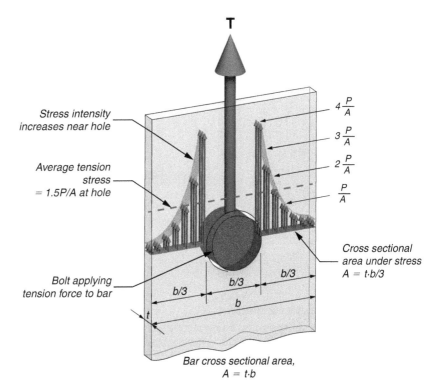

Stress intensity increases near hole

Average tension stress = 1.5P/A at hole

Bolt applying tension force to bar

$4\frac{P}{A}$

$3\frac{P}{A}$

$2\frac{P}{A}$

$\frac{P}{A}$

Cross sectional area under stress $A = t \cdot b/3$

$b/3$ $b/3$ $b/3$

b

t

Bar cross sectional area, $A = t \cdot b$

Such increases in stress due to holes, sharp notches, and other abrupt changes in cross section are called *stress concentrations*. Such concentrations occur not only at bolt and rivet holes but also at sharp corners and at the base of threads (their *roots*) in bolts and nuts. To reduce these increases in stress, sharp corners and thread roots are usually rounded. As seen in Fig 9.8, the increased stresses near the hole are as much as four times the value of the average stress in the bar further from the hole. Stress concentrations may increase the value of the stress above its average value by large factors, thus bringing the stress not only above the elastic limit, but above the ultimate strength of the material. For example, when the edge of a bolt of cloth is nicked, a high stress concentration is developed under tension, and the material tears as the stress concentration moves along the cut. Without the nick, it is practically impossible to tear the material. Similarly, to cut plate glass the glazier scratches it with a diamond and bends the plate at the scratch, creating stress concentration on the tensile side of the plate.

Figure 9.9a gives the stress lines for a simply supported beam under uniform load. The bending and shear stress pattern considered in previous sections is shown by the stress lines to be equivalent to a set of arch stresses, indicated by the red compression lines, combined with a set of cable stresses, indicated by the blue tension lines, at right angles to each other. The bottom fiber of the beam is in tension, and if a notch is cut out of the beam at midspan, stress concentration arises, because the tensile stress lines must go around the notch and crowd at its top (Figure 9.9b). When the load on a beam is increased beyond safe values, if the beam is made of a material stronger in compression than in tension, then compressive arch action carries more of the load to the supports as the fibers in tension fail. Thus, the beam at first carries the load by bending but changes its mechanism to compression as soon as failure in tension makes bending impossible. This readjustment of stress cannot be counted upon for normal loads, since it is accompanied by large displacements and the partial failure of the beam, but is useful in preventing or retarding the total collapse of a structure under exceptional loads.

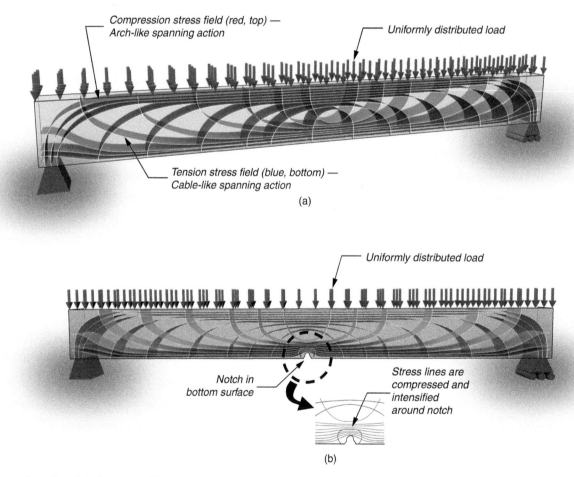

Figure 9.9 Isostatics of simply supported beam

A simply supported beam can be envisioned as a series of arches (red lines) and cables (blue lines), when the isostatic contours are constructed (9.9a). Shear forces at the supports interact with the bending forces to redirect the angle of the principal stress lines perpendicular to the surfaces. If a notch is made on the bottom (tensile) surface, then it can be shown that the isostatic lines crowd around it (indicating an increased stress level) in the same manner as the pure tensile stresses in Figure 9.7b.

9.3 THE IMPORTANCE OF PLASTIC FLOW

It was stated in Section 3.1 that most structural materials have an elastic stress range, followed by a plastic range, and that under normal loading conditions stresses are kept below the elastic limit in order to avoid the accumulation of permanent deformations, which may occur through plastic flow. On the other hand, plastic flow plays an essential role in the behavior of all structural elements.

In the post-and-lintel system, it is generally assumed that the lintel rests on the post, and that a uniform compression is developed between their contact surfaces. This assumption would be correct if the two surfaces were ideally smooth; in practice, instead, both are uneven and present small irregular bumps. If two bumps coincide (Figure 9.10), the contact is limited to a small area and the compressive stress on it may be many times greater than if the contact took place over the entire surface. But, if the high stress developing at the bumps is above the yield point, the material flows, the bumps flatten, the contact area increases, and the stress decreases to acceptable values. Thus, *plastic flow eliminates stress concentrations* due to the roughness of the surfaces and brings about a stress situation nearer the ideal one on which the design was based. Post-and-lintel systems of brittle materials (see Section 3.1) could not readjust and would fail owing to high stress concentrations.

Besides the unavoidable presence of small irregularities, another cause of stress concentration exists at the inner edge of the post. As the lintel bends under the applied loads, it tends to rest on the inner portion of the contact area with the post, and, at the limit, on its inner edge alone (Figure 9.11a). If the post were made of a brittle material, the partial contact would increase the contact stresses and the edge of the post would cut into the lintel or shear off. In practice, plastic flow comes to the rescue. The rotation of the lintel produces a stress concentration at the inner edge of the post, and the elastic-plastic material of the post flows together with the lintel material. Plastic flow increases the contact areas between these two elements until the stresses are reduced to acceptable values, and flow stops (Figure 9.11b).

The stress concentrations at the ends of the horizontal diameter of the hole in a tension bar (Figure 9.8) are similarly relieved by plastic flow. In the neighborhood of these points, flow spreads until the stress lines acquire greater radii of curvature and do not crowd as much as they did when compelled to go around a hole of small radius. The analogy with the flow of water may be carried one step further: In order to reduce the pressure on the piles of a bridge, piling guards are set up which divert the flow of water in a smoother fashion. The plastic areas (Figure 9.12) around the stress concentration points perform the same role and ease the stress flow of elastic stresses.

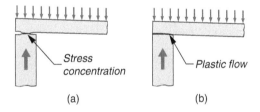

Figure 9.10 Stress concentration and plastic flow
When two structural surfaces contact one another to transfer a compressive force, the members are referred to as exhibiting *bearing stress* against one another, essentially the load divided by the contact area of the two components. When defects exist on the surface of the materials, however, the pressure distribution is not even and localized overstress will cause deformation of the materials (9.10a). Plastic flow allows the materials to redistribute the stress to a more uniform condition (9.10b).

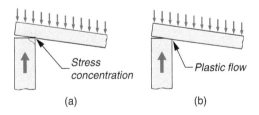

Figure 9.11 Redistribution of stress by flow
Another condition of plastic flow occurs when a member bearing upon another rotates under bending stress. Due to the angular change at the support, the edge nearest the beam span will encounter a far higher stress than the opposite side (9.11a). If the stress exceeds the yield point of the material, plastic flow will allow redistribution of the stress to acceptable levels (9.11b)

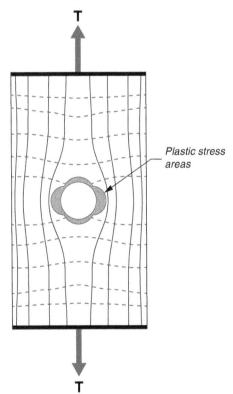

Figure 9.12 Plastic flow at edge of hole in tension bar
As illustrated in 9.7b and 9.8, a tension bar with a hole in it produces localized stress concentrations which dramatically increase in magnitude at the edge of the hole. Plastic flow at the areas of highest concentration will redistribute stress to lower values in these regions.

Stress redistribution due to plastic flow is responsible for the *reserve of strength* exhibited by beams, frames, and other structures in which loads are carried essentially in bending. It was found in Section 7.2 that the midspan section of a simply supported beam develops the highest bending stresses under symmetrical loads. The heaviest loads the beam can safely carry are those for which the upper and lower fibers at midspan are stressed up to the allowable stress for the given material, which is a fraction (usually 60 percent for steel) of its elastic limit. If under increased loads the extreme fiber stresses reach the elastic limit, the stress distribution through the beam depth is still linear and all but the extreme fibers are stressed below the elastic limit (Figure 9.13a).

If the loads on the beam are increased further, the extreme fiber stresses cannot grow any higher, since the elastic limit is

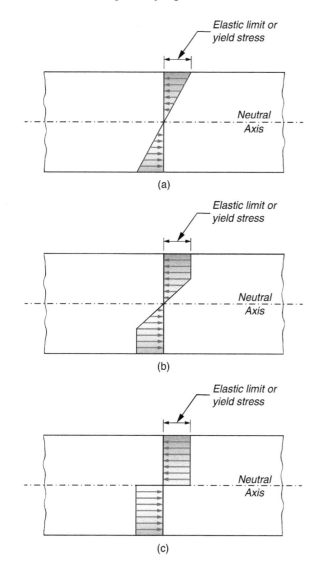

(a)

(b)

(c)

Figure 9.13   Plastic flow in beam section

As material reaches its yield point in the outer edges of a bending member, the maximum stress value will creep into the depth of the member, causing progressive yielding throughout the member depth. The beam will continue to carry load until the point where the entire depth of the member has reached its yield point.

practically the value of the stress at which plastic flow starts; but the other beam fibers may help support the additional loads by increasing *their* stress up to the elastic limit. The stress distribution in the beam stops being triangular and follows the diagram of Figure 9.13b. The loads may be increased until all fibers reach the elastic limit, at which point the stress distribution is rectangular (Figure 9.13c), the beam reaches its *ultimate load* capacity, and all fibers flow, the upper fibers in compression and the lower in tension. The beam fails at this point, because its two halves rotate about the middle section *as if the beam had a hinge at midspan*, and the *plastic hinge* makes the beam into a three-hinge moving mechanism (Figure 9.14). For a rectangular cross section, the ultimate load is 50 percent higher than the load at which flow starts; plastic flow introduces a 50 percent reserve of strength, which may prevent collapse in case of exceptional loads. (See Section 10.7 for the added strength reserve in beams with longitudinally restrained ends.)

A simply supported flanged I-beam is more efficient in bending than a rectangular beam, but does not have such high reserve of strength. The beam is efficient *because* most of its area is concentrated in the top and bottom flanges, where the stresses are highest. But as soon as the flanges flow, the only fibers that can help carry additional loads by increasing their stresses are those in the web, and the web has such a small area that the corresponding increase in load is small. Practically, the flanged beam develops a plastic hinge soon after the flanges reach the elastic limit. Wide-flange steel sections have a reserve of strength of about 15 percent against collapse, because their flanges have almost no reserve of strength.

The reserve of strength is greater for beams with fixed ends. It was seen in Section 7.3 that a fixed-end beam under uniform load develops the highest bending stresses at the supports. If the load is increased beyond the value that brings such stresses up to the elastic limit, the fibers of the end sections flow progressively until the stress distribution at the end sections is that shown in Figure 9.13c. At this point, the end sections develop plastic hinges (Figure 9.15). As the load keeps increasing, the beam, although originally fixed, behaves under the additional loads as if it were hinged at the ends, and the stresses at the middle section (b) start increasing until they progressively reach the elastic limit. At

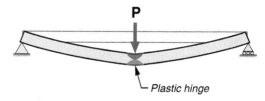

Figure 9.14   Plastic hinge in simply supported beam

In a simple beam of elastic material loaded to failure, plastic flow must occur throughout the cross section before ultimate collapse. The complete plastic flow through the cross section (9.13c) creates what is referred to as a *plastic hinge*, since the failure mechanism mimics the behavior of a true hinge that allows rotation. A single plastic hinge in a simply supported beam will lead to collapse.

this point, the beam develops *three plastic hinges*, reaches its ultimate capacity, and collapses (Figure 9.15). It is thus seen that a beam with fixed ends has a twofold source of strength reserve: the plastic flow of the end sections and the plastic flow of the middle section. The ultimate load for a fixed beam of rectangular cross section is twice the load at which flow starts at the extreme fibers of the end sections. Such a beam has a reserve of strength of 100 percent.

A flanged beam of constant cross section cannot substantially redistribute stresses *at a section*, but has a reserve of strength of more than 33 percent when its ends are fixed. As soon as the higher stresses at the flanges of the end sections reach the elastic limit and hinges are developed there, the lower stresses in the flanges of the midspan section grow until they develop a third hinge. *Redistribution* of stress occurs *between end and midspan sections*, rather than at any *one* section.

Stress redistribution due to plastic flow occurs in any framed structure, since the loads produce maximum stresses only at a few critical sections; as soon as one of these sections develops a plastic hinge, stresses start growing at other sections. Continuity between the various bars of a frame allows such redistributions to occur at a number of sections (Figure 9.16). Thus, continuity insures strength reserve and

develops in the frame a "democratic structural action"; understressed sections come to the help of more severely stressed ones as soon as these have reached their ultimate capacity. Continuity is one of the best guarantees against collapse, even if, under certain loads, it may produce higher elastic stresses at particular points.

Large amounts of flow should usually be avoided, since they introduce permanent, and often increasing, deformations in the structure. On the other hand, small amounts of plastic flow are necessary to smooth out local stress concentrations, which otherwise might endanger the structure. Plastic flow is the mechanism through which a structure redistributes stresses and is made safe against a sudden collapse. It is thus clear that no material may be properly used structurally unless it exhibits some plastic flow under high loads.

A structure is only capable of redistributing stresses *between* its elements if it can stand up when one or more of its elements are cut out or removed. For example, a fixed beam has such capacity because if we cut it through its middle section, it may still carry loads as two cantilevers (Figure 9.17a); a simply supported beam does not, because, if cut at midspan, it collapses (Figure 9.17b). A structure so designed as to be capable of stress redistribution between its elements is called *redundant*: It is *always* a statically indeterminate structure (see Section 4.2). Redundancy is an essential guarantee of safety against structural collapse and should be the basis of design for all important structures (see Section 13.3).

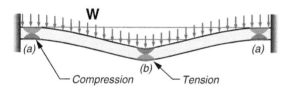

Figure 9.15   Plastic hinges in fixed (built-in) beam
A fixed end beam must have plastic flow occur at three locations before ultimate failure. First the ends (a) will yield, and then transfer load to the center (b), which must also yield. Three plastic hinges are thereby created that form a collapse mechanism.

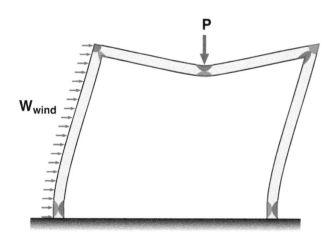

Figure 9.16   Plastic hinges in simple frame
Plastic hinges in a frame provide for multiple points of stress redistribution. For a fixed-base frame as illustrated, *two additional* plastic hinges beyond the fixed beam are required to form at the base before ultimate collapse.

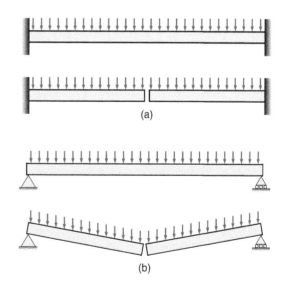

Figure 9.17   Results of stress redistribution for (a) fixed beam and (b) simply supported beam
Stress redistribution can *only* happen in members with redundancy (i.e., statically indeterminate). Thus, a fixed end beam (a) is capable of standing if cut at midspan, turning into two independent cantilevers. The same cannot be said of a simply supported beam (b). Cut in the middle, due to the hinged and roller supports permitting rotation to occur, the two sections have no ability to stand without a counter-rotating moment.

KEY IDEAS DEVELOPED IN THIS CHAPTER

- Stress distribution at and in the vicinity of connections or at points of load application are usually quite complicated, but at a short distance away they revert to simple stresses.
- A rectangular piece of material isolated in a structural component is usually subjected to a combination of tensile, compressive, and shear stresses. As the element is rotated, at some orientation the shear stresses vanish and direct stresses become maxima and minima, acting at right angles to each other.
- At holes, sharp corners and thread roots, stress concentrations increase three to four times the average stress values.
- Components made of elastic-plastic materials flow plastically if stresses exceed the yield stress. If this stress is concentrated over a small region, the flow alleviates the stresses.
- Plastic flow provides a reserve strength to continuous structures.

QUESTIONS AND EXERCISES

1. Cut a 2 inch (50 mm) wide, 1 foot (300 mm) long strip out of a plastic bag. Stretch the strip between two hands. What do you observe about the width of the strip? What happens to its shape? Can you explain relative to the concepts presented in this chapter?
2. Cut a hole in the middle of the strip. Stretch it again and describe the deformations. What happens to the strip as the load is increased?
3. Take a deck of playing cards. Clamp one end of the deck with a strong binder clip. Press down on the free end. Why do the ends slide past one another? Clip the other end too and press down again. Is it easier or harder to press down than before? What causes the difference?

FURTHER READING

Allen, Edward Zalewski, Waclaw, *Form and Forces: Designing Efficient, Expressive Structures.* John Wiley & Sons, Inc. 2009. (Chapter 17)

Millais Malcom. *Building Structures: From Concept to Design,* 2nd Edition. Spon Press. 2005. (Chapter 4)

BEYOND THE BASICS

Part III of this text builds upon the concepts and principles addressed in the first nine chapters of the text. Virtually all of engineering mechanics can be reduced to the elemental principles introduced in the earlier chapters, such as linear and rotational equilibrium, stability and axial stress, introduced in Chapters 4–6, and bending in Chapter 7.

Chapters 10 through 12 introduce the spanning of space by employing various forms of surface structures, whether planar, curved or reticulated. Many of these are funicular, spanning space in the most efficient manner in pure tension or compression. Chapters 13–15 look more closely at incidents of structural failure and the lessons learned, the aesthetics of structural systems, and future directions for structure.

Roof structure of the Papal Audience Hall in Vatican City, Rome (see Figure 10.36)

Photo courtesy of Mario Carrieri, Italicementi

GRIDS, PLATES, FOLDED PLATES, AND SPACE-FRAMES

10.1 LOAD TRANSFER IN TWO DIRECTIONS

The structural elements considered so far have in common the property of transferring loads in one direction. A load set on a cable or a beam is channeled to the supports along the cable line or the beam axis; an arch, a frame, and a continuous beam produce the same type of "one-directional load dispersal." These structures are labeled one-dimensional resisting structures, because they can be described by a straight or curved line, along which the stresses channel the loads. (Lines are said to have only one dimension, because a single number—for example, the distance from one end of the line—is sufficient to define the position of a point on the line.)

One-dimensional resisting elements may be used to cover a rectangular area, but such an arrangement is often impractical and inefficient. For example, a series of beams, all parallel to one of the sides of the rectangle, serves this purpose; but a concentrated load on such a system is carried entirely by the beam under the load, while all other beams are unstressed (Figure 10.1). The system is impractical because one beam deflects while the other beams remain horizontal, and inefficient because it does not work as a whole in carrying the load. The load transfer occurs always in the direction of the beams, and the loads are supported by

the two walls at their ends, while the walls parallel to the beams remain unloaded. This may be a proper solution when unloaded walls are needed for functional purposes, but it becomes inefficient when all the walls enclosing the space can be used to support loads.

These considerations suggest that in the example given above it would be structurally more efficient to have "in-plane two-way load dispersal." Such dispersal is obtained by means of grids and plates, two-dimensional resisting structures acting in a plane.

10.2 RECTANGULAR BEAM GRIDS

If two identical, simply supported beams at right angles to each other, placed one on top of the other, are connected at their intersection and a concentrated load is applied there, the load is transferred to the supports at the ends of both beams and thus dispersed in two directions (Figure 10.2). This mechanism involves the loading of a beam and its consequent deflection, the deflection of the first beam by the connection to the other, and the sharing of the load-carrying action between the two beams. Since the two beams are connected, they are bound to have the same deflection; and,

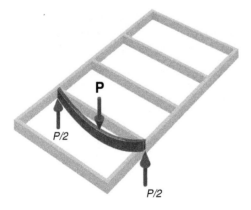

Figure 10.1 Concentrated load on one-dimensional system of beams
A one-dimensional structure consists of a parallel series of elements (such as cables, beams, or trusses) that each share a portion of the overall load. When a large concentrated load is applied to one of the beams, however, that beam will deflect and carry the load independently to the supports. All others are unaffected and do not contribute to carrying the load.

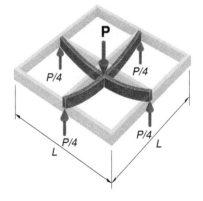

Figure 10.2 Two-way load dispersal by equal beams
If two beams of equal lengths, crossed at right angles to each other are loaded at their intersection, both will deflect the same amount and share the load equally. The load is therefore distributed along the span of each beam in two perpendicular directions, or what is referred to as *two-way load dispersal* or *two-way action*.

because they are identical, these equal deflections must be due to equal loads. Hence, each beam must carry half the load (Figure 10.2). Thus, each support reaction equals one-fourth of the load, and "two-way dispersal" reduces the loads on the supports to one-half the value they would have if only one-way dispersal took place.

Two beams at right angles must deflect by the same amount at their intersection even if they have different lengths or different cross sections. However, a greater load is required to deflect a stiffer beam by the same amount as a more flexible one. Hence, the stiffer beam will support a greater share of the load than the more flexible beam, and the loads on the two beams will not be equal. The stiffness of a beam under a concentrated load is inversely proportional to the cube of its length. Thus, if two beams of identical cross section have spans in the ratio of one to two, their stiffnesses are in the ratio of eight to one. Consequently, the short beam will carry eight-ninths of the load and the long beam one-ninth of the load (Figure 10.3).

This example shows that true two-way transfer takes place only if the two beams are of equal, or almost equal, stiffness. As soon as one beam is much stiffer than the other, the stiffer beam carries most of the load, and the transfer occurs essentially in its direction. To obtain an efficient two-way transfer in the case of unequal spans, the longer beam must have either a substantially stiffer cross section, (i.e., a greater moment of inertia—see Section 7.1), or "stiffer" support conditions. For example, if two identical beams have lengths in the ratio of 1 to 1.59 (which is the cube root of 4), their midspan deflections under the same load would be in the ratio of $(1.59)^3$ to 1 (i.e., 4 to 1). But if the longer beam's ends are fixed, its deflection becomes one-fourth as large and, because

the deflections of the two beams are now equal, the simply supported shorter beam and the longer fixed beam connected at midspan carry equally a load concentrated there.

The sharing of one concentrated load between two beams may be extended to a series of concentrated loads by setting beams in two or more perpendicular directions, one under each load. Once again, the intersection points must deflect by the same amounts. In the arrangement of Figure 10.4, the deflections of an unsupported long beam at the quarter points would be smaller than the deflection at midspan. Hence, if three, equally spaced, shorter cross-beams of equal stiffness support it, the middle beam deflects more and takes a greater share of the load than the side beams. The greater dispersal in the direction of the shorter, stiffer span is evident: The two long sides of the boundary support more than 90 percent of the total load. With a long beam eight times stiffer, the long sides still carry about 65 percent of the total load. The loads tend to move to the support along the shortest possible paths, and two-way action vanishes as soon as the ratio of the long to short sides of the rectangle (what is referred to as the *aspect ratio*) is greater than 1.5.

Finally, one may cover a rectangular area by a grid of beams at right angles and obtain the two-way dispersal of loads located at any grid intersection (Figure 10.5). If all the beams of the grid running in one direction are connected to the beams running in the other the, any one of the upper beams acts as a continuous beam on flexible supports provided by the lower beams. In this grid, an off center load may produce upward deflections of an upper beam (Figure 10.6).

A better interaction of the two beam systems is obtained by "weaving" the beams so as to have their relative positions change at each intersection (Figure 10.7). In this case, the grid

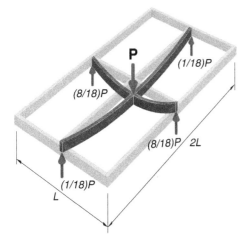

Figure 10.3 Two-way dispersal by different beams

If two beams, one twice as long as the other are laid on a rectangular frame and are loaded at their intersection, they will deflect the same amount because they are interconnected. Deflection of a beam of a given size and loading is proportional to the cube of its span (see Chapter 7.2), so a doubling of a span means an increase of deflection by eight times. Therefore, since the short beam is eight times stiffer, it carries a much greater portion of the load. This demonstrates that very little two-way action happens when one side is much longer than another.

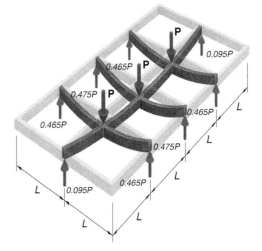

Figure 10.4 Two-way dispersal of three loads

When three loads are carried by three short beams in one direction and one twice as long in the long direction, the short beams again carry the largest portion of the loads, because the defection is greatest at the midpoint of the assembly. Thus, it is shown that the majority of the load is carried by one-way action in the short direction. True two-way behavior vanishes above a long-side to short-side ratio of about 1.5 to 1, or an *aspect ratio* of 1.5.

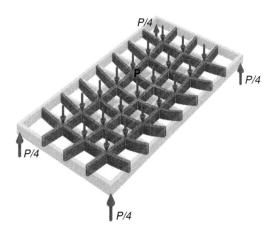

Figure 10.5 Rectangular grid

Two-way load dispersion takes place in a rectangular grid of beams. The beams are connected at the intersections either by welding or bolting.

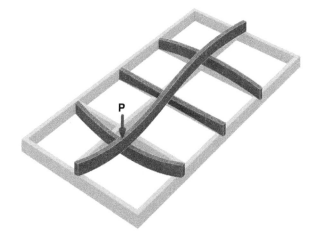

Figure 10.6 Uplift of upper beam of grid

If beams are laid on each other with firm connections, a beam on top of the other beams will act as a continuous beam on flexible supports. When a heavy load is placed on the top beam, some uplifting may occur. The magnitude of the support reactions will depend on the stiffness of the bottom beams.

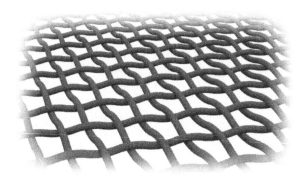

Figure 10.7 Woven grid

Fabric is a material held together by threads crossing over and under each other. Through weaving, multiple threads thus become linked into a surface structure. If an architectural structure can be similarly constructed by welding or bolting (or contiguous construction in the case of concrete), true two-way interaction of the elements can be obtained.

system is capable of giving support from both below and above: A beam moving up is pulled down by the cross-beam set above it, while a beam moving down is pushed up by the cross-beam under it. In practice, this behavior is obtained by means of welded or bolted connections between the beams of steel grids and occurs naturally in reinforced-concrete grids, due to their monolithic construction. An example of this is the roof of the National Gallery in Berlin, Germany, by the famous modernist architect Mies Van der Rohe (Figure 10.8). The substitution of trusses for beams transforms grids into rectangular space frames, which exhibit the same two-dimensional behavior (see Section 10.9).

Rigid connections between the beams of the two systems introduce another structural action in the grid. When two perpendicular beams connected at midspan are deflected by a concentrated load at their intersection, their midspan sections (Figure 10.9, point 'a'), move downward,

Figure 10.8 National Gallery in Berlin

The roof of the gallery employs a two-way grid of intersecting steel beams. The entire roof is supported by only eight columns around the perimeter. The columns placed two per side are located at a position creating corner cantilevers, the downward bending of which partially offsets the bending and deflection at the center of the structure.

Photo courtesy of Terri Meyer Boake

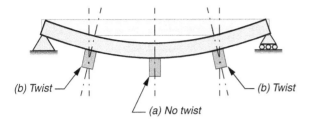

Figure 10.9 Grid action with twist
More complex interaction occurs when intersecting beams of a grid are firmly connected. Under load, members at right angles to each other are not only deflected symmetrically in the center (point 'a') but become twisted away from the center due to the deflection of the opposite-direction beams (b). The grid is made stiffer through this twisting action than if the members did not interconnect and could move independently. This efficiency of stiffness, however, comes at a higher fabrication cost due to the many connections.

but remain vertical because of symmetry. When the intersection of the two beams does not occur at midspan, the beam sections (Figure 10.9, point 'b') deflect and rotate, and the continuity introduced by the rigid connection transforms the bending rotation of one beam into a twisting rotation of the cross-beam. The stiffness of the grid is greater when the beams are rigidly connected than when they are simply supported over one another; this indicates that the twisting mechanism is capable of transferring part of the load to the supports, thus producing smaller displacements in the grid. The reader may use a simple grid model with "woven" beams to show that the partially rigid connections introduce a certain amount of twist in the beams, and that the woven grid is stiffer than the unwoven grid (see "Questions and Exercises" No. 1).

Grid systems are seen to be particularly efficient in transferring concentrated loads, and in having the entire structure participate in the load-carrying action. This efficiency is reflected not only in the more even transfer of the loads to the supports but also in the reduced depth-to-span ratio required in rectangular grids. Systems of parallel beams used in ordinary construction have depth-to-span ratios of the order of one-tenth to one-twentieth. This ratio varies somewhat with the beam material: Steel beams may be somewhat shallower than prestressed concrete beams; reinforced-concrete beams deeper; wood beams still deeper; but the depth-to-span ratio cannot go much below one-twentieth, if the beams are to be practically acceptable from the viewpoints of both strength and deflection. Sufficiently stiff rectangular grid systems with aspect ratios near unity may be designed with depth-to-span ratios of as little as one-thirtieth to one-fortieth. When accumulated over a large number of floors, this reduction in depth lowers structural and other costs, by reducing the height of the building.

Notwithstanding their interesting properties, it is well to remember that, in practice, two-way grid systems are seldom more economical than one-way beam systems when they require connections, because these are always costly.

10.3 SKEW GRIDS

An additional saving in floor depths may at times be achieved by the use of skew grids, in which the beams of the two systems do not intersect at right angles (Figure 10.10). The advantages thus obtained are twofold.

In the case of rectangular areas with one side much longer than the other, most of the beams of a skew grid span equal distances, and the loss of two-way action is substantially reduced. Moreover, the beams diagonally across the corners of the rectangle, shorter and stiffer than the other beams, give stronger support to the beams intersecting them. For loads concentrated around the center of the plate, because of these stiff supports the ends of the longer beams tend to lift up at the corners. Keeping them attached to the walls requires a downward force creating a reversal of curvature in the longer beams, which behave very much as if they were fixed rather than simply supported at those ends (Figure 10.11).

Figure 10.10 Skew grid
Because long beams carry less load than short ones, they are less efficient. To improve two-way efficiency, beams may be arranged in a skewed pattern in which most members are of the same length and disperse the load more evenly.

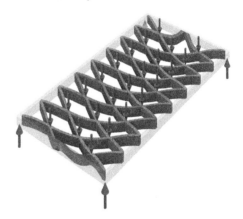

Figure 10.11 Reversal of curvature in skew grids
Members near the corners of the grid are shorter and consequently stiffer than those further from the corners. Because of their stiffness, these beams provide greater support for the long beams intersecting them. As a result, a load on a beam crossing the short corner beam tends to lift up from the support. The end must be firmly connected to prevent this uplift, which creates a reversal of curvature, and thus the corner behaves more like a fixed support. Because of this support condition, skew grids have a higher structural efficiency than orthogonal grids.

Figure 10.12   Seattle Public Library
The exterior façade of the library uses a skew grid as part of an overall spatial gridwork that wraps the core of the building in folded planes. The skew grid provides for lateral resistance to in-plane forces, while also making perpendicular (primarily wind) forces more effectively distributed.

Photo courtesy of Terri Meyer Boake

Since a beam with a fixed end has a load-carrying capacity for uniform loads 120 percent greater than an identical simply supported beam, skew grids have a somewhat greater structural efficiency. Their depth-to-span ratios may be as low as one-fortieth to one-sixtieth. The exterior façade of the Seattle Public Library (Figure 10.12) wraps the building with a folded set of planes, each surface constructed as skew grids.

10.4 PLATE ACTION

The two-way action of beam grids is due to the pointwise connection of the two beam systems at their intersections. Such action is even more pronounced in practice, because the openings in the grid are plugged with slabs or plates, thus making the roof or floor an almost monolithic structure. The advantages of an entirely monolithic structure having two-way action at all points would be even more pronounced.

A plate or slab is a monolithic structural element of relatively small depth covering an area which at first shall be assumed to be rectangular. Any strip of plate parallel to one side of the rectangle may be thought of as a beam acting in that direction; any strip at right angles to the first may also be considered as a beam, and since the two perpendicular beams act together, bending of one produces twisting in the other (Figures 10.13a and 10.9). The action of a plate is analogous to that of a welded beam grid with an infinite number of infinitesimally narrow beams near one another; but the beams of this grid may be considered to act in *any* direction, since the plate may be ideally divided into perpendicular or skew strips supported at any two points on its boundary (Figure 10.13b). Since, moreover, in any physical phenomenon nature follows the easiest path, structural elements tend to carry loads by their most efficient mechanism, and a plate acts as the set of welded grid beams

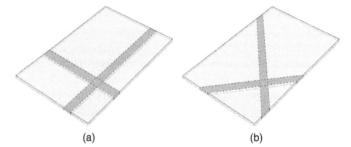

(a)   (b)

Figure 10.13   Equivalent grids in plate action
If members of a grid are assumed to be infinitely narrow and very close to each other, a solid plate results. The resulting monolithic action creates two-way action even more pronounced than in comparable beam systems. The internal slab "beams" may be considered orthogonal (a) or skewed (b).

transferring the load to the supports by means of the lowest possible stresses.

The analogy with a beam grid shows that a plate under load will bend and twist at every point. Bending produces beam-action, that is, bending and shear stresses; twisting produces torsional shears (see Section 5.3).

It is easy to visualize the role of bending, shear, and torsional stresses in the transmission of a concentrated load by plate action. If the plate consisted of a series of independent parallel beams, only the beam under load would deflect, thus transferring the load to the supports by bending and shear (Figure 10.14a). But the "beams" of a plate are all cemented together so that the loaded beam is partially supported by the two adjacent parallel beams and transfers part of the load to them. This transfer occurs through the shears developed between the vertical sides of the adjacent beams and produces bending and shear in the adjoining beams (Figure 10.14b). The two adjacent beams, in turn, transfer some of their load to the adjoining beams

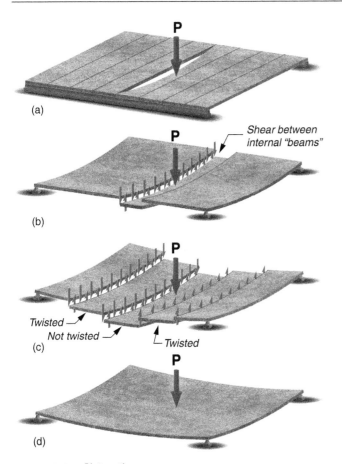

Figure 10.14 Plate action

A rectangular plate supported on four corners may be considered to consist of multiple parallel beam elements (a). If the beam elements act independently when loaded, a beam element bends and is subjected to bending and shear stresses (b). Because the elements are interconnected, though, as one beam element deflects downward, shear stresses are present between adjacent elements and some load is transferred through this shear(c). Conceptual beams at right angles to the first one also bend, and twisting takes place as indicated in fig. 10.9 creating additional shear stresses. A solid plate (d) is, therefore subjected to bending, vertical shears and twisting shear stresses simultaneously at every point. Load carrying by twisting is a characteristic of plates, which one-way beams and simple grids do not have.

through vertical shears; the difference between the downward shears coming from the loaded beam and those upward transferred to the adjoining beams produces the twisting action (Figure 10.14c). Thus, a combination of shearing and twisting transfers load at right angles to the loaded beam, while bending and shear transfer load in the direction of the loaded beam. Since a plate may be considered as a rectangular grid of beams, the transfer mechanisms just described take place not in one but in two directions, and plate action is equivalent to beam action in two perpendicular directions plus twisting action in these two directions (Figure 10.14d).

Load carrying by twisting action characterizes plates and differentiates them from beams and grids, since even welded grids cannot develop substantial twisting action. In this

connection, it is interesting to notice that the plate twist is responsible for a good percentage of its load-carrying capacity. For example, in a square plate uniformly loaded and simply supported along its four sides, twisting action is responsible for approximately 50 percent of the load transfer to the supports. In a rectangular grid weak in twist, 80 to 90 percent of the load is transferred to the supports by beam action.

It was noticed above that a plate may be considered capable of developing grid action in any direction. This means that any point in the plate may be considered as the intersection of two beams of a rectangular grid system, and that any number of rectangular grid systems may be thought of as passing through a plate point (see Figure 10.13). The bending stresses at the same point will vary depending on the direction of the conceptual grid system considered, because the beam spans to reach the boundaries change with the beam directions. An evaluation of the bending stresses in different directions at the same point of a plate shows that there are two perpendicular directions for which these stresses are, respectively, maximum and minimum, and that for these directions the torsional shears are zero. These are the *principal directions*, which were encountered in other stress situations earlier (see Section 9.2). Indicating the principal directions at various points in the plate by crosses, one may plot the *principal stress lines* or *isostatics*, which show the flow of bending stresses in the plate. Since no torsional shears are developed along the isostatics, the plate may also be thought of as a grid of beams curved in the plane of the plate, which meet at right angles, but do not transmit loads to the adjoining beams by twisting action. The isostatics for a simply supported square plate under uniform load are shown in Figure 10.15a and for a rectangular plate in Figure 10.15b.

The deflections of the plate of Figure 10.15a, shown in Figure 10.16, explain visually the structural behavior of the plate. The center strips of the plate deflect very much as simply supported beams. At the corners, the two sides of the plate are compelled to remain horizontal by the boundaries, assumed capable of reacting both up and down, and the entire corner areas are stiff; the ends of the diagonal strips are thus prevented from rotating, and these strips behave as fixed-end beams with a reversal of curvature at the ends similarly to beam grids shown in Figure 10.5 and 10.9.

A peculiarity of plate behavior occurs at the corners if a rectangular plate is simply supported over a hole with boundaries that can provide only upwardly-acting reactions: The plate corners curl up (Figure 10.17a). In order to have the entire plate boundary in contact with the supports, the corners must be pushed down by concentrated corner forces (Figure 10.17 b), which must be added to the load carried by the plate in computing its support reactions. For a square plate under a load concentrated at the center, the four corner forces add about 50 percent to the total plate reactions (Figure 10.16).

The stiffness of a simply supported, uniformly loaded square plate may be compared to the stiffness of simply

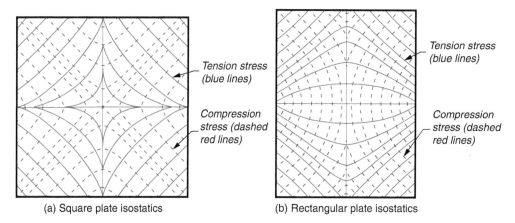

(a) Square plate isostatics   (b) Rectangular plate isostatics

Figure 10.15   Isostatics of simply supported plates under uniform load
Because a plate may be considered to consist of either a rectangular or a skew grid, the bending stresses will depend on the direction of the conceptual beam. There are two perpendicular directions in which these stresses are having maximum and minimum values. The principal stresses in these principal directions may be plotted as isostatics (see Chapter 9) for a square plate (a) and a rectangular plate (b). Along these, no twisting action occurs to transfer loads between the orthogonal lines.

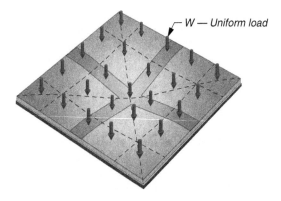

Figure 10.16   Deflections of a square plate, supported along its boundaries and under uniform load
As a plate may be considered as a set of rectangular or skew beams, the central strips bend as simple beams. Near the corners an imaginary skew beam is supported by a shorter imaginary beam and will have a reversed curvature as indicated in Figure 10.11.

supported beams having spans equal to the sides of the plate. The center deflection of the plate is 42 percent smaller than the corresponding beam deflections. The bending stresses in the plate are 29 percent smaller than in the beams. The reversed bending stresses at the corners of the plate are two and a half times smaller than the stresses in the fixed diagonal beams connecting opposite plate corners. Although the bending-stress distribution across the thickness of the plate is the linear distribution typical of beam bending, and as such is not ideally efficient, the two-way plate action reduces the maximum stresses by substantial amounts.

In locating the reinforcement in concrete slabs, it is essential to know the maximum and minimum stresses (or reversed stresses) and the directions in which the torsional shear stresses become maximum. Maximum shears always occur in directions at 45 degrees to the principal bending directions; in the corners of the plate, shear acts in the

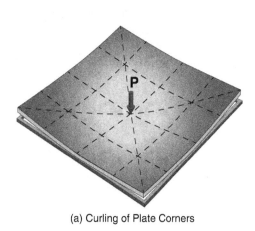

(a) Curling of Plate Corners

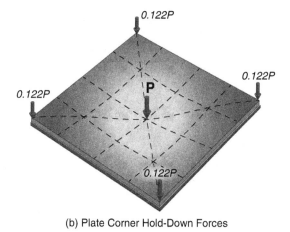

(b) Plate Corner Hold-Down Forces

Figure 10.17   Curling up of plate corners
If the square plate is simply laid on its boundaries and a load applied at the center, the corners will curl up (a) unless they are held down or pressed down by corner forces (b) to maintain full contact along the plate boundary.

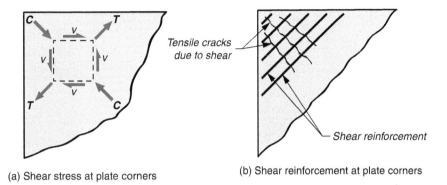

Figure 10.18 Shear reinforcement at plate corners

Shear stresses are present at 45° to the principal stress directions (a). In a concrete plate, reinforcing bars are needed near plate corners (b) to prevent diagonal shear cracking.

(a) Shear stress at plate corners

(b) Shear reinforcement at plate corners

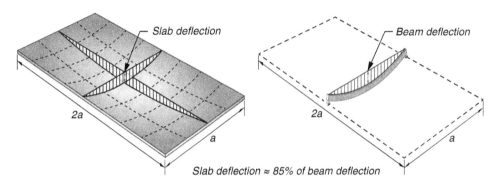

Figure 10.19 Action in simply supported rectangular plate

The rectangular plate with sides a and $2a$ carries a distributed load, p psi, (Pa). The total load on the plate is $P = p \times 2a^2$. The distributed reaction on the boundaries has maximum values of $0.31 \times P$ and $0.25 \times P$ on the long and short sides, respectively.

Slab deflection ≈ 85% of beam deflection

upper half of the plate as shown in Figure 10.18a, and diagonal reinforcing bars should be oriented as shown in Figure 10.18b in order to directly absorb the equivalent tensile stress.

It was shown in Section 10.2 that rectangular grid systems lose most of their two-way action as soon as one of the sides of the rectangle is much longer than the other. Plates exhibit the same behavior, as indicated in Figure 10.19 for the case of a uniformly loaded rectangular plate with sides in the ratio of two to one. The center deflection of the plate is only 15 percent smaller than the deflection of beams parallel to its short sides, while it was 42 percent smaller for a square plate. The largest stress in the plate is only 20 percent smaller than in the short equivalent beam. Since the short span is much stiffer than the long span, approximately two-thirds of the load is carried to the long-side supports.

What has been said of plates with simply supported sides is true of plates with fixed sides. Such plates are stiffer and behave like a rectangular, welded grid of fixed beams.

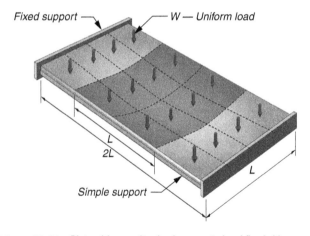

Figure 10.20 Plate with opposite simply supported and fixed sides

The central portion of the plate behaves as a simply supported square plate. The rigidity of the fixed end long span elements causes them to behave similarly to elements in the shorter (but simply supported) direction, thus equalizing the behavior in both directions.

10.5 PLATE STRUCTURES

Plates of many shapes and with different support conditions find application in construction, mostly as floor slabs.

The conditions of support may differ on the four sides of a plate. A plate may have two simply supported parallel sides and two fixed parallel sides. In this case, an increased rigidity of the long spans (due to their fixity) may compensate for a large aspect ratio: the plate exhibits two-way action comparable to that of a square plate with all sides simply supported (Figure 10.20). Similarly, a corner balcony

plate may have two adjacent sides simply supported and the other two sides completely unsupported or free. Obviously, no simply supported "unwoven" grid could be built in this fashion, since the beams of the lower system would be unsupported. A "woven" grid, instead, would be capable of carrying load mostly by twisting action. The same behavior is exhibited by a solid plate (Figure 10.21).

A stiffer corner balcony plate with two unsupported adjacent sides is obtained by fixing the other two sides (Figure 10.22); in this case, the analogy with a grid system

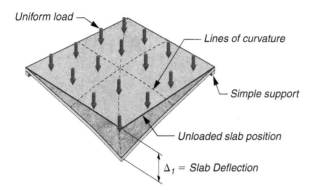

Figure 10.21 Plate with adjacent simply supported and free sides
A woven grid of beams could span between two simply supported and two free edges due to interactions between the members. Twisting action in the beams is responsible for much of the load carrying. By extension, a solid slab with the same support arrangement also carries some of the load by twisting.

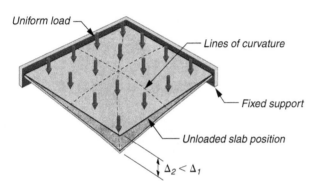

Figure 10.22 Corner balcony plate with fixed supports
Even greater stiffness may be obtained by providing fixity at the supports. In this case, the grid analogy is valid: Two conceptual beams perpendicular to the fixed sides behave as cantilever beams. In the equivalent slab, now bending action versus twisting is predominant.

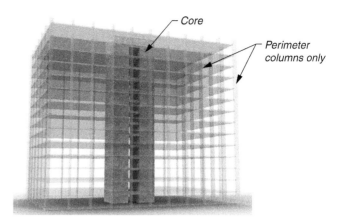

Figure 10.23 Floor slab of building with inner core
Rectangular plates may have many types of support conditions. Contemporary office buildings constructed with central cores that penetrate a floor plate. The floor plate is then supported around the building exterior perimeter as well as the interior core; a column-free interior space is thus created.

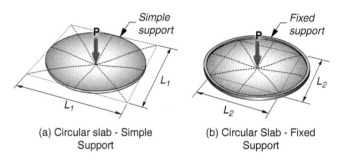

(a) Circular slab - Simple
Support

(b) Circular Slab - Fixed
Support

Figure 10.24 Circular plates
A circular-shaped plate, simply supported on its boundary, behaves essentially like a circumscribed rectangular plate (a). If the boundary is fixed, it behaves similarly to a square plate inscribed within the circle (b).

holds, since each beam of the equivalent system is cantilevered, and bending action prevails again.

Rectangular plates may be supported on more than one type of boundary. In modern office buildings, it is common to support the floor slabs on an outer wall or a series of columns, and on an inner "core," inside of which run the elevators, the air-conditioning ducts, and other parts of the mechanical, electrical, and plumbing systems (Figure 10.23). A completely column-free floor area is thus obtained.

The boundary of a slab may have a variety of shapes: Instead of covering a rectangle, it may cover a skew, a polygonal, or a circular area. The behavior of a circular slab simply supported at the boundary does not differ substantially from that of the circumscribed simply supported square slab (Figure 10.24a), since the corners of the square slab are rigid and only its central portion presents sizable deflections. The behavior of a circular slab fixed at the boundary is similar to that of a square slab inscribed within the circle (Figure 10.24b).

Ring slabs may be built to span the area limited by concentric outer and inner circular supports. As soon as the radial span of the slab is small in comparison to its average radius, two-way plate action under uniform load is considerably reduced, and the slab exhibits stresses of the same order of magnitude as those of beams spanning the radial distance between the two circular boundaries.

Slabs may also be "point-supported" on columns, either hinged or fixed to the slab. The connection between the columns and the slab must be designed to absorb the so-called punching shear from the columns and may require the use of capitals or of intermediate distributing plates called drop-panels (Figure 10.25). In order to avoid capitals, shear connectors of steel are used sometimes in reinforced-concrete design to guarantee the transmission of load from the column to the slab (Figure 10.26).

Slabs used as floors, or flat slabs, present the constructional advantage of smooth underside surfaces which permit the unimpeded running of pipes, ducts, and other

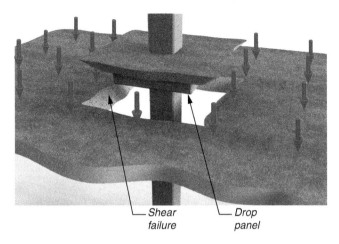

Figure 10.25 Punching shear in a column-supported slab
Punching shear is the propensity for a thin slab to be pierced by a supporting column. To reduce the tendency for punching shear to occur, two approaches are common. Drop panels can be used (which thicken the concrete and provide more shear area) and/or column capitols can be provided (which distribute the load to a greater area, reducing the stress level. Flat plates without capitols or drop panels will typically be used only for structures with lower loads.

Figure 10.26 Shear connector
A grillage of steel called a *shear head* may be used to transfer loads from columns to reinforced concrete slabs. The photo is a proprietary, prefabricated system, positioned at the top of a column prior to concrete slab placement.

Photo courtesy of Tuchschmid AG, CH-Frauenfeld

parts of the various mechanical systems required in a modern building. The savings achieved by the avoidance of pipe and duct bends around beams often justify the selection of a flat-slab system for floors and roofs. Flat-slab and column systems of reinforced concrete are among the most economical structures for buildings with limited spans and are commonly used in high-rise apartment buildings all over the world.

Reinforced concrete slabs supported by steel or concrete columns are easily poured on plane forms. An additional saving in formwork is achieved by pouring the slabs on the ground, one on top of the other, and lifting them to the top of the columns by means of hydraulic jacks. Buildings with a large number of floors have been built by this lift-slab technique, although a tragic construction failure at the L'Ambiance Plaza project in Connecticut in 1987 resulted in curtailed use of this technique. Another popular technique for walls only, called *tilt-up*, involves casting flat panels horizontally on the ground. The cured panels are then lifted and rotated by crane into a vertical position, serving as perimeter walls for buildings no higher than two or three stories.

An important application of reinforced-concrete slabs as both vertical walls and horizontal floors occurs in prefabricated panel construction. The slabs are fabricated and cured in a factory, and erected and joined on the site, with savings in labor and construction time, particularly when the electrical and plumbing systems are incorporated in the panels. The various panel systems invented and patented so far differ mostly in the method adopted for joining the plates. In box construction, boxes made out of panels are built integrally at the factory and erected at the site. Box construction has been used in Russia and other former Soviet states since the 1950s and was applied by Moshe Safdi to the construction of the apartment cluster called "Habitat" in Montreal in 1966 (Figure 10.27).

Inclined slabs are used in roof design as the sides of gabled roofs or the sloping sides of "north-light roofs," in which the higher boundary of the slab is supported on a truss (Figure 10.28). The vertical trussed elements are light, and permit the use of large glass surfaces, usually exposed to the uniform north light required in factories and artists' studios. Inclined slabs are often used in storage bins and other components of industrial buildings (Figure 10.29). Such inclined slabs carry some of the load by tension or compression down their slope, and some by plate action in a direction perpendicular to the plate.

10.6 RIBBED PLATES

The structural efficiency of plates is reduced by the linear stress distribution across their thickness. Only the top and bottom plate fibers at the most stressed point, in the most stressed direction develop allowable stresses. In the case of beams, this inefficiency is remedied by locating as much material as possible at a distance from the unstressed neutral axis, thus obtaining I and T sections; the same remedy can be adapted to plates. Some of the material may be located away from the neutral middle plane of the plate and used to create ribs, which run in one, two, or even three directions (Figure 10.30). A ribbed slab presents the advantages of continuity due to plate action and the advantages of depth due to its ribs. On the other hand, the underside of a ribbed plate is not smooth, and a soffit may have to be hung from it. Pipes and ducts are not bent around the ribs but, usually, hung from them.

Figure 10.27 Montreal *Habitat*
One of the most iconic images of Montreal's *Expo '67*, the *Habitat* condominium block remains a popular living destination. The structure is constructed of precast concrete modules, which included roof, floor and walls, and were outfitted with doors and windows as well as all utilities and appliances before being craned into place. A key feature of the complex is the multilevel dwelling units, each with an individual garden terrace on the roof of the unit below.

Photo courtesy of Terri Meyer Boake

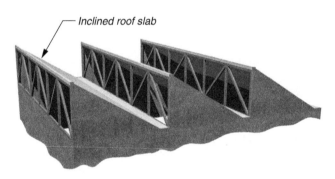

Figure 10.28 Plates in north-light roofs
Plates do not have to be horizontal. Inclined or vertical slabs may be used as walls or slanted roofs. In a north-light roof, a concrete slab may be supported on the bottom chord of one truss at the low end, and the top chord of the next truss at the high end, spanning between the two. The sawtooth design provides for abundant glass and daylight to the north, with the solid slab providing shading facing south.

Figure 10.29 Inclined slabs in storage bins
Concrete slabs may be arranged to form pyramidal storage bins. Loads in such inclined slabs are carried by tension or compression on the slope and by plate action perpendicular to the plate span.

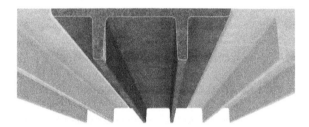

Figure 10.30 Ribbed slab
Slabs are relatively thin and hence have small moments of inertia. Similarly to T beams, the inertia may be increased with the addition of ribs in one or more directions.

Figure 10.31 Concrete and tile floor
A common means of constructing a ribbed slab outside the United States is to use hollow tiles as permanent forms. They are set on temporary shores, with space between parallel rows. Reinforcing steel is placed into the gap between parallel tile rows, and thus forms a solid rib when cast with the slab. The ribs increase the moment of inertia, and the hollow tiles can also provide space for piping and conduit, and the bottom tile surface makes for a smooth ceiling.

An economical solution of the slab problem for rectangular floors with relatively small spans is commonly obtained outside the United States by a mixed structure of reinforced concrete and tiles. The hollow tiles are set on a horizontal scaffolding of planks to create the formwork for parallel reinforced concrete ribs; steel bars are set between the tiles to reinforce the ribs and above the tiles to reinforce the slab (Figure 10.31); concrete is then poured in between the tiles and above them. Once the concrete has hardened, the floor acts as a reinforced concrete slab, ribbed in one or two directions, but with a smooth underside due to the tiles. For fairly large spans the portion of the tiles in the compressed area of the cross section is included in the evaluation of the ribbed slab strength. Cinder blocks or other insulating material may be used instead of tiles as formwork for the ribbed concrete slab.

In the United States, the floors of steel buildings are often built by pouring a concrete slab on a steel deck, which is supported on steel beams and acts as both the formwork for and the tensile reinforcement of the concrete slab (Figure 10.32). The steel deck is made of thin galvanized sheet steel and, through its wavy cross section,

introduces one-way ribs in the concrete slab. In composite action floors, a concrete slab is poured and rigidly connected to the top of steel beams; the beams stiffen the slab, which in turn acts as an upper compression flange for the beams (Figure 10.33).

A concrete slab ribbed in two perpendicular directions is called a waffle slab and is used to cover a rectangular area (Figure 10.34). When the sides of the rectangular area

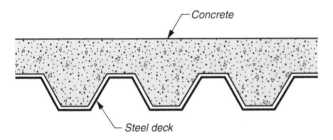

Figure 10.32   Steel deck concrete floor

Consisting of corrugated or formed sheet steel, the ribbed steel deck is a permanent formwork for supporting a concrete slab on top of it. The corrugations give the deck bending stiffness to support the wet concrete, but when the concrete has cured, it has also bonded to the steel due to dimples on the surface. The steel deck in the completed slab therefore acts compositely with the concrete to become external tensile reinforcement to the concrete slab.

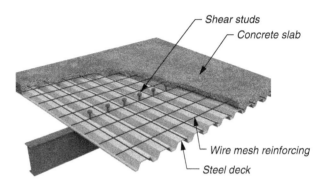

Figure 10.33   Composite steel beam-concrete deck floor

In a composite steel beam-concrete deck, steel *shear studs* are welded through the steel deck and on the supporting beam. These studs engage the concrete floor slab through horizontal shear transfer (see Chapters 5 and 7). As with the composite steel deck of figure 10.32, the composite steel beam-concrete floor assembly works as a unit. The composite assembly acts in bending like a Tee-Beam with the slab in compression and the steel beam in tension. This assembly is more efficient and lighter than a slab that is not composite with the beam.

Figure 10.34   Waffle Slab

To achieve two-way slab action and greater efficiency, a *waffle grid* of ribs may be created. Such slabs are economically formed by reusable steel or reinforced plastic domes set atop a temporary formwork, with reinforcing placed in the space between the domes. Shown here, the ceiling of the Cleo Rogers Memorial Library in Columbus, Indiana, by architect I.M Pei.

Photo courtesy of Ken McCown

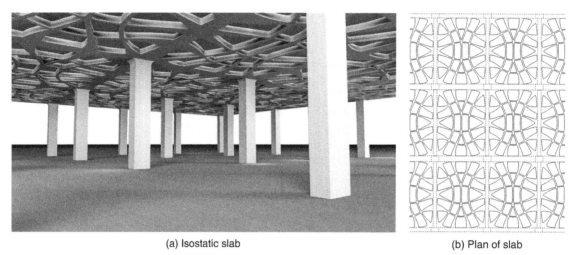

(a) Isostatic slab   (b) Plan of slab

Figure 10.35   Slab ribbed along isostatics

Italian engineer Pier Luigi Nervi pioneered new construction techniques in concrete, including slabs with ribs curved along isostatic lines. To create the ribs, Nervi fabricated permanent ferrocement shell forms from reusable master molds, into which reinforcement was placed and then filled with concrete.

are substantially different, deeper ribs may be used in one direction, or, in the case of reinforced concrete, the two sets of ribs may have the same depth and be reinforced differently. Waffle slabs are obtained by laying steel or reinforced plastic "domes" on a horizontal scaffold in a grid pattern, laying reinforcing bars between and above the domes and pouring concrete between the domes to create the ribs, and above the domes to create the slab. The domes can be reused a large number of times.

The appearance of a ribbed concrete slab may be enhanced by the suggestion of the Italian engineer Aldo Arcangeli to have curved ribs along the isostatic lines of the plate (see Section 10.4). The pattern depends on the support conditions at the plate boundary, and on the loads to be carried. For simply supported plates under uniform load the isostatic pattern is often similar to that shown in Figure 10.15, and was actually used by Nervi in floor design (Figure 10.35). Figure 10.36 presents the isostatically ribbed ceiling in the

Figure 10.36   Nervi's isostatic roof slab at the Papal Audience Hall

The ceiling in this corridor of the Papal Audience Hall at the Vatican is constructed of a ribbed slab oriented closely along the isostatic lines. The resemblance to magnetic field lines is apparent and is an apt metaphor: In a magnetic field, the lines represent constant intensity of magnetism. In the isostatic ribbed slab, the lines are constant intensity of stress with no twisting in the slab. The fact that such beautiful patterns emerge by the ideal stress patterns in a ribbed slab was not lost on the designer, Pier Luigi Nervi, although he did employ artistic license in a less than perfectly strict interpretation of the stress patterns in the ribbed form.

Photo courtesy of Mario Carrieri, Italicementi

Papal Audience Hall in the Vatican. Such curved ribs require expensive forms, and are economically acceptable only if the forms can be reused a number of times.

Two-layered concrete slab structures, analogous to I-beams, are obtained by pouring a concrete slab on a flat scaffold and then pouring on it a waffle slab. In this case, disposable boxes, used instead of reusable domes to create the ribs or webs, are lost in the process. Similar sandwich plates (with high strength-to-weight ratios) are obtained by inserting between two thin plates of plywood, gypsum, steel, aluminum, or plastic materials a plastic foam with good shear resistance. In such *stressed skin panels,* the plastic foam constitutes a lightweight two-dimensional web, while the two plates develop the bending resistance of the sandwich. Such plates have, in addition, good thermal and acoustical insulating properties. Sandwich plates of cardboard with honeycomb paper webs are used in many applications where a reduced dead load is essential, as in aircraft structures and packaging. Sandwich panels of graphite fiber are being used increasingly in aerospace applications where lightweight and high strength are at a premium. Styrofoam sandwiched between plywood or oriented strand board sheets (known as *structurally insulated panels,* or SIPs) can be used as roofs and walls in buildings.

Steel plates constitute the webs of plate-girders, deep beams used in bridges, or other large structures. Their buckling resistance is highly increased by vertical and longitudinal stiffeners, which essentially reduce the length of the plate strips that tend to buckle (see Section 7.1). Thus, steel and aluminum webs are often ribbed, not only to increase their direct load-carrying capacity but also to reduce their weakness in buckling (see Figure 7.30).

10.7 STRENGTH RESERVE IN PLATES

Plates are called two-dimensional resisting structures because two numbers are needed to define the location of one of their points. In the case of rectangular plates, these numbers are the distances of the point from each of the two sides, measured parallel to the other side.

One of the essential characteristics of two-dimensional structures is their high reserve of strength, which stems from two separate sources: stress redistribution and membrane action (see also Section 11.2).

It was shown in Section 9.3 that a simply supported rectangular beam can only redistribute stresses across its depth at the most stressed section, but that a fixed-end beam, after redistributing stresses first across the depth of the most stressed sections at the supports, proceeds to redistribute stresses across the depth of another less stressed section. In a plate, peak stresses (i.e., maximum principal stresses) occur only at one or at a few sections, and most sections are understressed. As the load increases, the most stressed sections redistribute stresses across their

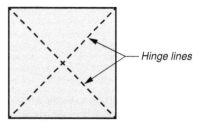

Figure 10.37 Hinge lines of simply supported square plate
In a beam, stress distribution varies from a maximum tension on one surface to maximum compression on the other. When the maximum stresses reach the yield strength of the material, on further loading, internal regions will yield until the whole cross section yields and a hinge develops (Figure 9.15). Similarly, as principal stresses reach the yield strength of the material in a plate, stresses are redistributed trough the depth of the plate and also to neighboring areas. When yielding takes place along a diagonal line, a hinge develops. If enough such hinge lines develop, the plate fails.

depth, while stresses increase at other sections. As soon as one of the sections yields and flows, the stresses at other sections increase progressively up to the yield point and then flow until entire lines of sections may flow, thus creating hinge lines (Figure 10.37).

When the plate has developed enough hinge lines to become a moving mechanism, it cannot carry additional loads. Whereas redistribution of stress in a beam can take place at only one or a few sections before failure occurs, in a plate redistribution takes place along an entire line of sections. Hence, the reserve of strength is generally greater in a two-dimensional element like a plate than in a one-dimensional structure. The reserve of strength in a simply supported square plate under uniform load is 80 percent as against 50 percent in a simply supported beam.

In the discussion of stress redistribution in beams (see Section 9.3), it was emphasized that the end sections of both simply supported and fixed beams are assumed free to move longitudinally relative to each other, so that the neutral beam axis does not change length (Figure 10.38a). When the end sections are restrained from moving longitudinally, the deflection of the beam under load requires a lengthening of the middle axis of the beam and introduces tension in the beam (Figure 10.38b). Thus, in a beam made of tensile material, the longitudinal restraint of the end sections develops a certain amount of cable action, due to the tension along the beam axis curved by the loads. A beam with ends free to move carries loads exclusively by bending and shear, that is, by beam action; a longitudinally restrained beam carries loads partly by beam, and partly by cable action. An additional reserve of strength thus appears in the restrained beam; as the beam yields at some sections, and finally becomes a mechanism incapable of carrying additional load in bending, the sag of the middle axis increases and additional load is carried by cable action. Such cable action cannot take place in longitudinally unrestrained (i.e., simply supported) beams, because, due to the movable roller support, they can deflect without stretching their middle axis.

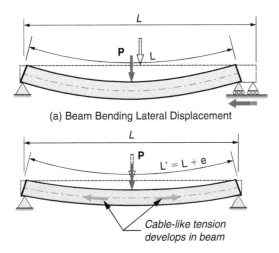

(a) Beam Bending Lateral Displacement

Cable-like tension develops in beam

e = Tensile elongation due to elastic deformation

(b) Tension in Restrained Beam

Figure 10.38   Cable action due to longitudinal restraint of beam ends
If the ends of a beam are free to move horizontally, the length of the originally straight neutral surface does not change when bent (a). On the other hand, if the beam is restrained from moving, the bent neutral surface must stretch and develop cable action (b). In the first case, the load is carried by bending alone; in the second case, it is carried by a combination of bending and cable action. Because of the cable action, additional reserve strength is present in the restrained beam. A plate supported on all sides is essentially restrained and develops membrane action in an analogous manner. See Chapter 11.

A rectangular plate, simply supported along two opposite sides and free along the other two, behaves very much like a series of parallel beams; if its simply supported sides are free to move relative to each other, the plate cannot develop cable stresses. Cable stresses will be developed only if the two opposite sides are restrained from moving.

A square plate, simply supported on all four sides, deflects under a uniform load into a shape that stretches its middle surface even if its sides are free to move (see Figure 10.16). Such shapes are called nondevelopable, because they cannot be "developed" or flattened into a plane without stretching them or introducing appropriate cuts. A sphere is a nondevelopable surface; it can't be flattened without being stretched only if it is cut radially at an infinite number of sections (see Figure 12.4); a cylinder, on the other hand, is a developable surface because it can be flattened without stretching or cutting (see also Section 12.2). The rectangular plate with two opposite sides simply supported and two sides free cannot develop cable stresses because it deflects under load into a cylindrical surface.

The surfaces of plates deflecting under loads are in almost all cases nondevelopable. Hence, the middle plane of the plate must stretch to acquire such deflections, and stresses are developed in the middle plane of the plate capable of carrying some load by membrane action. Membrane action is the two-dimensional equivalent of one-dimensional cable action; it is the action developed under load by a very thin sheet of flexible material, such as cloth (see also

Section 11.2). Membrane action develops even when the plate sides are free to move, but is substantially greater when the sides are restrained.

In general, whatever its support conditions, a plate develops membrane action as soon as it deflects under load; its load-carrying capacity is due to both plate action and membrane action. Under growing loads, a plate first redistributes bending stresses at various sections until it becomes incapable of carrying any additional load by plate action. But, at the same time, the plate develops membrane action and does not fail until the tensile membrane stresses reach the yield point. It is thus seen that, inasmuch as the plate deflects into a nondevelopable surface because of its "two-dimensionality," the continuity of the plate material in two directions gives the plate an inherently greater reserve of strength.

10.8 FOLDED PLATES

As previously shown, the structural efficiency of plates can be increased by stiffening them with ribs, thus removing some of the material from the neighborhood of the "middle plane" of the plate and increasing its moment of inertia. The same result may also be achieved by folding the plate. A sheet of paper, held along one side, cannot support its own weight; its minute thickness does not give a sufficient lever arm to develop an internal couple capable of resisting the bending stresses (Figure 10.39a). Folding the paper sheet brings the material of the cross section away from its middle plane and increases the lever arm of the bending stresses, which becomes comparable to the depth of the folded strips (Figure 10.39b). In fact, the moment of inertia of two plates at an angle is equivalent to that of a rectangular cross-section beam with a depth equal to the depth of the plates and a width equal to the combined horizontal widths of the two plates (Figure 10.40).

Folded plates spanning 100 feet (30.5 m) or more may be made of wood, steel, aluminum, or reinforced concrete. Reinforced-concrete folded plates are particularly economical because their formwork may be built of straight planks, or the concrete slabs may be prefabricated on the ground, lifted into place, and connected by welding the transverse bars at the folds and concreting the joints, thus eliminating most of the formwork altogether, Figure 10.41a.

Folded-plate action is a combination of transverse and longitudinal beam action. Since the length of the slabs is many times their width (Figure 10.41b), the single slabs behave like plates with large aspect ratios and develop only one-way beam action in the direction of their width—that is, transversely (see Section 10.4). This transverse beam action transfers the load on the slabs to the folds so that each transverse strip of slab behaves like a continuous beam with equal spans, elastically supported at the folds (Figure 10.42a). The folds are actually rigidly supported in the neighborhood of the plate ends, where the folded plate rests on rigid end frames (Figure 10.41a). Towards the middle of the span, instead, the fold supports are actually flexible; but since all

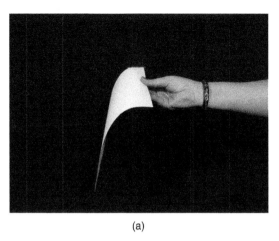

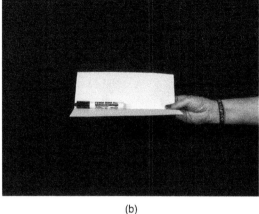

(a)   (b)

Figure 10.39 Strength of folded paper (folded plate)
A sheet of paper bends under its own weight (a) but the same paper bent into a V-shape can support a load in addition to its own dead load (b) due to the increased moment of inertia.

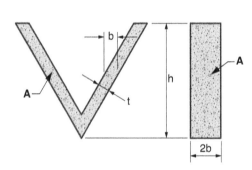

Figure 10.40 Rectangular beam equivalent to folded plate
When a thin sheet of material is bent into a V-shaped beam, its moment of inertia increases. It then behaves similarly to a rectangular beam of the same depth and twice the horizontal width of the plate.

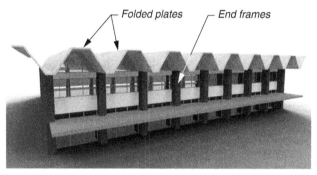

(a) Folded Plate Roof

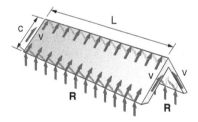

(b) Folded Plate Forces

Figure 10.41 Longitudinal beam action in folded plate
Folded plates first carry loads transversely to the ridge and valley of the plates, where they are then transferred in bending action along the length of the plate. The plate in this manner functions as a thin but deep beam.

folds deflect practically by the same amount, the transverse slab strip moves down by the same amount at each fold and the fold supports act everywhere as if they were rigid, as far as transverse bending is concerned. The reactions of the transverse strips at the folds may be split into components (as in a triangular truss), which load the slabs in their own planes (Figure 10.42b). These longitudinal reactions are carried to the end frames by the slabs acting as deep rectangular beams (Figure 10.41b). Thus, the load is first transferred to the folds by beam action of the slabs in the transverse direction and then to the end frames by longitudinal beam action of the slabs: Each unit strip of slab acts transversely as a fixed-end beam of unit width and depth h, and each slab acts longitudinally as a rectangular beam of width "b" and depth "a" (Figures 10.40–10.42).

A uniformly loaded roof, consisting of a large number of folded plates, develops the same deflections in all the slabs, except those near its external longitudinal boundaries. It was

seen that the transverse strips of interior slabs move down by the same amount at each fold and behave as continuous beams on rigid supports, since equal displacements of all the supports do not stress a continuous beam. The exterior slabs, instead, have differential displacements between the external and the internal fold supports, and carry more load by transverse bending than the interior slabs. To avoid overstressing, the exterior slabs may be stiffened by additional vertical deep

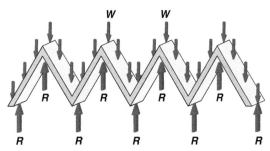

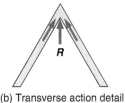

(a) Transverse continuous-beam action in folded plates

(b) Transverse action detail

Figure 10.42 Transverse continuous-beam action in folded plates
The slab action in the transverse direction is that of a folded continuous beam on supports at each of the ridge and valley locations—the folds of the plates act as supports (a). The end reactions of the plates can be looked at in a truss-like manner, breaking the forces into components (b).

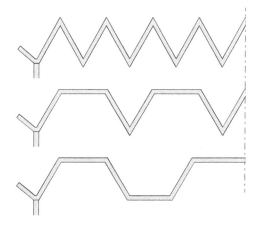

Figure 10.43 Folded plate cross sections
Folded plates can come in a variety of cross sections. Some of the more common are illustrated above. The exterior, final, fold of a plate will have a different transverse stiffness than its interior counterparts. To help equalize the transverse slab action at the outer edges, one approach is to add a vertical deep beam to function as a stiffener.

beams at the boundary; the one-sided lack of support of the exterior slabs is thus compensated for by the supporting action of the boundary beams (Figure 10.43).

Folded plates may be given a variety of cross-sectional shapes. Some of the commonly used arrangements are shown in Figure 10.43.

The increase in stiffness due to folding may be extended to other than longitudinal folds. Polygonal and circular folded plates may be used to cover circular areas; in this case, the plates are given a rise at the center, where the depth of the slabs peters out (Figure 10.44a). Each radial element of the folded plate behaves like a truss or arch; it develops a thrust, usually absorbed by a circumferential tie-rod, and may be considered hinged at the crown, since the small depth available there makes the development of bending stresses impossible (Figure 10.44b).

Folded plates are used mainly as roof structures, but their application to floors may become practical when their depth can be used to house mechanical systems. Folded plates may also be used as vertical walls to resist

both vertical and horizontal loads. Combinations of folded-plates have been used to enclose large spaces. A beautiful example of folded plates used as both walls and roofs is the Air Force Academy Chapel in Colorado, as shown in Figure 10.45. The reader may wish to experiment with this type of construction by following Exercise 2 at the end of the chapter to create a folded paper barrel vault. The barrel-type structure thus obtained is capable of supporting 300 or more times its dead load, although the paper may be only one-hundredth of an inch thick (¼ mm) (Figure 10.46). The creased-paper structure collapses if its boundary is allowed to move. The collapse is due mostly to buckling of the thin slabs in those areas where compression is developed.

10.9 SPACE FRAMES

The two-way action of a rectangular or a skew grid system can be used to span large areas by substituting trusses for beams. The rectangular (or skew) roofs thus obtained (Figure 10.47) exhibit a minor amount of load-carrying capacity through twisting action because of the small twisting resistance of the trusses.

Figure 10.44 Polygonal folded plates
Structural forms of folded plates can be shapes other than rectangular. In both radial (a) and linear (b) arrangements, they behave similarly to arches and must be restrained against lateral thrust. The circular form is compressed circumferentially by tie rods, the linear form is stabilized by a cross tie. At the center of the assemblage, a hinge develops because the small depth at that point can't support bending stresses.

(a)

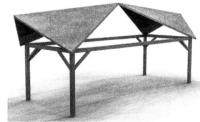

(b)

(a) Exterior: Photo: Michele Chiuini

(b) Interior: Photo: John Hoffman

Figure 10.45 U.S. Air Force Academy Chapel, Colorado Springs, Colorado
Folded plates constitute both the walls and roof of this beautiful structure. There is no other supporting armature beyond the visible structure. The building, designed by Walter Netsch of Skidmore, Owings and Merrill and completed in 1962 was designated a U.S. National Historic Landmark in 2004.

(a) Photo courtesy of Michele Chiuini; (b) Photo: John Hoffman/Shutterstock

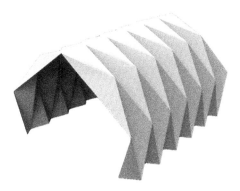

Figure 10.46 Creased-paper barrel
A folded plate structure of this type is capable of carrying a considerable load if the longitudinal borders are prevented from displacement, and the load is uniformly distributed along its surface.

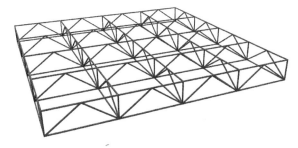

Figure 10.47 Rectangular space truss
Large areas may be covered by rectangular or skew grid by replacing beams with trusses, thereby increasing the strength without using heavy deep beams. Because the trusses have only a small resistance to twisting, such a roof carries very little load by twisting action.

A substantial increase in the twisting action of such roofs can be obtained if panel points of parallel and perpendicular trusses are connected by skew bars, as first suggested by Alexander Graham Bell (Figure 10.48). The truss systems thus become triangulated in space and their behavior is analogous to that of thick plates made out of a spongy material rather than to that of grids. Such triangulated space trusses, called *space frames*, are often used as roof structures to cover large areas because they consist of a large number of identical bars of a few different lengths and may be built in a variety of patterns whose laced appearance has high aesthetic content.

Although entire space frames have been assembled on the ground (by welding, bolting, or screwing their steel bar ends to joint connectors), and then jacked up in place, most of them consist of prefabricated sections built by the use of standard connectors (Figure 10.49) and erected in place without requiring costly scaffolds. The highly efficient Takenaka truss (Figure 10.50) has compressed bars on a square

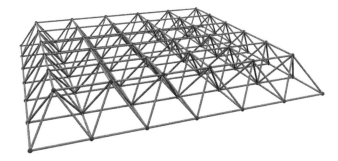

Figure 10.48 Space frame
The introduction of skew bars increases the twisting resistance of a rectangular space truss to create a *space frame*. The behavior of a space frame is analogous to a thick plate made from a spongy material.

grid in its upper chord, inclined compressed diagonals, and tensile bars in its lower chord on a square grid with sides the square root of 2, that is, 1.41 times the sides of the upper chord grid, skew to this grid. Stephan du Chateau has designed and built Takenaka-type space-frames by the use

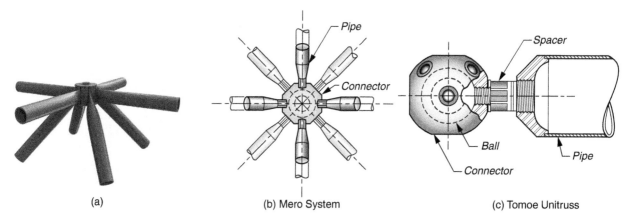

(a)   (b) Mero System   (c) Tomoe Unitruss

Figure 10.49   Slotted and screwed connectors
A variety of prefabricated connectors are used at the intersections of truss members in space frames. The intent of all of these systems is reduction in cost and increase in construction speed through simplicity and standardization of the connectors and bars. Each proprietary system manufacturer has its own particular approach to achieve these objectives.

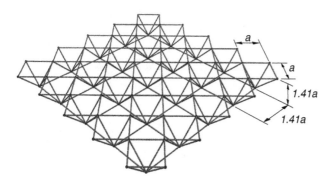

Figure 10.50   The Takenaka truss
The configuration uses standard connectors such as those shown in Figure 10.49 to assemble trusses ion place without the need for heavy lifting equipment.

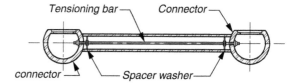

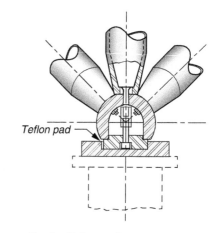

Figure 10.52   Details of PG space frame bar and support
The PG system consists of a tensioning rod in the center of the pipe. The pipes are connected to hollow spherical connectors, and the tensioning rods bolted from the inside. By pretensioning the rods in the entire assembly, the system can be constructed on the ground and then lifted into place by crane.

Figure 10.51   du Chateau space-frame element
A prefabricated element used in a Takenaka type space frame, connected to adjacent members by bolting at the corners.

of identical prefabricated pyramids connected at the corners by high-strength bolts (Figure 10.51). Lightweight planar space frames have also been built by means of steel pipe elements with crimped ends inserted in slotted connectors (Figure 10.49a).

In the PG system, developed by Gugliotta, the bars of steel pipe are connected to hollow spherical connectors by tensioning against the connectors' steel rods running inside the bars (Figure 10.52). The precompressed units of a PG frame can thus be assembled on the ground as a tinker toy and lifted in place. The PG system was used to build the 570,000 square foot space frame for the roof of the Jacob K. Javits Convention Center in New York City, the world's largest at its time of construction in 1986. Engineered by Weidlinger Associates, it consists of square units supported on concrete pylons 90 feet (27 m) on center (Figure 10.53).

It can be shown that a simply supported Takenaka truss under uniform load is statically determinate

Figure 10.53 Javits Convention Center, in New York City
At one time, the largest space frame structure in the world, the building was constructed using the PG system. It covers an area of 570,000 square feet (53,000 m^2). Because of their twisting capacity, space frames can be constructed with a span-to-depth ratio as high as 60 to 1.

Photo courtesy of Angie Allen

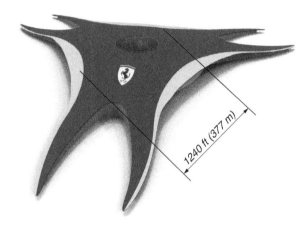

Figure 10.54 Ferrari World Abu Dhabi
The world's largest space frame structure, the structure covers more than 2.1 million square feet (200,000 m^2) of retail and entertainment space. It comprises approximately 172,000 members and 43,100 interconnecting The central enclosed space is nearly the length of four soccer fields laid end-to-end.

(a) Photo courtesy of Wajahat Mahmood

(see Sections 4.2 and 9.3) and hence is not redundant, while other types of space frames are statically indeterminate and may be designed with various degrees of redundancy (see Section 13.3). The research of Makowski shows that because of their twisting capacity space frames can be built with span-to-depth ratios as high as 60.

Increasingly, with improved computer analysis, design and fabrication technologies, space frames are being used for curvilinear and irregular surface structures. The current world's largest space frame is the roof over the Ferrari World amusement park in Abu Dhabi, enclosing an astonishing 2.1 million square feet (200,000 m^2) of retail and entertainment space, making it nearly four times the size of New York's Javits Center (Figure 10.54). The behavior of such curvilinear frame structures becomes more membrane-like (Chapter 11) or shell-like (Chapter 12) and has extended the possibilities for amorphic architectural shapes that were simply not possible until close to the end of the twentieth century. Owing to their lightweight and structural efficiency, their continued and expanded use is anticipated.

KEY IDEAS DEVELOPED IN THIS CHAPTER

- A set of beams laid parallel to each other may be used to cover a rectangular opening. A load applied to one of the beams is carried to its supports, and the other beams are not affected if the beams are not interconnected. If the beams are crisscrossed, load applied to any beam will be shared by neighboring beams, and the load is carried to the boundaries of the rectangle. Shorter beams in such a rectangular grid are stiffer and carry more loads than longer ones.

- To eliminate longer, less effective members in a grid structure, beams may be laid at angles in a skew grid. In this arrangement most beams have the same length and therefore distribute the load more efficiently.

- If the beams of a grid are assumed to be very close to each other, the grid becomes a solid plate.

- Every point on a plate is subjected to bending, shear, and also torsion. Since bending stresses consist of tension, compression, and shear, all these stresses are present in the plate.

- As discussed in Chapter 9, Tension and compression have principal directions and have maximum and minimum values without shear or torsion in those directions. They form lines of "*Isostatics*."

- Plates may cover areas of various shapes and may have a variety of supports. When a plate is loaded, depending on the support conditions on its boundaries, it deflects into a barrel or dish shape. In a dish shape, membrane action will develop (see Chapter 9) and help in carrying the load.
- Plates have relatively shallow depths and therefore have small moments of inertia. To increase the moment of inertia, ribs may be added to their undersides in a manner analogous to T beams, discussed in Chapter 7. Such ribbed plates have greater load-carrying capacities.
- To increase the moment of inertia of a plate, it may be folded into a V-shape that forms it into a beam. If several folds are made, a folded plate is obtained. Like ribbed plates, it has a higher moment of inertia than a flat plate and thus has greater load-carrying capacity.
- Folded plates can be made into vertical walls, radial domes, or even into barrel vaults.
- To cover large areas, instead of beams in a grid, trusses may be substituted for deep beams. Due to low twisting resistance, such a truss grid carries most load in bending action.
- Space trusses may be arranged three dimensionally to create space frames by providing interconnecting diagonal members. Such *space frames* are much stiffer than space trusses because they can develop substantial twisting resistance.

QUESTIONS AND EXERCISES

1. Obtain some slender wooden sticks such as tongue depressors or craft sticks from a hobby store. Create a platform of books on four sides to create a square opening. Perform two experiments using the sticks to span the opening:

   a. Lay the sticks across the opening side by side with about one-stick width between each, then at a 90° angle, span across the opposite direction. Press down on one of the top sticks: Notice how it moves down and also pushes down on any stick below it. Observe how much pressure it takes to push down.

   b. Make a weave of the sticks to create a rigid mat, and place the finished mat on the opening. Again, push down on one of the sticks. Can you notice the difference in behavior from the prior experiment? Now the entire mat moves together, and you may also observe a twisting action of some of the sticks closer to the edge of the opening. Can you explain the difference between this construction and that of the nonwoven sticks?

2. Following figure "a" above, create the barrel shell of Figure 10.46 from a sheet of copier or construction paper. Heavy paper (approx. 28 lb (105 g/m^2)) works very well, but lighter paper can be used, too. Draw light lines on the paper as illustrated (colors are for clarity here only).

   a. Fold on the blue parallel lines in one direction (either toward or away from yourself), and on the red diagonals in the other direction. It doesn't matter which folds are made toward or away from you, only that the diagonals *must* go in the opposite direction to the parallels.

   b. It helps to pre-crease all fold lines first and make them as crisp as possible, and also as close as you can to perfectly intersecting the parallel and diagonal lines at any one point. You will see that the paper has a great "memory" for these folds!

   c. Once all the creases are made, it will already be starting to naturally curl inward —the challenge is to coax it into the final shape. Gently tap the folds in their natural directions and begin to carefully collapse the two long sides together along the previously made fold lines. Be sure not to force anything or it may ruin a crease.

   d. Continued coaxing will transform the paper from a plane sheet into a 3D form. You can "gather up" the folds by holding on to one of the short ends of the paper while continuing to gently push the folds into their natural direction. It will eventually curl up and pancake into a flat six-sided circular donut after all folds have completely collapsed (see figure "b"). Press it flat to set the creases as tightly as you can.

   e. Expand the form into the barrel shell...the diagonal creases will pop upward on the outside surface and the linear creases will fold downward on the inside surface of the shell. You have created a folded-plate barrel shell, substantially stronger than the original flat paper sheet.

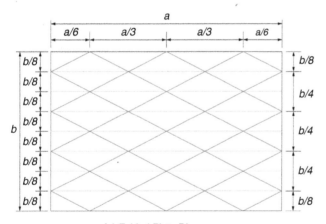

(a) Folded Plate Diagram

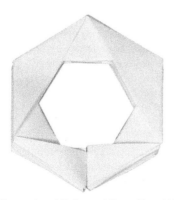

(b) Creased and Collapsed Paper Barrel Vault

3. Load the shell with a distributed weight and observe the behavior. See above figure for example.

   a. Buttress the sizes of the barrel so that it does not spread out at the base. Heavy books can be used for this, or a pair of pencils or short dowels taped to the table.

   b. Place a soft fluffy cloth on top, or make a small cardboard platform with legs to rest on the shell as shown. It's important that the load is not causing a local pressure point on any of the creases or this will cause a premature failure. Try to distribute the load as unifromly as possible.

   c. Gradually apply light loads on the shell until it collapses. Lightweight paperback books work well for this. Load it slowly and watch what happens at each stage.

   d. Observe the behavior: How much weight did it support? How did it collapse? If you have access to a kitchen scale, compare the ratio of the total load carried vs. the self-weight of the paper. How many more times larger is the total load supported than the weight of the paper itself? Are you surprised?

4. If you have access to "Tinker Toys" or a child's construction toys such as K'nex, you may build many of the structures described in this and foregoing chapters.

FURTHER READING

Schodek, Daniel and Bechthold, Martin. *Structures*, 7th Edition, Pearson. 2014. (Chapter 10)

Sandaker, Bjorn N, Eggen, Arne P. and Cruvellier, Mark R. *The Structural Basis of Architecture*, 2nd Edition, Routledge. 2011. (Chapters 6.9 and 6.10)

MEMBRANES

11.1 MEMBRANE ACTION

A membrane is a sheet of material so thin that, for all practical purposes, it can only develop tension. A piece of cloth or a sheet of rubber is a good example of a membrane. Soap films are among the thinnest membranes one can build: They are only a few thousandths of an inch thick, can span plane or skew closed boundaries, and have the remarkable property of being surfaces of minimum area among all those with a given boundary (Figure 11.1). By the same token, a soap film is also the smoothest surface connecting the points of the given boundary.

Although a membrane is a two-dimensional resisting structure, it cannot develop appreciable plate stresses (bending and shear) because its depth is very small in comparison with its span (see Section 10.4), neither can it stand compression without buckling. Therefore, the load-carrying capacity of membranes is due exclusively to their tensile strength.

It was noted in Section 6.1 that a cable can support loads in tension because it sags, and that it is structurally efficient because its tensile stresses are distributed uniformly across its sections. A membrane supports loads by a similar two-way mechanism and presents the same kind of structural efficiency. But just as a plate develops, in addition to beam actions, a twisting action which is due to its two-dimensional character, a membrane, due to its two-dimensionality, develops, in addition to cable actions, an in-plane "shear action" which increases its load-carrying capacity.

Figure 11.2 shows an element cutout of a curved membrane and acted upon by a load normal to it, that is, a pressure. Since the sag of the element produces in the membrane curvatures in two directions, the membrane may be considered as the intersection of two sets of cables; each set carries

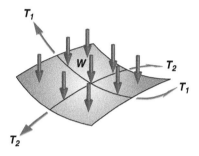

Figure 11.2 Cable actions in membrane

A membrane can be conceived of as a series of intersecting cables, each carrying a portion of the load in two-way action. Closely spaced lines create a closed surface such as a fabric membrane, and more widely spaced but heavier cables create surfaces known as cable nets (section 11.3).

a share of the load, and the total load carried by the membrane is the sum of the two sets of "cable-supported" loads. Thus, a membrane is seen to be capable of "cable action in two directions" due to its curved shape, that is, to the geometrical characteristic of its shape called curvature. It will now be shown that the two-dimensional resisting character of a membrane makes it also capable of a second load-carrying mechanism through the development of in-plane or tangential shears acting within the membrane surface.

One may prove that a thin piece of material can develop such shears by holding a vertical sheet of paper along one of its edges and by pulling down vertically along its opposite edge. The paper sheet carries the load acting in its own plane by tangential shears (Figure 11.3). But, just as the tension in a horizontal cable cannot carry vertical loads, the shears in the plane of the paper sheet cannot, by themselves, carry loads perpendicular to it. The load-carrying action of a membrane due to tangential shears is essentially connected with another geometrical characteristic of its shape: *twist*.

Figure 11.4 shows a square element cutout of a curved membrane and indicates that, in general, its four sides are not parallel, but askew in space; this means that one of its sides slopes more than the opposite side, and that a difference in slope occurs between these two sides. The figure also shows that a difference in slope between two opposite sides necessarily requires a difference in slope between the other two opposite sides, since the two more inclined sides must meet at a point. The difference in slope between two

Figure 11.1 Minimum surface membrane

A minimal surface membrane represents the smallest possible surface area that can be created about a specific boundary. For every unique boundary, there will be one minimal surface; it is akin to the funicular form of a cable. It is the shape that a soap film would take.

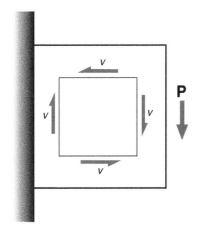

Figure 11.3 Shear development in plane of membrane
A sheet of paper held along its edge will carry a vertical load in its own plane by tangential shearing forces.

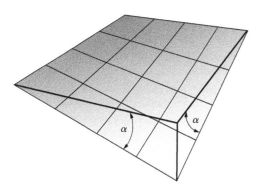

Figure 11.4 Geometrical twist
A square cutout of a curved membrane illustrates that the opposite sides are not parallel, The *geometrical twist* of the element is the difference in slope between two opposite sides spaced one unit apart. The curvature and geometrical twist define the surface at a given point.

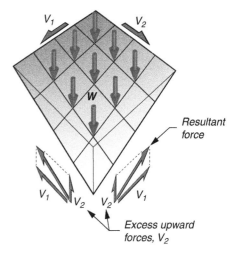

Figure 11.5 Portion of normal load carried by membrane shears
Membrane shears equilibrate in the same manner as shears in planar structures (see Figure 5.12). The difference here is that because of the twist, an *out of plane* component is generated (shown upward as here diagrammed, although the membrane can be oriented in any direction). It is this out of plane force that balances load such as wind pressure acting against the surface. Furthermore, the equivalent principal compression force due to shears must be lower than the equivalent principal tension force, or the membrane surface will buckle since it cannot support compression.

opposite sides one unit apart, that is, the unit change in slope at right angles to the direction of the slope, is called the geometrical twist of the membrane surface. Since, in addition, the slope of a membrane surface usually changes from point to point in the direction of the slope, membranes, in general, also have what we call curvatures. The curvatures and the twist characterize the geometrical behavior of the membrane surface at a given point (see Section 11.2).

The two pairs of equal shears on the opposite sides of any membrane element have directions that guarantee equilibrium in rotation (see Section 5.3), as shown in Figure 11.3; but in a twisted membrane the difference in slope between two equal shears on opposite sides produces an excess of upward forces (Figure 11.5). It is this excess that balances part of the load and gives the membrane its extra load-carrying capacity by shearing action within its own surface.

Inasmuch as shear is equivalent to tension and compression, this additional load-carrying capacity can be developed

by the membrane if and only if the equivalent compression due to shear is smaller than the tension due to cable action. Otherwise, the excess of compression would tend to buckle the thin membrane. If this happens, the membrane changes shape to carry the load by tension only.

Membranes that do not tend to buckle are thus seen to carry normal loads without changing shape by three separate mechanisms: Cable action due to curvature in one direction; cable action due to curvature at right angles to the first; and shear action due to twist. Any one of these actions vanishes if the corresponding curvature, or the twist, vanishes.

11.2 PRINCIPAL CURVATURES AND PRINCIPAL MEMBRANE STRESSES

The results obtained in the preceding section indicate the two essential characteristics of membrane action: (a) Membrane stresses, both tensile and shearing, are always developed within the membrane surface and never at right angles to it; (b) membrane action depends essentially on the geometrical characteristics of the membrane shape, that is, on its curvatures and its twist.

In order to visualize the curvatures of a surface, one must cut the surface with a plane perpendicular to it. For example, Figure 11.6 shows that, in a half circular cylinder, a perpendicular plane parallel to the axis cuts the cylinder along a straight line, indicating a lack of curvature in this direction. A cut at right angles to the axis shows a half circle.

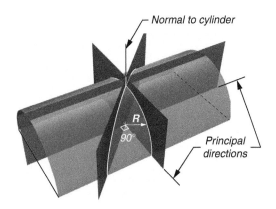

Figure 11.6   Cylindrical curvatures

To facilitate understanding and visualizing the curvatures of a surface, a plane cut through a half cylinder highlights the curvature at the intersection between the cylinder and plane. A plane parallel to the cylinder's axis will intersect along a straight line. A plane cut perpendicular to the cylinder's axis will be a semi-circle, and represents the maximum radius of curvature of the cylinder. Any other angle of cut will have a varying degree of curvature, from zero when parallel to the cylinder's axis to maximum when perpendicular. These minimum and maximum curvatures are called the *principal directions of curvature*.

In this direction, the surface has a large curvature, while cuts in any other direction show smaller curvatures. It is thus seen that as the cutting plane rotates around the normal to the cylinder, the sections acquire curvatures that vary from zero in one direction to a maximum value at right angles to it. The two perpendicular directions in which the curvatures become respectively maximum and minimum (in the present example the minimum is zero) are called the *principal directions of curvature* of the membrane surface.

For an element of the cylinder with sides parallel to these directions, it is seen that the element has no twist: The horizontal slope remains horizontal as one moves at right angles to it (Figure 11.7a). But an element cut out of the cylinder with sides not parallel to the principal directions has twist (Figure 11.7b). Thus, a surface may have twist in

certain directions and no twist in others. (The only surface without twist in any direction is the sphere, since an element cut by any two pairs of parallel sides is identical to an element cut by any other two pairs of sides.)

The properties exhibited by the cylinder in relation to curvature and twists are not peculiar to this surface, but completely general. Any surface cut by planes passing through its normal at a point exhibits two directions at right angles to each other for which the curvature becomes, respectively, maximum and minimum; for these two perpendicular directions, the surface has no twist. It is often convenient to spot principal curvature directions by the lack of twist—that is, of slope change—in these directions (Figure 11.8).

One may mark by means of small crosses the directions of principal curvature on a surface, as was done for the case of principal stresses, and obtain a pattern of lines of principal curvature. For a cylinder, this pattern is a rectangular mesh with sides parallel to the axis and at right angles to it. Figure 11.8 indicates the principal curvature lines for a more general surface.

The tensile direct stress at a point of a membrane, just like the stresses in a plate, has different values in different directions. As the orientation of the tensile stress rotates around a point, the stress becomes maximum and minimum in two directions at right angles to each other, the directions of principal stress. Elements oriented in these directions do not develop shear; shear instead becomes greatest at 45 degrees to the directions of principal stress. The perpendicular directions of principal stress are those in which the membrane acts as two sets of cables at right angles; one set develops the maximum stress, and the other set the minimum stress at each point. Shear stress develops between the cables in any other direction.

Indicating by small crosses the directions of principal stress at a number of points, one may obtain a pattern of principal stress lines for the membrane. Whereas the lines of principal curvature are a geometrical characteristic of the membrane surface, the lines of principal stress depend not

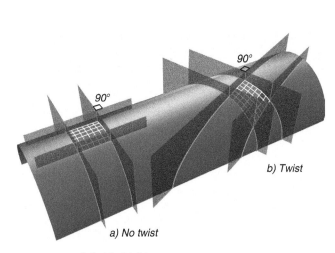

Figure 11.7   Cylindrical twist

Elements cut from a cylindrical surface in either principal direction exhibit no twist. At any other angle between the principal directions, however there will be twist.

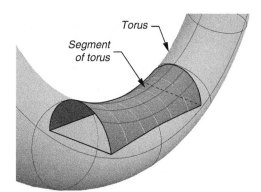

Figure 11.8   Principal curvature lines

Not only cylinders but any curving surface will exhibit maximum and minimum principal curvature directions. Since the direction of principal curvature will change from point to point, a plot of the curvature at regular intervals can be drawn on the surface. Doubly-cuved surfaces can be seen as segments of a torus (a donut shaped ring). In this example, horizontal and vertical slices are taken, but many other forms are possible with cuts at different angles.

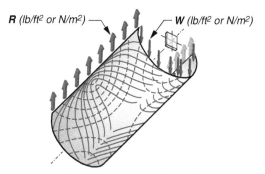

(a) Cylindrical Membrane Dead Load Stress

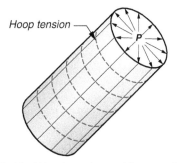

(b) Cylindrical Membrane Internal Pressure Stress

Figure 11.9 Principal stress lines of cylindrical membranes (a) under dead load and (b) under internal pressure

Lines of principal stress may be plotted on surfaces in a manner similar to that of principal curvatures. Unlike the lines of principal curvature, which depend only on the surface shape, the lines of principal stress also depend significantly on the load applied and character of the supports. A side-supported half cylinder with a downward curvature and load creates the pattern shown in (a). In certain symmetric geometries such as a full cylinder under internal pressure (b), the lines are all parallel. The entire load here is carried by *hoop stresses* that run circumferentially along the axis. It is the principle of hoop stresses that is at work in traditional wooden barrels wrapped with iron or steel rings.

only on the shape of the surface but also on the character of the load and the conditions of support (figs. 11.9a and b). In certain cases of symmetry, the two patterns may exceptionally coincide, as for a sphere or a cylinder under internal pressure (Figure 11.9b).

The analysis of cable action showed that the stresses developed in a cable are related to its span-to-sag ratio (see Section 6.1). Conversely, for a given allowable stress, the greater the sag, the greater is the load the cable can safely carry. Considering the cable action of a membrane along its principal stress directions, it is thus found that the direction with greater sag ratio—that is, greater curvature—carries more load than the direction with smaller sag ratio, smaller curvature. In the case of a sphere, the curvature is the same in all directions, and a spherical membrane under uniform pressure carries half of it by cable action in one direction and half of it by cable action at right angles to it. In the case of a cylinder under uniform normal pressure, the straight lines carry no load, since they have no curvature, and the entire load is carried by stresses along the curvature lines at right angles to the cylinder axis. These so-called hoop stresses are twice as large as those in a sphere of the same radius under the same pressure.

When the load distribution on a membrane changes, stresses in the membrane also change, but—provided the new stresses do not buckle it—the membrane does not have to change shape to support the new load. While a cable is funicular for only one set of loads, a membrane is funicular for a variety of load distributions because it can distribute the load between its two-dimensional tensile and shear mechanisms in a variety of ways. Thus, a membrane is inherently more stable than a cable, although it is unstable under loads that tend to buckle it. Under such loads the membrane changes shape so as to support them by purely membrane stresses.

In general, membranes must be stabilized, mostly because their funicular shape for horizontal loads differs from that for vertical loads. Stabilization is obtained either by means of an inner skeleton or by prestressing produced by external forces or by internal pressure. Prestressing allows a loaded membrane to develop compressive stresses up to values capable of wiping out the tensile stresses locked in the membrane by the prestressing; it adds to the advantage of greater aerodynamic stability that of an increased carrying capacity through the shear mechanism.

It must be noted that the tensile stresses in a membrane are uniformly distributed across its thickness; the utilization of the material in membranes is optimal. Moreover, tensile strains are always small compared to bending strains, so that the membrane deflections due to loads are usually small. (These deflections should not be confused with the displacements of the membrane due to load variations, which require changes in its shape.) Thus, by the nature of their load-carrying action, membranes are light and stiff under steady loads. As noted above, their use in permanent structures must take into account their mobility, which can be greatly reduced, but not totally eliminated, by prestressing.

11.3 TENTS AND BALLOONS

Notwithstanding their instability, from time immemorial human ingenuity has found ways and means of using membranes for structural purposes, mainly because of their lightness. The nomad's tent is a membrane capable of spanning tens of feet, provided its skins are properly supported by compressive struts and stabilized by tensioned guy ropes. The circus tent spans hundreds of feet using the same technology (Figure 11.10). The tent withstands the pressure of the wind, but even under ideal circumstances presents the drawback of moving under variable loads. Moreover, because of its light weight, the tent vibrates or "flutters" under the action of a variable, or even a steady, wind. Tents are useful as temporary covers, and acceptable as permanent roofs only if highly prestressed.

Large tents supported by complicated cable nets on vertical or inclined compression masts have been used by Frei Otto to build temporary or permanent pavilions for fairs and other uses. One of the largest to date (2015) covers 808,000 square feet (75,000 m^2) at the site of the 1972 Olympics in Munich, Germany (Figure 11.11). It is supported by nine

Figure 11.10 Circus tent
Tents are one of the oldest forms of human shelter, with examples of stretched animal hide dating back to antiquity. A contemporary tent, such as used in a traveling circus, derives stability through double curvature of its surface. The downward curving direction is responsible for load carrying; the upward curing direction provides a stabilizing force against flutter and uplift in wind.

Photo: Creativenature.nl/Fotolia

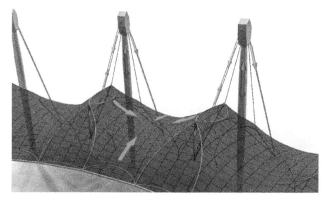

Figure 11.11 Portion of cable net structure of the 1972 Olympic Stadium in Munich, West Germany
The cable net can be thought of as a fabric membrane with large diameter cables for the material, spaced widely apart. And as with any membrane structure, there are primary downward curving load-carrying cables and cross-directional upward curving stabilizing cables. The Munich Olympic stadium by Frei Otto and Günter Behnisch remains one of the largest and most famous of this genre of structure. It is a popular tourist destination, and for a fee one can walk on the roof itself. (See also Figure 14.13b.)

masts up to 260 feet (79 m) tall, and has prestressing cables of up to 5000 tons (45 MN) capacity, which give the roof the appearance of a series of interconnected saddle surfaces. The cables support translucent Plexiglass slabs, lightly tinted a neutral gray-brown.

The largest tent built to date (2015) is the prestressed tent for the Hajj terminal of the Jeddah Airport in Saudi Arabia, engineered by Geiger-Berger for the pilgrims to Mecca. It consists of 210 square modules, 151 feet by 151 feet (46 m by 46 m), covering an area of 23 acres (9.3 hectares or more than 93,000 m^2) (Figure 11.12).

The tent of the Franklin Zoo in Boston, Massachusetts, engineered by Weidlinger Associates, consists of three cable-reinforced panels of plastic fabric, supported by a tripod of three inclined steel arches resting on a reinforced-concrete ring that is compressed by the tension in the cables (Figure 11.13). It covers a circular area of 7 acres (2.8 hectares) with a radius of 176 feet (54 m). Its membrane is partly stabilized by a minor internal suction. Although designed to support a snow load of 30 pounds per square foot (1,400 Pa), under normal conditions it will not carry such load because warm air of the conditioned interior will melt the snow.

(a) Single Module

(b) As Constructed Roof

Figure 11.12 Hajj Terminal roof
The largest membrane structure in the world to date in terms of surface area covered is the Hajj Terminal, located at the King Abdulaziz International Airport in Jiddah, Saudi Arabia. It was constructed to shelter the millions of the faithful in the yearly Hajj pilgrimage. It is constructed of 210 square modules, each 151 feet (46 m) on a side. Unlike the Munich Olympic Stadium, the roof is a fabric membrane, and the direction of stabilizing curvature is circumferential as opposed to downward curvature. Tensioning is through uplift of the central tension ring of each module, up to supporting towers nearly 150 feet (45 m) tall.

(b) Photo courtesy of Birdair, Inc.

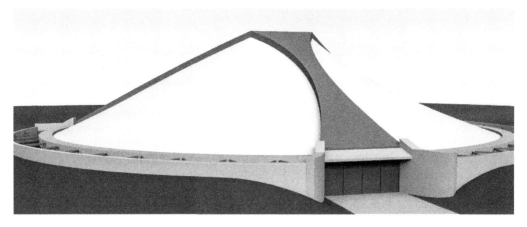

Figure 11.13 Boston's Franklin Park Zoo Tropical Rainforest Pavilion
The structure is supported by three large steel arches that support fabric panels. The panels span between the arches and the perimeter ring. The membrane is stabilized by an internal air pressure lower than the external ambient pressure.

The structural action of a membrane is greatly improved by tensioning it before it is loaded. Firefighters use a round sheet of cloth, tensioned on an outer ring, to catch people jumping from great heights. The tension in the sheet and the compression in the ring are produced by the radial pull of the rope connecting the sheet to the ring. The thin sheet takes the impact of the falling person, deflects elastically, and saves the jumper by its flexibility and strength. The combination of sheet and ring, even though unloaded, has "locked-in" stresses of the type described in Section 2.5. Firefighter's membranes are typical prestressed elements.

The umbrella is another example of a prestressed membrane with locked-in stresses. The steel ribs, pushed out and supported by the compressive struts connected to the stick, tension the cloth and give it a curved shape suited to resisting loads. Within limits, the membrane of an umbrella can take pressures from above and from below: The supporting steel skeleton reverses its stresses with the reversal of wind action, but the membrane is under tension in both cases (Figure 11.14).

The Eskimo kayak is built by means of a membrane of seal skins tensioned by a skeleton of compressed longitudinal bars and circumferential rings pushed into it. A modern version of the kayak is built by means of a membrane of rubberized cloth, stretched by an internal frame. Such is the strength of these little boats that they have been sailed across the Atlantic by solitary navigators, although only 17 feet long and 3 feet wide (5 m by 1 m).

The interesting roof shapes of membranes prestressed by external forces are always saddle shapes. Figure 11.15 shows a tent in the form of a saddle, designed by Frei Otto, in which the tension of the ropes anchored to the ground stabilizes the tent suspended from two poles (see also Section 6.2). Figure 11.16 illustrates a circular tent roof,

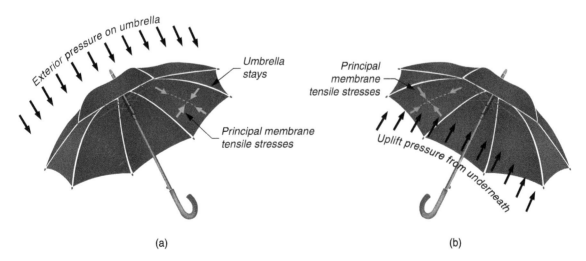

(a)   (b)

Figure 11.14 Prestressed membrane of an umbrella
A membrane that is pretensioned (or *prestressed*) will see an improvement in its load-carrying capacity. A classic example is an umbrella: The open framework stretches the fabric creating locked-in stresses, enabling it to carry both downward forces (a) and a modest amount of upward (or outward) force (b). The supporting framework will reverse its stresses upon a load reversal; however, the membrane will *always* be in tension. Clearly, the outward force-carrying capacity is substantially lower than the inward directed, as the many images of inside-out umbrellas in heavy winds attest.

Figure 11.15   Prestressed membrane tent

One of the earliest membrane tent structures constructed by architect-engineer Frei Otto was a simple hyperbolic paraboloid, or saddle-shaped roof, tensioned to the ground at the two low points and uplifted with two masts at the high points. The fabric was only 0.04 inches (1 mm) thick but spanned nearly 60 ft (18 m). This faithful recreation was made on the occasion of Frei Otto's posthumous award of the 2015 Pritzker Prize, one of the highest honors an architect may receive.

Photo courtesy of Birdair, Inc.

designed by Horacio Caminos, in which a circular cloth attached to a compression ring is tensioned by the action of compressive struts. Prestressed membranes are stiffer and more stable than unstressed membranes: They do not flutter or move as easily under variable loads.

The hulls of the German Zeppelin dirigibles were stabilized by an internal skeleton of aluminum girders and rings, which supported the external fabric membrane. Hydrogen contained in multiple internal sealed gasbags provided the lift. Its rigidity allowed the hull to take the severe dynamic loads imposed by air pockets and turbulence. While the Zeppelins relied on the frame for structural integrity, *blimps* (such as those made famous by the Goodyear Company) are like balloons and rely on the internal pressure of helium to support the shape using no frame at all. Since they have no structural strength without a frame, however, blimps cannot be as large as Zeppelin-type airships that employ structural frames to carry increasing forces with inceasing scale.

Membranes can be prestressed by internal pressure alone when they completely enclose a volume, or a number of

Figure 11.16   Prestressed circular membrane tent

Many shapes are possible with fabric membranes, by combining multiples of basic shapes. A circular form elevated on a structural frame creates a self-stabilized structure with no cable anchorages to the ground.

Photo courtesy of Birdair, Inc.

separate volumes. They then constitute *pneumatic structures*. Hot air balloons consist of an air bag open at the bottom where hot air from a burner enters and, being lighter than the surrounding cold air, makes the balloon rise. The air bag is reinforced by a series of ropes surrounding it. Membrane structures consisting of an enclosed volume are used, for example, in rubber rafts: The outer inflated ring is stiff enough to be used as a compression ring for the unstressed membrane forming the bottom of the raft. An inflatable airplane has been built on the same principle, with inflated wings and an inflated fuselage divided into separate compartments; it was flown by using a light engine. Paraglider canopies and parachutes are also simple membrane structures.

The long plastic balloons used at children's parties become so stiff when inflated that they can develop compression and bending action. The internal pressure locks into the cylindrical balloons tensile stresses in the circumferential and in the longitudinal directions. (The circumferential stress is often referred to as the "hoop stress" by analogy with the tensile stresses locked into the hoops of a barrel.) The inflated cylinder supports compressive loads, as a column, up to the point where the compression due to the load equals the longitudinal tension due to the internal pressure. When the compression nears this value, the membrane buckles.

Similarly, a balloon used as a horizontal simply supported beam carries vertical loads up to the point where the compressive stresses in the upper fibers of the beam equal the tensile stresses produced by the internal pressure. The reader may easily prove that when the compressive stress nears this value, the beam buckles at the top.

Based on this principle, the Fuji pavilion at the Osaka World's Fair of 1970, designed by Murata, had a curved roof spanning 164 feet (50 m) obtained by means of inflated curved tubes of plastic material, in which the internal pressure could be varied to increase the stability of the structure with increasing wind velocities (Figure 11.17). Enclosures for athletic fields, warehouses, and meeting halls have been built on the same principle.

Figure 11.17   Fuji pavilion at Osaka World's Fair

An early air-supported structure was the Fuji Pavilion of the 1970 Osaka World's Fair. It consisted of 16 air-filled fabric tubes bundled together in parallel rows. A somewhat amorphic shape was generated by arranging the ends of each tube along half circles on each side of the structure, to make a circular plan. The internal pressure of the structure could be increased or decreased in response to wind loading.

Photo courtesy of Birdair, Inc.

Inflated plastic "mattresses," similar to those used to lie on the beach or to float on water, have been used structurally as horizontal roofs, vertical walls, and inclined floors, for example, for theatre floors.

All the pneumatic structures mentioned so far are inflated, closed membranes. Some of the most interesting and impressive examples of pneumatic structures belong, instead, in the category of air-supported roofs, the so-called balloon roofs or bubbles.

Plastic fabric balloons covering swimming pools, tennis courts, and other temporary installations may be blown up by a small pressure to create stable barrels (Figure 11.18). An overpressure of only one to two tenths of a pound per square inch is sufficient to hold up such structures, which can be entered through revolving doors, since the loss of

pressure in the large enclosed volume is negligible even when doors are used frequently; such losses as do occur are replaced intermittently under control of a pressure gauge. Large spherical domes over radar installations (radomes) are built on the same principle by means of extremely thin plastic membranes, which do not interfere with the reception and transmission of electromagnetic beams. Radomes are built up to diameters of 150 feet (46 m), using high-strength membranes capable of developing allowable tensions of 300 to 800 pounds per inch (2 kPa to 5.5 kPa).

New material developments in the form of tough, thin fluoropolymers such as ETFE are spurring application in building skins. The thin films have been fabricated into air-inflated "pillows" forming the envelope of some buildings, such as the Eden Project in Cornwall, England (Figure 11.19),

Figure 11.18 Space Shuttle museum pneumatic enclosure

Upon retiring the Space Shuttle fleet, the flight deck of the Intrepid Museum in New York City became home to one of the vehicles. It was initially housed in this inflatable fabric membrane structure. Barely more than three months after the exhibit opened, however, 2012's Hurricane Sandy knocked out power to the fans providing pressure to the roof structure. With the loss of its tension and consequently loss of the structural stability, fierce winds ripped apart the fabric, destroying the structure. This was never intended to be a permanent enclosure, though, and it was subsequently replaced with a rigid steel frame structure.

(a)

(b)

Figure 11.19 Eden Project

A contemporary development of the inflated structure is to use a rigid structure that supports a façade of air-inflated "pillows." The Eden Project by Grimshaw Architects in Cornwall, England, was among the first projects of this type. The skin "pillows" are made of ETFE plastic, a very durable yet translucent material. Built on the site of an abandoned quarry, the project is an environmental education center, featuring "Biomes" containing plants from several distinct climactic regions of the world. A noteworthy feature of this building is that because the skin dead load is so low, the structural frame has to do little more than support its own weight and withstand lateral wind forces. The resulting total structure weight is so efficient that in total it weighs no more than the volume of air that it encloses.

Photo courtesy of Terri Meyer Boake

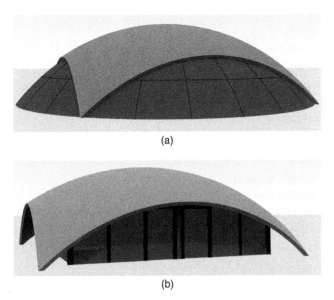

(a)

(b)

Figure 11.20   Igloo (balloon) houses
A balloon-formed house is created by applying concrete to the outer surface of an inflated membrane. When the concrete has cured, the membrane is deflated and removed, leaving a rigid concrete shell in its place.

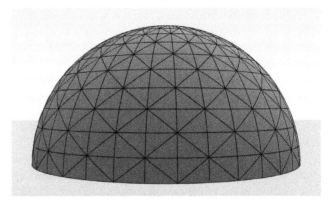

Figure 11.21   Birdair pneumatic dome concept
With increasing size of a dome, the pressures exceed that of the fabric, which must be reinforced by steel cables. The fabric then only spans between surrounding cables, which in turn serve as the primary load-carrying members.

Photo courtesy of Birdair, Inc.

and the natatorium for the 2008 Beijing Summer Olympics games (better known as the "water cube"). Given the material's light weight, high strength, and excellent thermal properties, the success of these projects will undoubtedly spur on continued investigation and use of lightweight pillow membranes for façade applications.

Balloon forms for concrete were used in the construction of the so-called igloo houses, invented by Neff and designed by Noyes and Salvadori (Figure 11.20). The inflated balloon supports a reinforcing steel mesh, over which a layer of a few inches of Gunite concrete is sprayed by a concrete gun. Once the concrete has hardened, the form is deflated and pulled out of the house through the door opening.

The technique invented by Dante Bini to build his "Binishells" is an improvement on the Neff idea. A special type of sliding reinforcement is set on the deflated balloon, on which is also poured very fluid concrete; the inflation of the balloon lifts both the reinforcement and the concrete, which is kept in place by the helical steel springs sheathing the reinforcement and vibrated after inflation. Over 1,000 Binishells with spans of up to 120 feet (37m) have been built and larger spans are technically feasible.

As the span of the air-supported balloon roof increases, the tension in the membrane exceeds its strength and the membrane must be reinforced, usually by steel cables. In such systems, the membrane spans only between cables and can be thinner, lighter, and more economical. In the Birdair domes, which can span up to one thousand feet (305 m), the cables are laid on a triangular mesh about 10 feet (3 m) on a side (Figure 11.21). In the U.S. Pavilion at the Osaka World's Fair of 1970, engineered by David Geiger, the cables were laid on a skew grid 20 feet (6 m) by the side and anchored to a

funicular compression ring of concrete in the shape of a rectangle with rounded corners 460 by 262 feet (140 m by 80 m). The internal overpressure was only 4.3 pounds per square foot (three hundredths of a pound per square inch (206 Pa)) because the low rise of the roof (23 feet (7 m)) and the earth berm surrounding it induced wind suction over almost the entire roof. Air-supported roofs based on this scheme have been built to date covering stadiums with areas of over 300 thousand square feet (28,000 square meters) and seating over 80,000 spectators, like the Silver Dome stadium in Pontiac, Michigan, or the Minneapolis Metrodome (Figure 11.22), both engineered by Geiger.

Studies by Weidlinger Associates and others show that air-supported, cable-reinforced roofs, spanning over 6000 feet (1800 m) and covering over 600 acres (243 hectares) are feasible with present-day technology. The recent development of stronger and longer lasting membrane fabrics allow the building of roofs capable of covering entire towns.

Figure 11.22   Minneapolis Metrodome
This air-supported stadium roof structure consists of a skew grid of cables supporting a membrane. Only a very slight overpressure is needed to keep the roof inflated. Many large-scale arenas have used this principle, the fabric membrane providing both weather enclosure and abundant natural light compared to other roof options.

Photo courtesy of Birdair, Inc.

The cost per square foot of air-supported bubbles is among the lowest for large span roofs. Their behavior in case of fire is safer than originally predicted because their membranes do not support flames, and superior to that of other large structures because the membranes are light and deflation takes hours even if large holes open in them. Transparent, translucent, and opaque membranes, glass-reinforced and vinyl-, hypalon-, or teflon-coated, permit a variety of natural lighting, different temperature distributions, and various degrees of privacy. They create artificial environments adaptable to human use in practically any part of the world and have good energy conservation characteristics. Since, moreover, under such roofs the architect has great freedom in "landscaping" the space with structures that are independent of the outer environment, their use is bound to increase in popularity. Air-supported, cable-reinforced bubbles are one of the modern breakthroughs in large-span roof construction.

Membrane roofs have been built not only of cloth and plastics, but of steel, aluminum, and reinforced concrete. Metal membranes, well adapted to carrying loads by tensile stresses, have been used to build stiff, permanent, circular roofs in Austria and Russia. Reinforced concrete may be so reinforced, or prestressed, as to make it suitable to the development of tensile stresses: The Viera circular roof may be considered a concrete prestressed membrane (see Figure 6.20). The cable-supported roof of the Nowicki arena may be considered a cable-supported membrane made up of metal sheet elements (see Figure 6.19). It was indicated in Section 6.2 that light metal structures have a tendency to flutter, while the heavier concrete membranes do not. Thus, the thickness of a structural membrane may be increased for reasons other than strength, and the membrane may acquire some of the characteristics of a plate. Just as plates are bound to develop some membrane action in deflecting under applied loads (see Section 10.7), membranes are bound to develop some plate action due to their finite thickness. In view of the greater efficiency of membrane action (see Section 11.1), the thickness of a membrane should be kept minimal, provided its shape and displacements are functionally acceptable.

KEY IDEAS DEVELOPED IN THIS CHAPTER

- A membrane is a thin sheet of material that has practically no moment of inertia. It can therefore only carry tensile stresses and will buckle under compression.
- A curved membrane behaves essentially likes a series of closely placed cables connected in two directions. Though its principal load-carrying mechanism is tension in two directions, there is a small amount of torsion, and hence in plane shear stress also present to help in carrying part of the load.
- A curved membrane usually has two principal curvatures (a maximum and a minimum). Along these lines there is no twist or shear stress.
- Membranes, like cables, are unstable structures. They change their shape under varying loads: They may be stabilized by internal or external ribs such as umbrellas, by two-directional opposing curvature, or by internal pressure (such as in balloons and air-supported structures).

QUESTIONS AND EXERCISES

1. Inflate a long skinny toy balloon. Use it as a beam and as a column. Describe the mechanism of failure for each.
2. What kinds of stresses are present in a trampoline?
3. Aside from differences in size, what are the differences between a child's balloon and a dirigible?

FURTHER READING

Allen, Edward Zalewski, Waclaw, *Form and Forces: Designing Efficient, Expressive Structures.* John Wiley & Sons, Inc. 2009. (Chapter 12)

Berger, Horst, *Light Structures—Structures of Light: The Art and Engineering of Tensile Architecture.* Birkhäuser Basel. 1996.

Sandaker, Bjorn N, Eggen, Arne P. and Cruvellier, Mark R. *The Structural Basis of Architecture*, 2nd Edition. Routledge. 2011. (Chapter 10)

Schodek, Daniel and Bechthold, Martin. *Structures*, 7th Edition. Pearson. 2014. (Chapter 11)

THIN SHELLS AND RETICULATED DOMES

12.1 FORM-RESISTANT STRUCTURES

A sheet of paper held in the hand bends limply and cannot support its own weight. The same sheet of paper, if pinched and given a slight upward curvature, is capable of supporting its own weight and some additional load (Figure 12.1). The new carrying capacity is obtained not by increasing the amount of material used, but by giving it proper form. The upward curvature increases the stiffness and the load-carrying capacity of the cantilevered paper sheet because it locates some material away from the "neutral axis," so that the bending rigidity of the sheet, acting as a cantilevered beam, is substantially increased. The same result is achieved by creasing the paper, that is, by giving it sudden changes in slope or concentrated curvatures (see Section 10.8).

Structures in which strength is obtained by shaping the material according to the loads they must carry are called *form-resistant* structures. Membranes (see Chapter 11) depend on curvature and twist to carry loads; they belong to the category of form-resistant, purely tensile structures. A membrane turned upside down, frozen, and carrying the same loads for which it was originally shaped is the two-dimensional antifunicular for those loads (see Section 6.1): It is a form-resistant structure developing only compression.

Since membranes resist loads entirely through tension, no form is structurally more efficient. But their flexibility and incapacity to resist compressive stresses restricts the use of membranes (see Section 11.1). In thin shells, however, all the disadvantages of membrane action are avoided, and most advantages maintained. Thin shells are form-resistant structures thin enough to not develop appreciable bending stresses, but thick enough to carry loads by compression, tension, and shear. Although thin shells have been built of wood, steel, and plastics, they are ideally suited to reinforced concrete construction. Thin shells allow the construction of domes and other curved roofs of varied form and exceptional strength. They are among the most sophisticated expressions of modern structural design.

The work of the Swiss engineer, Heinz Isler, clearly demonstrates the antifunicular relationship between tension membranes and shell structures. To create his

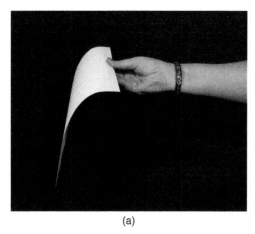
(a)

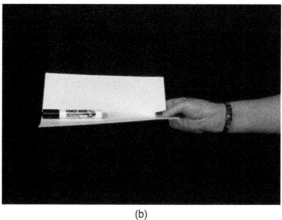
(b)

Figure 12.1a & b Form-resistant structure

A sheet of paper has such little resistance to bending that it cannot even support its own weight (a). Provided with a very slight curl, though, a moment arm is created between the high point at the edges and the low point in the center, which allows for a resisting moment to develop. The formed sheet can then hold not only the paper's self-weight but also considerably more (b). Another way of conceptualizing this is to recognize that the moment of inertia is significantly increased in bending the sheet vis-à-vis the thickness of the sheet itself. See also Figure 10.39 for similar behavior in a folded plate.

(a)   (b)

Figure 12.2 Isler gravity-shaped shell

The working method developed by the Swiss engineer Heinz Isler was to let gravity form a resin-coated fabric (a). Once cured, careful scaled measurements were used to create full-size formwork to support casting an inverted shell. The shell form is a perfect inverted compressive funicular form (b).

Photo courtesy of Princeton University Billington Archive

mathematically complex shells, Isler relied on a simple technique of letting gravity shape a resin-coated fabric scale model. The fabric membrane would drape naturally, and then cure into the draped shape (12.2a). The inverted shape then was the funicular compressive form, from which detailed testing and analysis would further refine the design. Finally, taking careful measurements, Isler would create full-scale formwork to match the model profile, and then cast concrete shells on top. The subsequent Isler shells are only a few inches (8 – 9 cm) thick, but span from 100 to130 feet (30 – 40 m) (Figure 12.2), providing a remarkable span to thickness ratio of up to nearly 400 to 1.

Unlike Isler's mathematically complex gravity formed shells, a great many thin-shell structures can be described and understood geometrically, which is where the discussion will begin. Thin shells, like membranes, owe their efficiency to curvature and twist. In order to understand their structural action, one must first become familiar with their purely geometrical characteristics.

12.2 CURVATURES

It was seen in Section 11.2 that the curvatures of a surface at a point are exhibited by cutting it with a plane pivoted around the normal to the surface at that point. The curvature

varies as the plane rotates; it may be up or down in all directions, or up in some and down in others. Moreover, a surface has, in general, a twist, since its slope changes as one moves at right angles to the direction of the slope.

All the intersections of a dome with a plane passing through its normal have downward curvatures. For spherical domes, the curvatures are all identical (Figure 12.3); for any other type of dome, they change from a maximum to a minimum as the plane rotates (Figure 12.4). On the other hand, the curvatures of a dish are all up. Surfaces like domes or dishes, in which the curvature changes value around a point but is always up or down, are called *synclastic* (from the Greek words *syn* = with, and *klastein* = to cut). It is convenient to call downward curvatures positive and upward curvatures negative; with this convention, at any point on their surface domes have positive curvatures and dishes negative curvatures in all directions. Surfaces with positive or negative curvature at all points in all directions are said to be nondevelopable, because they cannot be flattened without stretching them, unless they are cut at a number (usually infinite) of sections (Figure 12.5).

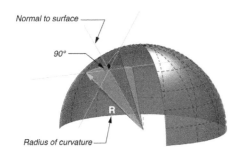

Figure 12.3 Curvatures of spherical dome

In a spherical dome, all curvatures along intersecting planes produce the same downward curvature anywhere on the surface.

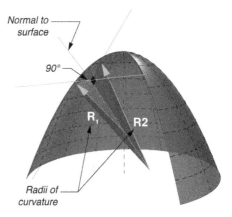

Figure 12.4 Curvatures of parabolic dome

In a parabolic dome, the degree of curvature depends not only on the position on the surface, but the angle and position of the intersecting plane. An infinite number of curves exist on this surface, varying from a maximum to minimum as the plane is rotated across the surface.

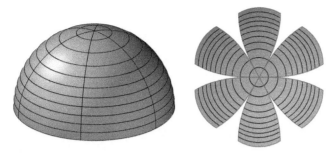

Figure 12.5   Development of synclastic surface
One characteristic of a synclastic surface is that it cannot be created, or *developed*, from a flat sheet. A hemispherical dome, for example (a), can only be created from a flat sheet if that sheet is cut into petal shapes (b).

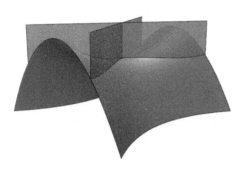

Figure 12.6   Saddle surface
A saddle surface has two directions of zero curvature, which can be defined by straight lines directly on the surface.

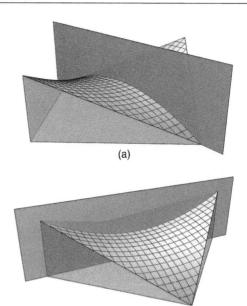

(a)

(b)

Figure 12.7   Principal curvature lines of saddle surface
A saddle-shaped surface has two principal directions of curvature: maximum downward (positive) curvature (a) and maximum upward (negative) curvature (b).

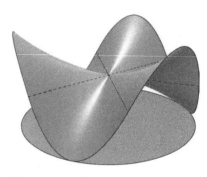

Figure 12.8   The monkey saddle
A surface with zero curvature in three directions is referred to as a *monkey saddle*.

Their stiffness and strength stems, in large part, from their resistance to those deformations which tend to flatten them, that is, to reduce their curvatures.

As the curvature of a synclastic surface becomes smaller in a given direction, the surface approaches the shape of a cylinder. The curvature in the direction of the cylinder axis, one of the principal curvatures, is zero (see Figure 11.7). Surfaces, like those of barrel roofs or gutter channels, with curvatures always positive or always negative but vanishing in one direction, are called *developable* surfaces; they can be flattened without stretching them or introducing cuts. Developable surfaces are obviously less stiff and strong than synclastic surfaces.

A horse saddle has a downward curvature "across the horse" and an upward curvature "along the horse": As one cuts the saddle with a vertical rotating plane, its curvature changes not only in value but also in sign (Figure 12.6); a mountain pass presents the same behavior. As the cutting plane rotates around its axis, the curvature of the saddle changes gradually from positive to negative values, and back to positive values again; therefore, the curvature must become equal to zero, that is, vanish, in two directions. Saddle or anticlastic surfaces have, in general, two directions of zero curvature, that is, two directions along which straight lines lie on the surface (Figure 12.6); they are also nondevelopable. The lines of principal curvature of a saddle surface are shown in Figure 12.7. One may talk of the "top of the dome," or even the "top of the cylinder," which is a ridge of

tops, but one cannot really define the top of a saddle surface. The "top of the mountain pass" is a real top for those coming from the valley, but is a "bottom" for those coming down the ridges of the adjoining mountains.

Surfaces exhibiting zero curvatures at a point in more than two directions are easily constructed. A surface with no curvature in three directions is called a monkey saddle, since a monkey can straddle it with its legs and have its tail down on it (Figure 12.8). A surface with more than three lines of zero curvature has a scalloped shape.

12.3 ROTATIONAL SURFACES

Shell surfaces were classified in Section 12.2 according to their curvatures at a point. A different classification, depending on their general shape, is also of great significance in thin-shell design, and will be used in later sections to

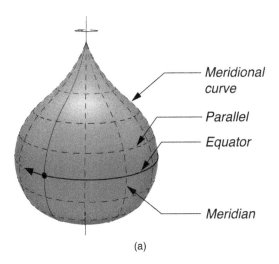

(a) (b)

Figure 12.9 Rotational surface

A rotational surface is generated when a planar curve (*meridional curve*) is rotated around a vertical axis (a). A variety of shapes are possible and have long been used architecturally, such as the onion domes of the Cathedral of the Annunciation in Moscow, Russia (b).

(b) Photo courtesy of Petar Milošević

Figure 12.10 Shell tank for liquids

A spherical tank is frequently used for high-pressure storage, such as this liquid natural gas (LNG) storage tank. Because the sphere has a perfect distribution of stress around its surface, it is the ideal shape to use for storage. A sphere is the shape that also encloses the greatest volume with the least amount of material, and so enjoys further economy of construction.

Photo: Muratart/Shutterstock

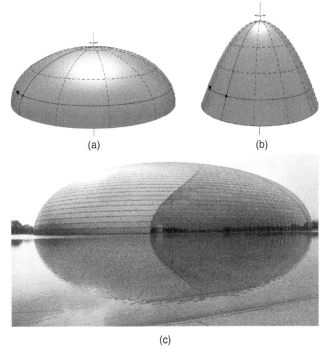

(a) (b)

(c)

Figure 12.11 Elliptical and parabolic surfaces of revolution

An ellipse (a) and a parabola (b) are two more shapes that can generate surfaces of revolution. An elliptically revolved surface was used to define the Beijing National Center for the Performing Arts (c). The project is a pure elipsoid dome with a façade of titanium and glass, sited in an artificial lake.

(c) Photo courtesy of Terri Meyer Boake

analyze the overall behavior of the most commonly encountered thin-shell roofs.

Rotational surfaces are described by the rotation of a plane curve around a vertical axis (Figure 12.9a). The plane or meridional curve may have a variety of shapes, thus giving rise to a variety of dome forms, well suited to the roofing of a circular area, which have historically found wide application in architecture (Figure 12.9b). The most commonly used dome is spherical: Its surface is obtained by rotating an arc of circle around a vertical axis. The vertical sections through the axis of a rotational shell are called its meridians and its horizontal sections, which are all circles, its parallels; the largest parallel is called the equator.

Complete spheres, supported on a ring of stilts, are often used as metal tanks for liquid chemicals (Figure 12.10). Liquid containers are also built as rotational surfaces in the shape of a drop of water. Since a drop of water is kept together by the constant capillary tension of its surface, these containers are designed to develop a constant tension and are most efficient in storing liquids.

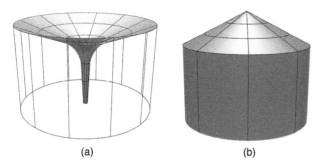

Figure 12.12   Conical surfaces

By rotating a straight line around a vertical axis a cone is created. It may be supported at a point as an umbrella form (a) or a dome-like structure when supported around the perimeter. Such structures are readily formed of concrete due to the lack of radial curvature.

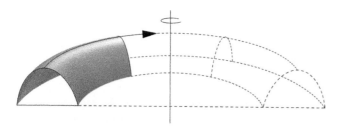

Figure 12.13   Toroidal Surface

A half-circle rotated about a central axis creates a doughnut-shaped surface known as a *toroid*.

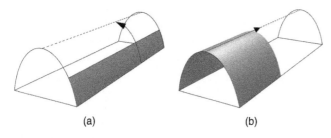

Figure 12.14   Cylindrical Surfaces

A cylindrical surface is a translational surface that can be generated by either (a) translating a straight line along the arc of a curve or (b) translating a curve along a straight line. The resulting cylinder can be circular, parabolic, or elliptic.

Elliptical domes are described by half an ellipse rotating around its vertical axis (Figure 12.11a); their action is not as efficient as the action of a spherical dome because the top of the shell is flatter, and the reduction in curvature introduces a tendency to buckle. On the other hand, the parabolic dome (Figure 12.11b) has a sharper curvature at its top and presents structural advantages even in comparison with the sphere. An elliptical shell forms the contiguous shape of the Beijing National Center for the Performing Arts in China (Figure 12.11c).

The surface described by rotating a straight line around a vertical axis it intersects is a cone. Conical surfaces are commonly used as umbrella roofs when supported at a point, and as domes when supported on their circular boundary (Figure 12.12). The lack of curvature in the radial direction of the cone makes the forming of cones in reinforced concrete somewhat simpler than the forming of curved domes.

Circular barrels are obtained by rotating around a vertical axis the upper half of a circle or any other curve not intersecting the axis; such doughnut-like surfaces, called toroidal, are well suited to roof ring areas (Figure 12.13).

12.4 TRANSLATIONAL SURFACES

The surface of a translational shell is obtained by "translating," that is, sliding a plane curve on another plane curve, usually at right angles to the first, while maintaining it vertical. A cylinder is obtained by translating a horizontal straight line along a vertical curve or translating a vertical curve along a horizontal straight line, perpendicular to it; depending on the curve, the cylinder may be circular, parabolic, or elliptic (Figure 12.14). When the straight line is inclined, the cylinder also becomes inclined and is well suited to roof a staircase by means of a barrel-like vault.

The translation of a vertical parabola with downward curvature on a perpendicular parabola also with downward curvature generates a surface called an elliptic paraboloid, which covers a rectangular area; its horizontal sections are ellipses, its vertical sections are parabolas (Figure 12.15). When the two parabolas are identical, the paraboloid covers a square area, and its horizontal sections become circles. The elliptic paraboloid was the first shape ever used to build a thin concrete shell (in 1907).

Figure 12.15   Elliptic paraboloid

Translational surfaces can be generated by two curved lines, as well as by a curve and straight line. When a downward parabolic curve is translated on a downward parabolic surface, an elliptic paraboloid is generated (a). Vertical sections through the surface generate parabolas, and horizontal sections generate ellipses (b)

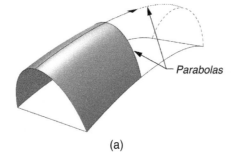

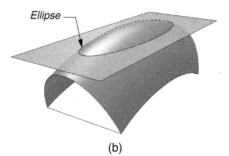

(a)

(b)

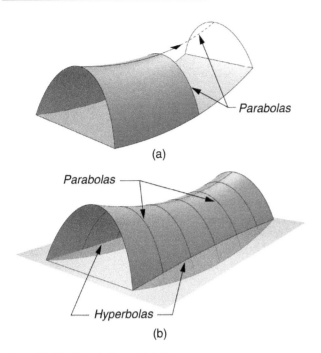

Figure 12.16   **Hyperbolic paraboloid**
A hyperbolic paraboloid is generated by the translation of a parabola with downward curvature on a parabola with upward curvature. Like the elliptic paraboloid, vertical sections generate parabolas. Horizontal sections, however, generate hyperbolas (which graphically appear as two parabolas on a plane opening away from one another. The surface has the shape of a saddle.

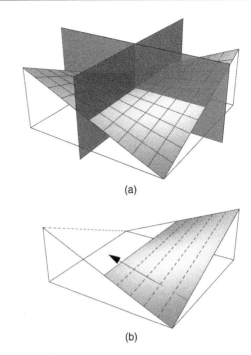

Figure 12.17   **Straight-line generatrices of hyperbolic paraboloid**
The hyperbolic paraboloid has two intersecting sets of straight lines (called *generatricies*) on its surface (a). One property of the hyperbolic paraboloid is that it may be constructed by linear translation of a straight line against two oppositely sloped lines (b). Because the curving surface can be described by straight lines, it is a natural form to create with cast concrete, placed atop straight wood formwork set along the lines of zero curvature.

A hyperbolic paraboloid ("*hypar*" for short) is obtained by translating a parabola with downward curvature on a parabola with upward curvature (Figure 12.16a). This surface has the shape of a saddle; its horizontal sections are two separate branches of a curve called a hyperbola, while its vertical principal sections are parabolas (Figure 12.16b). As for all other saddle surfaces, the curvature of the hyperbolic paraboloid vanishes in two directions; but for a hyperbolic paraboloid these two directions are the same at all points. This means that the vertical sections parallel to these two directions are all straight lines, called *generatrices*, and that the hyperbolic paraboloid has a double set of straight lines lying on its surface. Hence, this surface may also be generated by a straight-line segment sliding with its two ends on two straight-line segments askew in space (Figure 12.17). The forming of such a surface in reinforced concrete is relatively simple, since it involves the use of straight planks in the direction of one set of lines of zero curvature.

When the two curves used in translation are not at right angles to each other, the translational surface may cover a skewed rectangle.

A great variety of surfaces can be obtained by translation, since any curve may be translated on any other curve for this purpose, but not all translational surfaces are well suited to carry loads.

12.5 RULED SURFACES

The hyperbolic paraboloid was shown in Section 12.4 to have two sets of straight lines lying on its surface. Any surface generated by sliding the two ends of a straight-line segment on two separate curves is called a *ruled surface*. When the two curves are two straight line segments askew in space, the ruled surface is a hyperbolic paraboloid (Figure 12.17). The cylinder is a ruled surface generated by sliding a horizontal line segment on two identical vertical curves (see Figure 12.14a).

A different family of ruled surfaces called conoidal surfaces, is obtained by sliding a straight-line segment on two different curves lying in parallel planes. Conoidal roofs cover a rectangular area and allow light to enter from a given direction, usually from the north (Figure 12.18).

Conoids are conoidal surfaces obtained by sliding a straight-line segment with one of its ends on a curve and the other on a straight line (Figure 12.19). The conoid is called circular, parabolic, or elliptic depending on whether its end curve is an arc of a circle, a parabola, or an ellipse. The conoid is a saddle surface; its principal curvature lines show that the curvature in the direction of the two lines connecting the top of the curve to the ends of the straight-line segments is upward, while the other principal curvature is downward. Conoids may be used as cantilevered shells with their curved

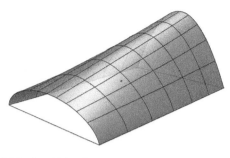

Figure 12.18 Conoidal surface
Sweeping a straight line across two different curves on parallel planes creates this ruled surface. In this case, the generating curves are a half-ellipse on one end and a parabola on the other end.

Figure 12.19 Conoid
A conoid is generated when a straight line is swept along a curve on one end and a straight line on the other end. The curved end may be circular, parabolic, or elliptic. The conoid is a saddle shape.

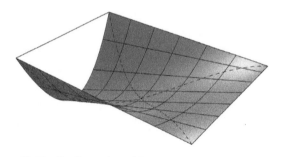

Figure 12.20 Cantilevered conoid
Inverted conoids can be used as cantilevered shells.

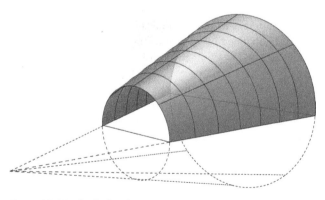

Figure 12.21 Conical sector
Sweeping a straight line along a curve at one end and a single point at the other end generates a cone. A portion of a cone is called a *conoidal sector*, and can be used to cover a trapezoidal roof area.

end at the root of the cantilever; in this case, their transverse curvature is directed upward (Figure 12.20).

Cones are ruled surfaces in which one end of the segment rotates about a point and the other slides on a curve. Conical sectors are conoidal surfaces and may be used to cover trapezoidal areas or as cantilevered roofs, much as conoids can be used (Figure 12.21).

An inclined segment, sliding on two horizontal circles lying one above the other, describes a surface called a hyperboloid of one sheet, whose vertical sections are the two branches of a hyperbola (Figure 12.22a). These easily formed surfaces are extensively used to build cooling towers in power plants, in cement factories, and other industrial plants, as well as in architectural applications (12.22b).

When the sliding line segment is vertical, the hyperboloid becomes a vertical cylinder; when it connects diametrically opposite points of the two circles, the hyperboloid becomes two inverted half-cones. It is thus seen that some surfaces, say the cone, may be generated in a variety of ways and be considered rotational, translational, conoidal, or ruled.

12.6 COMPLEX SURFACES

The elementary, mathematically defined surfaces of the previous section may be combined in any number of ways to obtain more complex surfaces. Two cylindrical shells intersecting at right angles cover a square or a rectangular area with a so-called groined vault (Figure 12.23). Parallel cylinders with curvatures alternately upward and downward create an undulated roof similar to a folded plate (Figure 12.24).

Scalloped roofs over curved boundaries may be obtained by joining sectors of cones with curvatures alternately upward and downward (Figure 12.25). Moreover, any of the elementary forms may be scalloped to obtain more playful and, at times, more efficient structural thin shells. Parabolic surfaces were used in concrete shell of the chapel of the St. Louis Abby Priory (Figure 12.26). An ellipsoid may be scalloped to give it transverse curvatures toward the supported edge (Figure 12.27). Spherical domes may be undulated for the same purpose (Figure 12.28). A parabolic cylinder may be undulated to transform it into a surface with curvatures in two directions rather than in one, thus stiffening it. Uruguayan engineer Eladio Dieste constructed numerous undulated complex surface structures in his native country, uniquely of reinforced brick masonry (Figure 12.29). The largest concrete spherical dome built to date (2015), the Kingdome designed by Jack Christiansen, covered the King County Stadium in Seattle, Washington. The scalloped shell spanned 661 feet (201 m) and was only five inches (13 cm) thick—the structure was demolished in 2000 after only 20 years, but for economic and political resons, not structural.

Hyperbolic paraboloids may be used in a variety of combinations. Four identical paraboloidal segments form a corner-supported roof covering a rectangular area, one of the most commonly used combinations of this surface (Figure 12.30). Two hyperbolic paraboloidal segments may

Figure 12.22 Hyperboloid of one sheet

A diagonal line connecting two circles of the same or different diameter when rotated about those circles generates a hyperboloid (a). Vertical sections through the surface produce the two branches of a hyperbola. The hyperboloid is commonly used as thin shell concrete structures for cooling towers, and also in architectural applications. The McDonald Planetarium at the St. Louis Science Center is a concrete hyperboloid shell (b).

(b) Photo courtesy of Hanneorla, Flickr. com https://www.flickr.com/photos/hanneorla/sets/

(a)

(b)

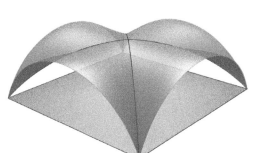

Figure 12.23 Groined vault

The groined vault is constructed by the intersection of two half-cylindrical shells. It is a complex surface generated by the combination of two simple surfaces.

be combined to form a northlight conoidal roof (Figure 12.31) or a cantilevered conoid shell (Figure 12.32). A series of such shells was used by Nervi as a roof for the stands of one of his stadiums in Rome, Italy. Four paraboloidal segments supported on a central column form an umbrella roof (Figure 12.33), a scheme frequently employed by the acclaimed architect-engineer Felix Candela in his works of the mid-twentieth century. When the angle between the planes of the two translating parabolas is not a right angle, the hyperbolic paraboloid is called skew and may be used to cover nonrectangular areas (Figure 12.34).

There is no reason to limit the shape of a thin shell to forms definable by geometrical formulas. Structurally sound, "free" shapes may be invented. But the imagination of the designer becomes mere fancy unless he or she is

Figure 12.24 Undulated cylindrical surface

Alternating half cylinders with downward and upward curvature generates an undulated cylindrical surface (a). The undulations (or *corrugations*) provide an internal moment arm and increase the moment of inertia in the same manner as the folded sheet of Figure 12.1b. A cantilevering undulated cylindrical concrete shell roof covers the drive-through teller of this bank in Oklahoma (b).

(b) Photo courtesy of Michael Brown

(a)

(b)

Figure 12.25 Undulated conical surface

Joining conical sections of alternating up and down curvatures generates this surface form. The curvature profile for shapes of this type, as with any conoid, can be circular, parabolic, or elliptical.

Figure 12.26   Undulated parabolic surface

The St. Louis Abby Priory Chapel is a parabolic surface in two levels revolved about a central point. In the interior view (b), it can be seen that the valleys of the lower and upper tiers coincide and thus act like continuous arches spanning between the foundation and a central compression ring.

Photo courtesy of Tyler Sprague

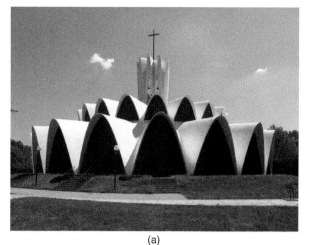

(a)

(b) Interior view

(a) Exterior view from above

(b) Interior view

Figure 12.27   Scalloped ellipsoid

The scalloped form of this elliptical shell provides stiffening for the roof of this seaside hotel restaurant.

Photo courtesy of Stephanie Rogers, http://www.stephanierogers.info

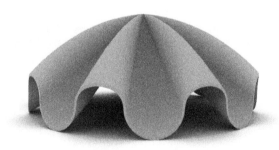

Figure 12.28   Scalloped spherical dome

This surface structure is similar to the scalloped ellipsoid (Figure 12.17), except it is based on a spherical segment.

Figure 12.29   Undulated parabolic cylinder

Uruguayan engineer Elasio Dieste constructed numerous thin shell structures such as this storage facility, made of doubly-curved reinforced brick (as opposed to concrete).

Photo courtesy of Hamilton Hadden

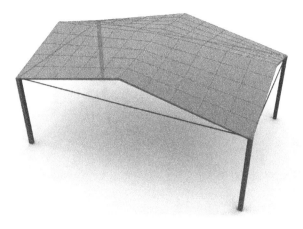

Figure 12.30 Hyperbolic paraboloidal roof
Combining four-corner–supported hypar shells is a common way of using multiple hypars together to cover an area.

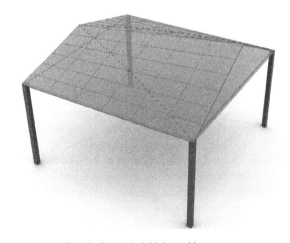

Figure 12.31 Hyperbolic paraboloidal conoid
Two hypars connected can form a northlight conoidal roof.

Figure 12.32 Cantilevered hyperbolic paraboloidal conoid
The inversion of a hyperbolic paraboloid conoid has been used to create cantilever roof structures.

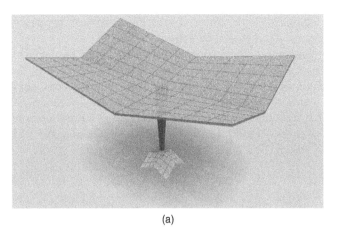

(a)

(b)

Figure 12.33 Hyperbolic paraboloidal umbrella
Many structures by architect-engineer Felix Candela were designed using the inverted hypar umbrella form (a). Candela further extended the concept by making the foundation also as a thin-shell hypar buried in the ground. This was done as a material- and cost-saving design approach that allowed Candela to win contracts by constructing them less expensively than more conventional construction. Like Candela's work, the Cesar E. Chavez Student Center at the University of California, Berkeley (12.33b), is constructed of a series of inverted hypar umbrellas. The staggering of alternate column rows means that the high point of two units joins at the low point of the next adjacent two units to form a diamond-shaped clerestory window at their juncture.

Photo courtesy of Deborah Oakley

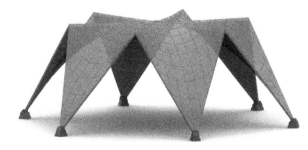

Figure 12.34 Skew hyperbolic paraboloidal combination
When the corner angles of a hypar are not 90 degrees, it becomes a *skew hypar*. Such elements can be used in combination just like a square hypar. By connecting them at a central point, a radial arrangement can be created that has self-stable dome-like qualities.

familiar with the structural behavior of the basic geometric forms. Thin shells are extraordinarily efficient structural elements, and their shapes should be dictated primarily by structural considerations. The behavior under load of the most commonly used thin-shell forms must be considered first in terms of their membrane action, following the basic ideas presented in Chapter 11. More refined considerations of stress, as well as additional practical requirements, will be taken up in some detail in later sections.

12.7 MEMBRANE ACTION IN CIRCULAR DOMES

The structural action of a rotational and, in particular, of a circular dome, supported around its entire boundary and acted upon by vertical loads symmetrical with respect to its axis (as the dead load), is a simple consequence of its geometrical characteristics. In these *axisymmetrical shells*, the meridional sections and the sections at right angles to the meridians are both principal curvature and principal stress sections (see Section 11.2). The stresses on these sections are simple tension or simple compression evenly distributed across their small thicknesses.

Figure 12.35 shows the stresses developed at a parallel of an axisymmetrical shell; they are compressive stresses in the direction of the meridian and constant along the parallel because the shell and the loads are symmetrical about the axis. Each meridian behaves as if it were a funicular arch for the applied loads; that is, it carries the loads without developing bending stresses.

Arches were shown in Section 6.4 to be funicular for only one set of loads. The meridians of a dome, instead, are funicular for any set of symmetrical loads. This essential difference in structural behavior is due to the fact that while isolated arches have no lateral support, the meridians of the dome are supported by the parallels, which restrain their lateral displacement by developing hoop stresses. As a consequence of its funicular behavior under any set of symmetrical loads, a dome does not change shape to adapt itself to a change in such loads; therefore, it is a stable structure for such loads. (It will

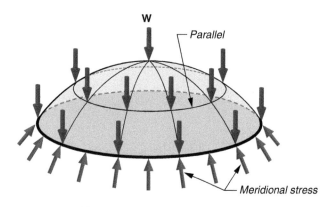

Figure 12.35 Meridional stresses at a parallel of a spherical dome
The meridians and parallels of a spherical dome are both its principal curvatures and also its principal stresses. Each meridian behaves like a funicular arch, carrying the load purely in compression without bending. Unlike an arch, which is funicular for only one loading, the meridians of a dome are funicular for all symmetric loadings due to the restraint (hoop stresses) of the parallels.

be seen later in this section that a dome is stable under any set of loads, symmetrical or unsymmetrical.)

The contribution of the parallels to the funicular behavior of the dome is indicated by the deformations of the meridians under load. In a shallow or small-rise dome, the meridians, shortening in compression, move down under load, and in so doing move inward, that is, towards the dome axis (Figure 12.36). In the two-dimensional structural dome, this motion is accompanied by a shortening of the parallels, because of their radii decrease. Hence, the parallels also are compressed, and their compressive stiffness reduces substantially the freedom of the meridians to move inward. In other words, an axisymmetric shallow dome may be considered to act as a series of meridional funicular arches elastically supported by the parallels. It develops compressive stresses along both the meridians and the parallels and could be built of materials incapable of developing tensile stresses, such as masonry or bricks.

When the dome has a high rise, it flattens at the top and opens up at the bottom; the points of its top portion

Figure 12.36 Deformation of shallow spherical dome under load

When a dome is shallow or low-rise, the arch action of the meridians compresses both the meridians and parallels. Being entirely in compression, such domes are suitable for construction in materials that are best suited for compression only, such as stone or brick masonry.

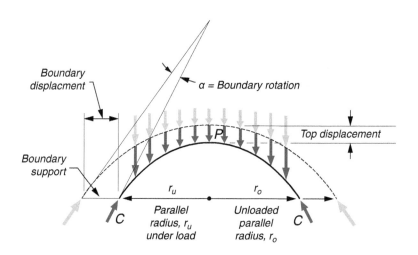

Figure 12.37 Deformation of high-rise spherical dome under load

The behavior of a high-rise dome differs from a shallow or low-rise dome. Here, the top tends to flatten out, compressing the parallels, but spreads out at the base, stretching the parallels. Under dead load, the transition point for the parallels in compression vs. tension is at an angle of 52° measured from the vertical. The parallel at this point is under zero stress. Under a snow load uniformly distributed on a horizontal projection of the dome, the unstressed parallel is at 45°.

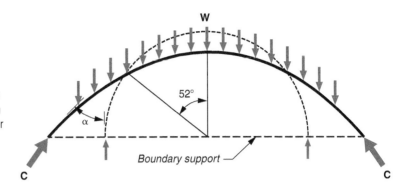

move inward under the action of the loads, but the points of its lower portion move outward, that is, away from the axis (Figure 12.37). The parallels of the upper portion shorten, but those of the lower portion lengthen and develop tensile stresses, again restraining the motion of the meridians. Depending upon the type of load, one specific parallel remains unchanged in length, while all those above develop compression and all those below develop tension. In a spherical dome under dead load, the parallel at an angle of 52 degrees from the axis does not change length (Figure 12.38). Under a snow load, uniformly distributed on the horizontal projection of the dome, the unstressed parallel is at 45 degrees to the axis.

Since the stresses developed by a dome are purely compressive and tensile, and the corresponding strains are very small (see Sections 5.1 and 5.2), the stiffness of circular domes is exceptionally high. The top of a concrete dome in the shape of a half sphere, 100 feet in diameter (30.5 m) and only 3 inches thick (8 cm), acted upon by a very heavy load of snow and its own dead load, deflects less than one tenth of an inch (3 mm). The ratio of span to deflection is 12 thousand and should be compared with an acceptable ratio of between 300 and 800 for a bending structure such as a beam. The outward displacement of the dome at its boundary is even smaller than its vertical deflection: only three one-hundredths of an inch (1 mm); the rotation of the meridians at the boundary, that is, their change in direction, is only one hundredth of a degree (Figure 12.36). A similar behavior is exhibited by rotational shells having other meridional sections.

The stiffness of circular domes explains why the thickness of reinforced-concrete domes may be reduced to such small values; ratios of span-to-thickness of the order of 300

or more are customary. For example, a reinforced-concrete dome spanning 100 feet (30.5 m) with a thickness of 3 inches (8 cm) has a ratio of span-to-thickness of 400. The steel domes used as containment spheres in nuclear reactors have ratios of the order of one thousand. Such ratios may be compared with the ratio of span-to-thickness in an eggshell, which is only 30, or in bending structures, which is of the order of 20. A shell thickness is limited mostly by the requirements of buckling resistance.

It was shown in Section 11.2 that membranes may change shape under varying loads in order not to develop compressive stresses. In a thin-shell dome, instead, compression can be developed, and, moreover, even if direct stresses along the meridians and the parallels cannot carry the entire load, a third mechanism is always available to equilibrate the unbalanced difference: the shear mechanism considered in Section 11.1. Provided safe stresses are not exceeded, direct stresses (compression and tension) on one hand, and shear within the shell surface on the other, will always share and equilibrate the total load on a spherical dome element without requiring a change in its shape. Taking into account the shear mechanism, a dome may be said to be funicular for all (smooth) loads and, hence, to be a stable structure under practically any circumstance. For example, a dome resists lateral loads, such as wind pressures and suctions, by developing all three types of membrane stresses (Figure 12.38). The shear stresses due to wind are also of modest intensity: A 100 mile-an-hour wind (161 km/hr) on a spherical concrete dome 3 inches thick (8 cm) spanning 100 feet (30.5 m) produces membrane shear stresses of only 14 pounds per square inch (96 kPA), while the allowable shear stress in concrete is at least 75 pounds per square inch (517 kPa).

Figure 12.38 Shear mechanism in dome under wind load

Unlike membranes, which will reshape under a compression stress, a shell is capable of resisting compression as well. A dome is stable and capable of carrying nearly all smooth loads with no change in shape. Under wind load, the dome will resist pressures and suctions by a combination of tension, compression, and shear.

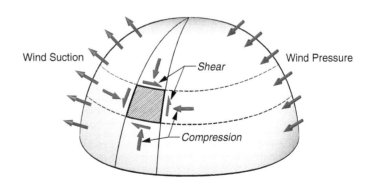

12.8 BENDING STRESSES IN DOMES

A dome was shown to carry loads essentially by membrane stresses (compression, tension, and shear), because the development of shear makes it funicular for all loads. On the other hand, because a dome is usually very thin, if there were a tendency to develop bending stresses anywhere in the dome, these could easily exceed allowable values. It is therefore necessary to investigate the possibility of such a tendency.

In the analysis of the carrying capacity of thin domes, it was tacitly assumed that domes are free to develop the minute displacements required by their membrane state of stress. Thus, the dome considered in Section 12.7 develops membrane stresses under load, and the corresponding strains give its top a small vertical displacement. Since this top deflection is not prevented, a pure state of membrane stress exists in its neighborhood. An entirely different situation may develop, instead, at the boundary of the shell.

A hemi-spherical dome under load opens up at the equator; its boundary displaces outward, even though by a very small amount (see Figs. 12.36 and 12.37). Moreover, the reactions supporting the dome must be in the direction of the slightly rotating meridians, since the meridional arches are funiculars of the load; reactions in any other direction would produce bending in the dome. In other words, in order to have a pure state of membrane stress in the shell, the boundary should be free to move outward, and the reactions should rotate in the direction of the meridians at the boundary. In practice this is impossible. A moving boundary presents practical difficulties and disadvantages and it is impossible to allow the reactions to rotate in order to remain tangent to the deformed meridians when the boundary rotates because of the deformation of the shell under load (see Figs. 12.36 and 12.37). In common practice, instead, the equator of the shell is reinforced by a stiff ring, which prevents almost entirely the outward motion of the boundary and its rotation, and hence introduces an additional inward radial thrust and bending at the equator. The shell, which would open up and rotate at the equator under the load-induced membrane stresses, develops a kink or sudden change in curvature and, hence, bending stresses around the boundary (Figure 12.39a).

The bending disturbance thus introduced at the boundary does not penetrate deeply into the shell, but is limited to a narrow band in the neighborhood of the boundary. This "damping out" of boundary disturbances, another useful characteristic of thin shells, occurs because the bending displacements of the meridians, which could be large in view of the small bending rigidity of the shell, are restrained by the parallels. Large meridional bending displacements imply large changes in the radius of the parallels, and, hence, large tensile or compressive strains in the parallels. The parallels, instead, are stiff in tension and compression, and do not allow such large displacements; they permit a small amount of bulging, in and out, of the meridian and, hence, a small amount of bending, which

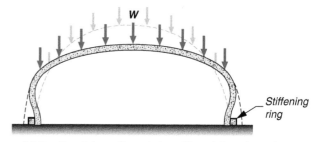

(a) Bending deformations at ring-stiffened dome boundary.

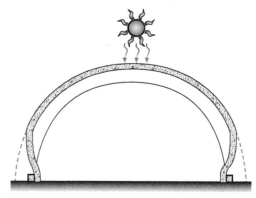

(b) Bending deformations at ring-stiffened dome boundary.

Figure 12.39 Bending deformations at ring-stiffened dome boundary
Hemispherical domes under distributed loads (a) or thermal loads (b) need to be prevented from expanding at the supports (equatorial boundaries). Stiffening rings are used to prevent outward displacement at the supports; these displacements however are required to keep the dome in funicular compression. The stiffening ring thereby introduces bending stresses around the circumference that must be designed for.

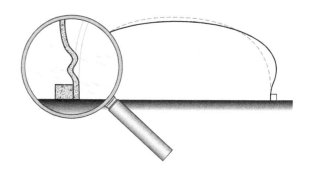

Figure 12.40 Bending disturbance at dome boundary
The bending stresses introduce some disturbance near the boundary, but this disappears a short distance away.

peters out as one moves away from the shell boundary (Figure 12.40). This petering-out of bending stresses is analogous to the phenomenon occurring in continuous beams (see Section 7.3), where bending vanishes rapidly away from the loaded span due to continuity over the supports. In the shell the meridians are continuous over the "supports" of the parallels, and bending vanishes away from the boundary with the same wavy displacement typical of continuous beams (see Figure 7.42a).

Figure 12.41   Principal stress lines of dome under load

As long as the boundary reactions are tangent to the meridians of a dome all around its circumference, the principal stress lines act along the meridians and along concentric parallels (a). If the dome is supported on columns, the pattern of membrane stresses changes, and localized bending stresses develop near the supports.

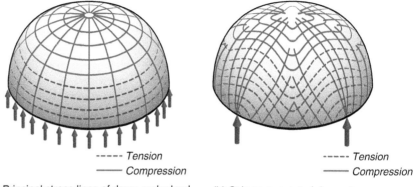

----- Tension
——— Compression

----- Tension
——— Compression

(a) Principal stress lines of dome under load

(b) Column-supported dome stress

The width of the area affected by the bending boundary disturbance is proportional to the square root of the ratio of thickness to radius of the dome; in order to reduce the width of the disturbed zone, the shell must be made thinner. For a thickness equal to one four-hundredth of the radius, the arc of the disturbed zone is only one-tenth of the radius (Figure 12.40). Thus, most of the shell is actually in a state of undisturbed membrane stress.

Bending disturbances, usually more severe than those due to the loads, are produced by thermal conditions. When exposure to the sun increases the shell temperature, the entire dome changes shape, uniformly increasing its radius (see Figure 12.39b). If an underground ring prevents the boundary displacement, the shell once again presents a sudden change in curvature and develops high bending stresses of the same nature as those discussed above. Since boundary displacements due to thermal changes are usually larger than those due to the loads, the bending stresses due to thermal changes are also usually larger than those due to the loads. If the temperature of a dome spanning 100 feet (30.5m) increases uniformly by 30 degrees Fahrenheit (17 °C) with respect to the foundation ring, the dome boundary displaces radially by one-tenth of an inch (3 mm). This displacement is three times larger than that produced by the dead load and a heavy snow load; hence, the thermal bending stresses at the boundary, when the thermal displacement is prevented, are three times as large as those due to the loads.

Whenever the reactions are not tangent to the meridians all along the boundary, bending stresses occur in the neighborhood of the boundary. Thus, if a dome is supported on evenly spaced columns rather than all around its boundary, not only does the membrane stress pattern change (Figure 12.41), but, moreover, the columns introduce horizontal reactions at the dome boundary, and bending stresses develop there in the dome. Similarly, if portions of the dome are cut out by vertical or inclined planes and the dome rests on a few points, the support conditions differ substantially from those ideally required by membrane action, and bending stresses are to be expected. Finally, any load capable of producing a kink or sudden change of curvature in a thin shell is bound to produce bending stresses: thus, concentrated loads cannot be carried by membrane stresses (Figure 12.42). The thickness of the

Figure 12.42   Concentrated load on thin shell

A concentrated load applied to a thin shell produces localized deformations in the vicinity of the load that peters out in a short distance.

shell is often dictated by bending disturbances rather than by the membrane stresses due to the loads.

Two additional conditions may call for a thickening of shells above the modest requirements of membrane stresses. One concerns reinforced-concrete shells and is of a purely practical character: Enough thickness must be provided to cover the reinforcing bars on both the outside and the inside of the shell. The exact location of the bars in the shell thickness is a delicate and expensive matter; in countries with high labor costs, it is often found less expensive to increase the shell thickness than to locate the steel carefully. Shells thinner than 2 or 3 inches (5 – 8 cm) are seldom economical in the United States; shells as thin as half an inch (13 mm) have been built in other countries.

Shells must often be thickened to prevent buckling. Any thin structural element subjected to compressive stresses may buckle, and thin shells are no exception. The buckling load for a thin-shell dome is proportional to the modulus of elasticity of the material and to the square of the thickness-to-radius ratio. With ratios often as small as one three-hundredth or one four-hundredth, the buckling load may be exceptionally low. The buckling load for a spherical dome 3 inches thick, spanning 100 feet (30.5 m) is about 150 pounds per square foot (7.2 kPa); with a factor of safety of 2½, the maximum uniform load on the shell cannot exceed 60 pounds per square foot (2.9 kPa). This is

Figure 12.43   The C.N.I T Dome

At 715 ft (218 m), the C.N.I.T. dome at the la Défense complex in Paris, France, is the world's longest unsupported concrete span. It is actually a double-layered shell, with internal stiffening ribs connecting the layers.

Photo courtesy of Terri Meyer Boake

equivalent to the dead load of the shell, including the roofing or insulating materials, and a small snow load. A shell of this sort could be built in the tropics, but not in the northern part of the United States.

The buckling resistance of a dome may be substantially increased, without increasing its thickness uniformly, by using meridional and parallel ribs. This practice is well suited to the stiffening of steel domes, in which the thickness required by membrane stresses may be quite small in view of the tensile and compressive strength of the material. Concrete domes are seldom stiffened by ribs because of the cost of forms, except in the case of very large spans. In order to ensure buckling resistance the largest concrete thin-shell roof built to date (2015), the C.N.I.T. dome in Paris, spanning 715 feet (218 m), consists of two separate shells connected by vertical diaphragms (Figure 12.43).

A shell acts "properly" if it develops membrane stresses almost everywhere; it is then said to carry loads by thin-shell action. It was shown above that the following three conditions must be satisfied for a dome to develop thin-shell action:

1. The dome must be thin; it will thus be incapable of developing substantial bending action.

2. It must be properly curved; it will thus be strong and stiff because of its form resistance.

3. It must be properly supported; it will thus develop a small amount of bending in a limited portion of the shell.

These three conditions are essential to thin-shell action, whatever the shape of the shell and the loads on it. Whenever these conditions are not met because of construction difficulties, aesthetic considerations or architectural requirements, bending action becomes important and the structural efficiency of the shell is reduced.

12.9 MEMBRANE ACTION IN CYLINDERS

Cylindrical or barrel shells are used to cover rectangular areas and are commonly supported on end frames, stiff in their own vertical plane and flexible at right angles to it (Figure 12.44). Their action may be considered a combination of beam action in the longitudinal direction and a particular funicular-arch action in the transverse direction.

The longitudinal membrane stresses in a "long barrel"—a barrel in which the length is at least three times the width—are similar to those developed in a beam. The thin shell obtains its strength from its curved shape, and one may think of the barrel as a beam with a curved cross section. The longitudinal membrane stresses are linearly distributed across the depth of the barrel; the upper fibers are compressed, the lower fibers are tensed (Figure 12.45).

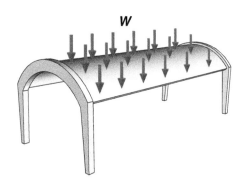

Figure 12.44   Barrel shell on end stiffeners

Structures of this type span space using a combination of beam action in the long direction and arch-type action transversely. They can be used to cover rectangular areas.

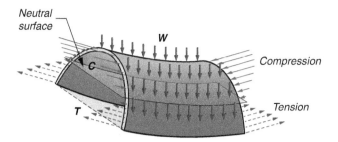

Figure 12.45 Beam (membrane) stresses in long barrels

A barrel shell is considered "long" when its length is at least three times its width. Away from the end supports, these shells exhibit beam action with tension stresses at the bottom and compression on top.

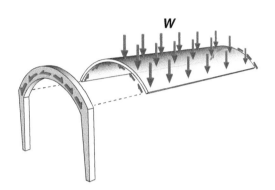

Figure 12.46 Shear reactions on barrel ends

At the end stiffeners, the load is transferred to the support by shear stresses, the vertical components of which add up to the shell vertical end reaction.

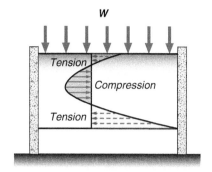

(a) Longitudinal stresses in short barrels.

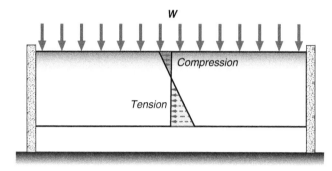

(b) Longitudinal stresses in long barrels.

Figure 12.47 Longitudinal stresses in short and in long barrels

Short barrels (a) deviate from the simple beam action in long barrels (b). The shape off the short barrel may become deformed and develop a curved stress pattern through its depth, notably with tension in *both* extreme upper and lower fibers.

Beam action carries the load to the frames. These, because of their horizontal flexibility, cannot react longitudinally; the longitudinal stresses vanish at the end of the shell, just as they vanish at the end of a simply supported beam. The load is transferred to the end frames or stiffeners by shear action; tangential shear reactions develop on the cylinder end sections, and their vertical components add up to the value of the load (Figure 12.46). On the other hand, the action of the barrel differs slightly from beam action because the cross section of a beam is solid and not deformable, whereas the thin cross section of a shell deforms under load. If a thin, curved piece of material (paper or plastic) is supported on two grooved stiffeners, and loaded, the longitudinal boundaries of the cylinder will be seen to move inward, indicating the absence of the outward thrust typical of arch action, and showing the deformability of the cross section.

As the barrel becomes shorter, the influence of the deformability of its cross section is more felt and the longitudinal stresses deviate from the straight-line beam distribution. There comes a point where the stresses acquire the curved distribution of Figure 12.47a), in which both the top and bottom shell fibers are in tension, while the intermediate fibers are in compression. If the cross section of the barrel is prevented from deforming, either by intermediate stiffeners

or by the continuity with adjacent barrels, the stress distribution becomes linear once again, indicating a return to the beam behavior of long barrels (Figure 12.47b).

12.10 BENDING STRESSES IN CYLINDERS

As the barrel becomes shorter, arch action and eventually longitudinal plate action become preponderant. One may visualize the behavior of short barrels as due to the action of arches supported by longitudinal beams consisting of barrel strips near the boundaries (Figure 12.48). Transverse, compressive arch stresses develop in the upper part of the barrel, petering out toward its longitudinal edges.

The compressive arch stresses introduce bending in the neighborhood of the barrel stiffeners. The arch compression produces a shortening of the transverse fibers of the barrel and an inward displacement of the barrel section (Figure 12.49). This displacement takes place freely at midspan, but at the barrel ends it is prevented by the stiffeners, which are rigid in their own plane. In order to restrain the inward displacement at the stiffener while allowing it at midspan, the shell must bend; thus, it develops some bending stresses at the stiffeners (Figure 12.49). This bending disturbance is of the same kind as that encountered at the boundary of a circular dome when

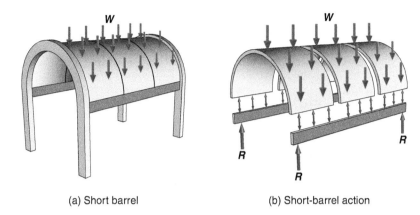

Figure 12.48 Short-barrel action
Short barrels supported along their longitudinal boundaries behave as a series of arches and carry loads by arch action.

(a) Short barrel   (b) Short-barrel action

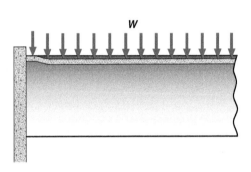

Figure 12.49 Bending at barrel stiffener
At end stiffeners, the barrel is subjected to localized bending similarly to the bending at the boundaries of spherical domes (fig. 12.40). This rapidly diminishes with distance from the end stiffener.

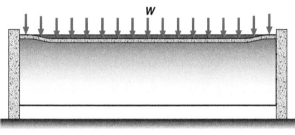

(a) Bending in long barrel

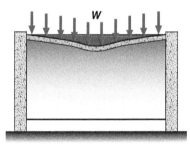

(b) Bending in short barrel

Figure 12.50 Bending in long and short barrels
For long barrels (those with lengths greater then twice the radius), (a) the end disturbances are very short compared to the length of the barrel, and cause only a small portion of the barrel to experience bending stresses. For short barrels, (b) bending disturbance may affect the full length.

its outward displacement is prevented by a stiff ring. The bending disturbance at the stiffeners peters out rapidly and is felt only in a short portion of the barrel. The width of the longitudinal band affected by the bending disturbance depends again on the square root of the thickness-to-radius ratio; the bending displacements bulge in and out and are rapidly "damped." In concrete barrels, the bending stresses at the stiffeners' edges are often absorbed by a local reinforcement in the shell thickness.

In any thin barrel, the disturbed area at the end stiffeners is narrow in comparison with the shell radius. When the barrel is long—that is, when its length is substantially greater than its radius—only a small fraction of its length is under bending stresses (Figure 12.50a). But when the barrel is very short—that is, when the barrel length is less than twice the radius—the width of the bending stress band, even if small compared to the radius, may become comparable to the length of the barrel (Figure 12.50b). Thus, in short barrels the bending disturbance may affect a large portion of the total shell area. This result could be foreseen intuitively: in short barrels the load is carried directly to the stiffeners by the bending action of longitudinal shell strips, because the thickness of the shell even if small relative to the radius is large compared to the short span. Actually, since the short barrel is a two-dimensional resisting structure, it acts very

much like a plate in carrying the load directly to the two stiffeners, and necessarily develops bending stresses over most of its surface.

Thermal displacements and shrinkage of the concrete during setting may produce bending disturbances at the stiffeners substantially greater than those produced by the loads. They often require an increase in thickness and reinforcement at the stiffeners' edges.

How essential it is for a shell to be "properly supported" in order to develop almost exclusively membrane stresses is clearly seen in barrel shells. A long barrel, supported on end stiffeners, carries loads by longitudinal tensile and compressive stresses and transverse shears, that is, by beam action

Figure 12.51 Barrel acting as vault

In contrast to long, end-supported barrels where beam action prevails (Figures 12.44 and 12.45), for barrels supported along their longitudinal boundaries (a), arch action carries the load. If the shape is funicular, the stresses will be purely compressive. Compared to a thin barrel supported on its ends, an edge-supported vault must be thicker to absorb the horizontal thrust (b), especially if the form is not funicular and bending stresses develop.

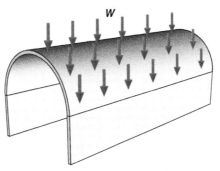

(a) Barrel acting as a vault

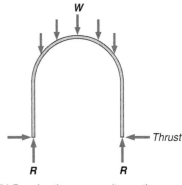

(b) Barrel acting as a vault - section

(Figure 12.45). The same barrel, supported along its longitudinal edges, develops essentially arch action, since each transverse strip may be considered as an arch supported at its ends and capable of developing the necessary thrust (Figure 12.51). If the shape of the cylinder section is the funicular of the loads, each strip will develop only compressive transverse stresses and no longitudinal stresses. This is also a membrane state of stress, but totally different from that developed by the stiffener-supported barrel, which consists essentially of longitudinal stresses. If the barrel cross section is not the funicular of the loads, bending stresses will develop in the arched strips (see Section 6.4); these bending stresses do not have the character of boundary disturbances, but pervade the entire shell. Although a stiffener-supported barrel may be relatively thin, a barrel supported on its longitudinal edges and acting as a vault must be fairly thick to resist the bending stresses in the arched strips.

The stiffener-supported barrel also develops some boundary disturbances along its longitudinal, free edges. These are minor disturbances, but have a tendency to penetrate more into the shell than the disturbances along the stiffener boundaries. It is not difficult to realize why. If a moment bends the edge of a thin flat plate, one must apply an equal and opposite bending moment to the opposite edge in order to equilibrate it. The entire plate bends and transfers the bending action from one edge to the other, unchanged (Figure 12.52). If the thin plate is curved into a

cylinder, bending applied at one of its curved boundaries is not transmitted to the opposite boundary, but dies out quickly (Figure 12.53a). It is thus seen that the curvature acquired by the plate "damps out" bending in a direction at right angles to the curvature. The bending disturbance at the stiffener ends of the barrel is damped out rapidly by the cylinder curvature at right angles to it. A curvature in the direction of the bending disturbance produces a lesser amount of damping, so that a longitudinal-edge disturbance dies out slowly, although usually it is not transmitted all the way to the other longitudinal edge (Figure 12.53b). The longitudinal-edge disturbance in the barrel is dampened less rapidly than in a shell curved at right angles to

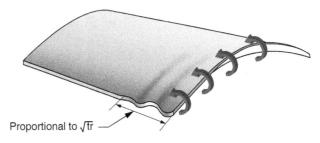

Proportional to $\sqrt{tr}$

(a) Bending penetration at edge of curved plate

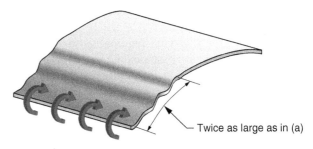

Twice as large as in (a)

(b) Bending penetration at edge of curved plate

Figure 12.53 Bending penetration at edge of curved plate

If bending is applied to the curved edge of a cylinder (a), the effect dies out rapidly. The curvature acts a damper for the disturbance at right angles to the curvature. Damping is less if the bending is applied in the direction of the curvature. Bending in the longitudinal axis extends approximately twice as far into the shell as that perpendicular to the curvature.

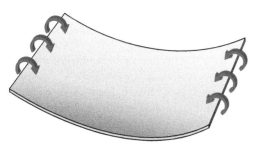

Figure 12.52 Bending of flat plate

When a bending moment is applied to one end of a flat plate, rotational equilibrium requires an equal and opposite moment at the other end. The effect of the bending is thus felt throughout the whole length of the plate.

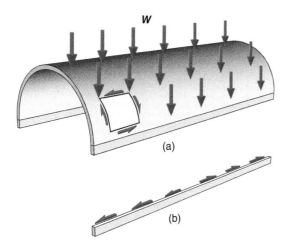

Figure 12.55   Ribbed barrel
In order to avoid buckling of thin shells due to compression, it may be necessary to increase the thickness of the shell. In such cases, it may be more economical to apply stiffening ribs to the shell. Stiffeners are commonly used in steel shell; in concrete, the required formwork increases cost and complexity.

Figure 12.54   Shears on longitudinal edge beams
Long barrels act similarly to beams with compression on top and tension on the bottom. To increase overall bending stiffness, edge beams are frequently used to absorb some of the tension generated by longitudinal shears.

the disturbances; for most shells the width of the disturbed band along the longitudinal edges is about twice as large as that along the curved boundary.

Barrel shells are not as stiff as domes, since their single curvature makes them behave very much like beams. Whenever it is necessary to increase their stiffness, it is customary to add longitudinal beams along their edges. This adds area on the tensile side of the cross section, so that a larger part of the barrel shell is under compression and the tensile stresses are absorbed mostly by the edge beams, acting as flanges. The edge beams, moreover, are used to absorb longitudinal shears developing at the edge of the shell, which introduce tension in the edge beams (Figure 12.54).

Just as the upper flange of an I-beam may buckle in compression, the upper fibers of a long barrel shell may buckle under longitudinal compression. The resistance of a long barrel shell to buckling is higher than that of domes, since compressive stresses are developed only in a portion of the shell and only in the longitudinal direction.

Notwithstanding the greater buckling strength of a long barrel shell, the danger of buckling may often demand an increase in the barrel thickness above the value required by a state of membrane stress.

Whenever the thickness required to avoid buckling makes the bending resistance of the shell so large as to reduce its membrane action, it becomes more economical to rib the shell transversely (Figure 12.55). Stiffeners are commonly used in steel shells; they are used less often in concrete shells because they increase the cost of the forms.

12.11 STRESSES IN SYNCLASTIC TRANSLATIONAL SHELLS

A variety of curves may be used to generate translational shells with double-positive curvature; arcs of circles and arcs of parabolas are the most common among them.

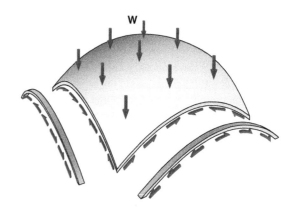

Figure 12.56   Boundary shears in translational shell
Translational shells transmit the load to rigid end frames by boundary shears similarly to barrel shells (Figure 12.43).

Translational shells are supported by shears on boundary arches, just as barrels are supported by shears on end stiffeners (Figure 12.56). The supporting arches are rigid in their own plane and flexible at right angles to it, so that compression or tension perpendicular to the arches is not developed at the shell boundary. The action of translational shells may be considered as barrel action in two directions, since the shells have curvature in two directions, but the central portion of the shells acts very much like a shallow dome, and develops compressive stresses along both generating curves. The portion of the shell near the supporting arches develops a certain amount of bending disturbances, and although these peter out rapidly into the shell in view of the curvature at right angles to it, a thickening of the shell is often required at the boundary.

A particular type of translational shell, engineered by Salvadori, is commonly used in Europe. These elliptic paraboloidal shells are built of hollow tile and concrete, with a minimal amount of reinforcing steel (0.6 pound per square foot) (29 Pa), on a movable form a few feet wide, shaped as one of the two generating curves and sliding on wooden arches having the shape of the other generating curve (Figure 12.57). The shell is built in slices, which act as separate arches and

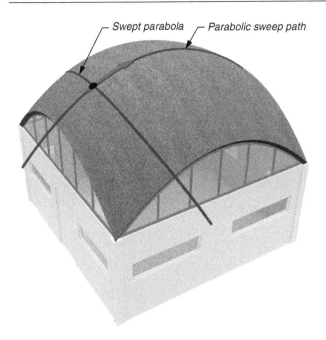

— Swept parabola *— Parabolic sweep path*

Figure 12.57   Tile and concrete elliptic paraboloid
This translational shell is constructed by using a wooden form a few feet wide and shaped as a parabola. Hollow tiles and reinforced concrete are laid on the form. After the concrete is partially cured, the form is moved over a similar parabolic support at right angles to the first and the process is repeated. Tie-rods are required to absorb thrust until all segments are completed.

Figure 12.58   Hyperbolic paraboloid supported on end stiffeners
The downward curving segments of this shell are mostly in compression and act as arches wile the upward curving parts are mostly in tension with membrane action. The double-curvature further stiffens the structure against bending disturbance at the ends. This project was among the first thin shell structures by Felix Candela, called the "Cosmic Rays Pavilion"; it was designed as a space for sensitive scientific apparatus at the National Autonomous University of Mexico in Mexico City.

Photo courtesy of Benjamin Ibarra-Sevilla

require tie-rods until they are made monolithic by the pouring and setting of concrete over the entire surface. These are among the few thin shells ever tested under total and partial loads. The tests proved that all the basic assumptions made about their action are met in practice. The bending disturbance at the boundary penetrates only a short distance into the shell; the shell does not thrust out along its sides and is supported exclusively by boundary shears. The rigidity of the shell proved to be exceptionally high: Under a uniform live load of 30 pounds per square foot(1.44 kPa) over its entire surface, a 60-by-90 foot (18–by-27 m) shell presented a top deflection of less than one-eighth of an inch (3 mm), or a span-to-deflection ratio of 9000.

12.12 SADDLE-SHELL ACTION

The behavior of the most common saddle shell, the hyperbolic paraboloid, shows the dependence of its structural action on support conditions. This surface was shown in Section 12.4 to be generated either by a vertical parabola with downward curvature sliding on a vertical parabola with upward curvature or by a straight-line segment the ends of which slide on two straight-line segments askew in space (see Figs. 12.16 and 12.17).

When the shell is supported on two parabolic arches or stiffeners, it transfers the load to the support by shears (Figure 12.58). Its action is similar to that of a stiffener-supported barrel, but the upward curvature in the longitudinal direction gives the shell additional strength, particularly against buckling. If the shell tends to buckle, the parabolas with downward curvature tend to buckle downward and flatten; this deformation is resisted by the parabolas with upward curvature, since their tension stabilizes the compressed parabolas.

If the paraboloid reaches the ground, its intersections with the ground consist of two outward-curved boundaries (the two branches of a hyperbola), so that the area covered by the shell has two straight sides and two curved sides (see Figure 12.16). When the paraboloid does not reach the ground, the shell may be supported at its four corners, but the end stiffeners must then carry the vertical and horizontal loads on the shell as arches. Bending disturbances, similar to those at the curved boundaries of barrels, are encountered at the intersection of the paraboloid with the stiffeners. Their values and the width of the disturbed band are of the same magnitude as those in cylindrical shells.

It is easy to see that even when the hyperbolic paraboloid is supported on its straight-line generatrices, the directions of principal stress coincide with those of principal curvature, that is, the parabolas (see Figure 12.7). Since the straight lines have no curvature, no cable or arch stresses can be developed along these lines by the shell that can only react in shear. Hence, tension along the upward parabolas must combine with compression along the downward parabolas to create a state of pure shear along the straight lines (Figure 12.59). The load is thus carried to the supporting boundaries by pure shear directed along the straight lines, and these shears accumulate along the boundary supports. The supporting elements are usually "beams," but these elements, rather than being loaded vertically, are loaded by the shears accumulating along their lengths, so that they behave like compressive struts (Figure 12.60) or tensile bars (Figure 12.61), except for the action of their own dead load.

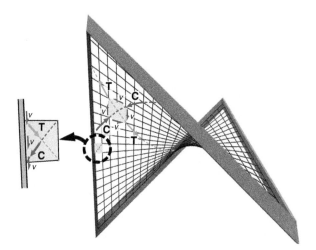

Figure 12.59 Hyperbolic paraboloid supported on straight generatrices

The lines of principal stress are not parallel to the straight-line generatrices. As a consequence, shear develops along the supporting boundary, acting somewhat like compressive struts or tensile bars.

When the hyperbolic paraboloid is shallow, not only are the principal membrane stresses identical tensions and compressions producing shears of the same intensity (Figure 12.59), but all three membrane stresses have the same value all over the shell. Inasmuch as membrane stresses are uniformly distributed across the depth of the shell, the hyperbolic paraboloid may be said to have a 100 percent overall efficiency; the material of the shell is equally stressed at any point of any cross section. The membrane stresses in the paraboloid are proportional to the total uniform load on it and inversely proportional to the shell rise and shell thickness. The shell rise should not be less than one-sixth to one-tenth the span in order to avoid high compressive stresses, which may buckle the shell.

It may be proved that the membrane stresses considered above arise in the hyperbolic paraboloid only if its straight-line boundary stiffeners are very rigid vertically and very flexible horizontally. Such conditions are seldom encountered in practice. If, for example, the boundary stiffeners are cantilevered from the two low points of the shell (Figure 12.59), the boundary of the shell is not rigidly supported vertically, as required by the theoretical assumptions. Actually, some of the stiffeners' weight hangs from the shell, and the two halves of the paraboloid behave very much like cantilevered beams with the variable, curved cross section of the downward parabolas.

Boundary disturbances are always present at the intersection with the boundary stiffeners; they penetrate into the shell more than the disturbances at the curved boundary of a cylinder, but less than those encountered along a cylindrical straight boundary.

Roofs are often built by combinations of hyperbolic paraboloid segments. One of the most commonly used involves four identical segments, the outer corners of

which are supported on columns (Figure 12.60). In this case, the inclined outer stiffeners and the horizontal inner stiffeners are all compressed by the accumulation of boundary shears. The thrust of the compressed outer stiffeners is absorbed by tie-rods, while the inner beams are in horizontal equilibrium under the equal and opposite compressions of each pair of paraboloidal segments. The roof has the appearance of a regular gabled roof, and its reactions at the corners are purely vertical, once the tie-rods are tensioned.

The same combination of four segments may be used as an umbrella supported on a central column (Figure 12.61).

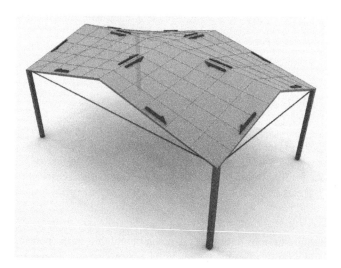

Figure 12.60 Shears on boundary beams of hyperbolic paraboloidal roof supported on outer columns

A combination of four hyperbolic paraboloids forms a roof. The outer edges as well as the inner ridges are compressed by boundary shears. The outer edge acts in compression, which is absorbed by tie-rods.

Figure 12.61 Shears on boundary beams of hyperbolic paraboloidal umbrella

The combination of four *inverted* hypars forms an umbrella supported on a central column. Here, the accumulated boundary shears produce tension at the outer edges that are equilibrated in the troughs of adjacent panels. The outer boundary edges act like tie-rods.

In this case, the outer horizontal stiffeners are tensioned by the accumulation of shears, and the tension in one of them balances the tension in the stiffener of the adjoining segment. The roof is innerly equilibrated and its boundary stiffeners are its tie-rods (Figure 12.61). In either combination, the weight of the horizontal stiffeners (the boundary stiffeners for the umbrella, and the inner cross for the gabled roof) is carried in part by the shell itself; the shell, therefore, develops some bending stresses. The roof supported at the corners may also buckle in the central region where it is flat, that is, where no appreciable curvature increases the buckling resistance of the thin shell.

A variety of other combinations of hyperbolic paraboloids may be used to cover areas of varied shape. Each one of these combinations must be carefully analyzed to determine whether it is innerly balanced. A typically unbalanced combination, consisting of two segments, is used as a northlight roof (see Figure 12.31). The thrust of the inclined boundary struts is equilibrated by a tie-rod, but the two compressive forces accumulating along the inclined strut common to the two segments are not equilibrated (Figure 12.62). Hence, the horizontal boundary beam is acted upon by a concentrated load at midspan and is subject to bending. The shell behavior, in this case, is a bending action in horizontal projection, with compression at the upper-boundary segments and tension at the lower horizontal boundary, due to a load concentrated at midspan; the two tensile forces at the outer lateral boundaries of the shell equilibrate the concentrated load, just like beam reactions (Figure 12.62).

Hyperbolic paraboloids with a very high rise have also been used as roof elements; others as almost vertical elements. In such cases, their behavior is totally different from

Figure 12.62 Shears on boundary beams of hyperbolic paraboloidal conoid

In this combination of two hyperbolic paraboloids, the inclined boundaries are compressed by the accumulated shears and are equilibrated by the tie-rod. The ridge is also in compression. This compressive load acts as a concentrated load on the horizontal boundary, which acts like beam on two supports with the two horizontal edges providing this support.

that of the shallow paraboloids considered above and similar to the behavior of a thin plate loaded in its own plane. It may be often approximated by the behavior of deep beams (see Section 7.1).

The different mechanisms by which the hyperbolic paraboloid carries loads are one more example of the dependence of shell behavior on support conditions.

12.13 STRESSES IN SCALLOPED AND OTHER TYPES OF SHELLS

Interesting shell shapes may be obtained by adding local curvatures to a simple surface, according to structural needs or aesthetic requirements.

An ellipsoid intersected by a horizontal plane is a smooth surface covering an elliptic area. In order to enhance its appearance and stiffen its boundary, it may be scalloped, thus creating local curvatures that peter out toward the interior of the shell (see Figure 12.27). The introduction of such curvatures substantially changes the shell behavior. The smooth ellipsoid carries load by stresses similar to those encountered in a rotational shell: compression along the meridians and compression or tension along the elliptical parallels. The introduction of local curvatures destroys the parallels' stiffness typical of membrane behavior. Under the action of loads, the shell tends to open up, since its undulated parallels cannot develop the hoop stresses that play such an important role in stiffening rotational shells. The action of the shell is similar to that of a series of arches with variable cross section, hinged at the supports and at the crown, where the lack of curvature reduces the bending resistance of the shell (see Section 8.5). Circular domes of large diameter have also been built by scalloping a spherical surface with the purpose of increasing its buckling resistance (see Figure 12.28). The waving of the parallels very much reduces their stiffening action; such shells behave as a series of arches crossing at the top of the dome and hinged there.

The membrane stresses of conoidal shells also depend on support conditions. If the straight boundaries are supported vertically and connected by tie-rods, each slice of the shell acts as a tied arch, and substantial bending stresses are to be expected for all loads for which the cross section of the shell is not a funicular curve. When the longitudinal boundaries are not supported, the conoid behaves as a cylinder with a variable rise and develops bending stresses in the neighborhood of the end arches only. If such a shell is scalloped, the transverse arch action is destroyed, since the conoid tends to open up; in order to restore its rigidity, the undulations must be stiffened transversely by stiffeners at regular intervals.

Northlight roofs for industrial buildings are often built by means of thin shells. Their structural behavior is similar to that of a beam with a shallow, curved cross section, whose neutral axis is not horizontal, but inclined toward the upper shell boundary (Figure 12.63). The shell fibers above the inclined neutral axis are compressed and those below are tensed. When the beam stiffening the shell at its upper

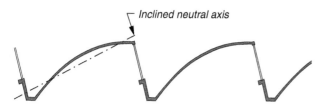

Figure 12.63 Cross section of northlight barrel shell
A series of thin shells are often used as roofs. They act as shallow inclined beams with curved cross-sections. The neutral axis here is inclined: Stresses above it are tensile and compressive below.

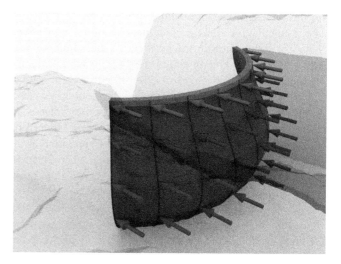

Figure 12.65 Thin-shell dam
Large dams over 1000 feet (300 m) high have been built with curvatures in both the horizontal and vertical planes. Their thickness of several feet (meters) is only about 1/400th of the radius of curvature, so they are therefore still thin shells.

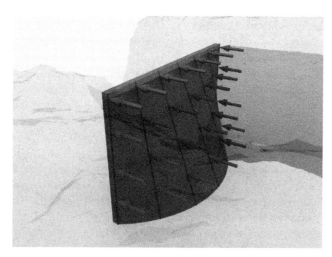

Figure 12.64 Conoidal retaining wall
Conoidal shells may be used as retaining walls to withstand horizontal pressures. Their behavior is similar to the structure of Figure 12.62.

boundary is large, the neutral axis may cross it, and most shell fibers may be in tension.

Some of the largest thin shells in the world have been built to resist horizontal rather than vertical loads. Retaining walls loaded by the thrust of the earth may be built as vertical cylinders or conoids (Figure 12.64). Cylindrical tanks containing water, oil, or other liquids are built as thin steel or prestressed concrete shells to resist the outward pressure of the liquid. Dams, with heights of over 1000 feet (300m), have been built with curvatures both in the vertical and the horizontal sections and a thickness of only a few feet (meters) (Figure 12.65). The ratio of radius of curvature to thickness in these dams is of the order of 400 or more, so that such shells are to be considered thin even though they may be several feet thick at their foot.

12.14 THIN-SHELL FORMWORK

The aesthetic, architectural, and structural possibilities of thin-shell construction are practically unlimited; only their cost may at times hamper their diffusion. The problem of building expensive curved forms on which to pour reinforced-concrete shells has made their use uneconomical in countries with high labor costs. The solution of the forming problem has been attacked in a variety of ways. Roofs

consisting of identical shell elements are built by pouring an element on a movable form, which is then lowered, shifted, and reused again to pour the next. Prefabrication of shell elements and their erection on a light scaffold has become a standard procedure, which eliminates expensive forms. The joining of the prefabricated elements must be carried out with great care, and often requires welding the reinforcing bars sticking out of the elements and pouring the joints at a later date. Some of the exceptional thin-shell structures designed by Nervi were built by this procedure.

Prefabrication of shells by elements is often used in conjunction with posttensioning. It was observed in previous sections that certain types of shells develop tensile stresses over large portions of their area. These tensile stresses may be eliminated by introducing in the thickness of the shell steel cables or tendons which are tensioned after the concrete has hardened (posttensioned); they compress the shell so as to eliminate altogether the tensile stresses due to the loads (see Section 3.3). The thin shells of cylindrical tanks are usually posttensioned by spirals of steel cables, thus guaranteeing the waterproofness of the tanks under all conditions. Northlight shells with large spans may also be posttensioned by curved tendons so as to eliminate tensile stresses in the shell portions below the inclined neutral axis (see Figure 12.63).

The use of a concrete gun allows the spraying of layers of concrete (Shotcrete or Gunite) on the reinforcing bars with a minor amount of formwork. The bars are supported on simple scaffolding, in which the insulation panels constitute the actual form surface against which the concrete is shot. Finally, balloons have been used as inflated forms to support the steel and against which to spray the concrete or, as in the Bini process, as inflatable forms to lift both the reinforcing steel and the concrete (see Section 11.3).

12.15 RETICULATED DOMES

Space frames (see Section 10.9) behave like structures in which the distributed material of a plate and, hence the stresses in it, are lumped in the frame bars. Their load

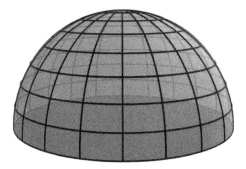

Figure 12.66   Ribbed spherical dome

The behavior of ribbed or reticulated domes is analogous to that of thin shells if it assumed that the shell material between the ribs is lumped into the ribs. The meridional ribs carry the load in compression to their base, and ribs along the parallels are hoop stresses in tension or compression depending on the opening angle (see Section 12.7).

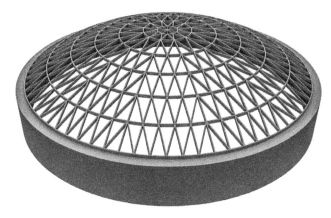

Figure 12.67   Houston Astrodome

The 642 ft (195m) diameter dome is supported by a structure consisting of bars arranged in merdional and parallel directions, interconnected by elements creating regular triangular regions. The lengths of the bars vary dependent on their location in the structure and hence are more difficult to assemble. When it opened in 1965, it was the world's first domed sports stadium.

carrying mechanisms can be understood by analogy with those of plates. Similarly, the behavior of reticulated or ribbed domes may be grasped readily, by analogy, with that of thin-shell domes.

For example, if the material in a spherical dome is lumped in bars lying along the two families of curves created by the meridians and the parallels (Figure 12.66), the meridional bars will absorb the compressive membrane stresses from the top to the boundary of the dome, and the parallel bars will be in tension or in compression depending on the value of the opening angle (see Section 12.7). The continuous thin shell has been transformed into a discrete structure of the same shape.

In order to use bars that do not differ too much in area, the meridional bars often branch off from the top of the dome down. The steel structures of some of the largest domes in the United States, like those of the 642-foot-diameter (195 m) Astrodome in Houston, Texas (Figure 12.67), and of the 680-foot-diameter (207 m) Louisiana Superdome in New Orleans, Louisiana, follow this scheme (Figure 12.68).

The meridional-parallel lattice presents the disadvantage of using bars of different lengths. Other schemes have been invented to avoid this difficulty and simplify the connections between the bars, but since the surface of a sphere cannot be entirely covered by a lattice of regular polygons, the bars of a spherical dome cannot all be identical. In the geodesic domes, designed by Buckminster Fuller, triangles and hexagons are used to obtain a subdivision in terms of bars of equal length (Figure 12.69). An irregular triangular lattice of ribs is used instead in the classical Schwedler dome (Figure 12.70), which presents the design advantages of being statically determinate (see Section 4.2), and in the Zeiss-Dywidag dome (Figure 12.71).

Ribbed domes spanning up to three or four hundred feet have been built in Europe with standard connectors, using bars made out of steel pipes on a triangular grid. These structures require only 4 to 5 pounds of steel per square foot (192–239 Pa), as against 10 to 20 pounds per square foot (479–958 Pa) for domes built of rolled sections, but are very sensitive to local buckling. Their behavior in this context is analogous to

Figure 12.68   Louisiana Superdome

The 680 ft. (207m.) diameter superdome is also constructed using a branching pattern very similar to the Astrodome. Since opening in 1975, it has remained the world's largest nonretractable steel-framed dome (as of 2015).

Photo couresy of Corey Seeman

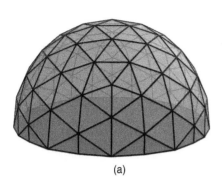

(a)　　　　　　　　　　　　　　　　　　(b)

Figure 12.69　Geodesic dome

The disadvantages of bars of varying length in the assembly of the structures of Figures 12.64 and 12.65 can be overcome in the construction of geodesic domes (a). In these bars of equal length are arranged as equilateral triangles and hexagons to approximate a dome shape. Here, the bars do not lie along the meridians or the parallels. The Nagoya Dome in Japan (b) is an indoor baseball field covered by a shallow geodesic dome spanning 328 ft (100m).

(b) Photo courtesy of Jeff A. Boyd

Figure 12.70　Schwedler dome

In the Schwedler dome, most bars are of equal length. These do follow the meridians and parallels. It has the design advantage of being a statically-determinate structure. Both the Astrodome and Superdome are double-layered Swhwedler domes.

Figure 12.71　Zeiss-Dywidag dome

Equilateral triangles are used in this statically determinate structure. Here, bars follow the parallel lines but not the meridians. Steel pipes and standard connectors are used to assemble both the Schwedler and the Zeiss-Dywidag domes.

that of a thin shell in which the steel of the bars, uniformly distributed over the dome surface, has a much smaller apparent modulus of elasticity and a smaller apparent density. In other words, these domes behave like a thin shell made out of a thicker, spongy material, just as space frames do in relation to plates. As the buckling load is proportional to the elastic modulus of the material, the lower apparent modulus lowers the buckling capacity of these latticed domes.

The analogy between latticed roofs and thin shells persists for roofs of other than spherical shape. Latticed barrel roofs can be built with bars parallel and perpendicular to the barrel axis (Figure 12.72) and can be supported on end diaphragms. The forces in their discrete bars are, to a good approximation, the resultants of the stresses in the barrel shell areas "contributory" to the bars, and can be obtained by considering the barrel as a beam (see Section 12.9). When the barrel roof springs from the ground and, hence, behaves like a series of parallel arches, a skew grid of discrete arches leads to a so-called Lamella roof (see Figure 12.73), which

is used to span hundreds of feet in wood, concrete or steel (see Section 8.5). Zeman built domed Lamella roofs spanning up to 300 feet (91 m), by erecting prefabricated steel-pipe sections with standard bolted connections. The structure of these roofs, also, weighs not more than 5 pounds per square foot (239 Pa).

Regular and irregular doubly curved surfaces called *gridshells* consist of a latticework of light members that follow the shell surface curvature. These have been constructed of wood or steel, such as the Savill Building outside of London, England (Figure 12.74). Such structures have shell-like behavior, but forces are restricted to axial tension and compression along the member length. They have been sheathed in wood, glass and fabric membranes. One of the first such structures was by Frei Otto, the 1974 Multihalle in Manneheim, Germany. It was constructed with the gridwork horizontally on the ground, and then hoisted into final position as a complete unit, and then secured at the foundation. Steel cables were used diagonally to laterally stiffen the structure.

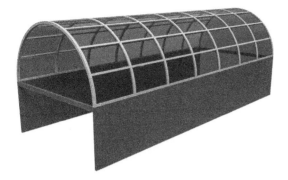

Figure 12.72 Latticed barrel roof
Large rectangular areas may be covered by barrels constructed of a rectangular mesh of bars parallel and perpendicular to the barrel axis. The bars following the curved surface have arch action, and the longitudinal bars carry loads by the bending stresses of the whole vault. Compressed bars are prone to buckling but are restrained by the covering material.

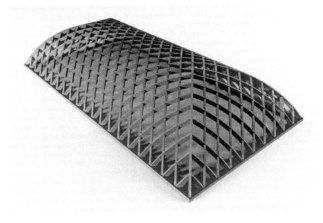

Figure 12.73 Lamella Roof
The bars of a barrel roof may be arranged in a skewed pattern forming intersecting grids of parallel arches called a *lamellae*. The bars may be made of steel or wood using bolted connections, or of concrete.

(a) Exterior view

(b) Interior view

Figure 12.74 Gridshell
A gridshell is a regular or irregular doubly-curved surface that is defined by a lattice-type structure, most frequently constructed of wood or steel. The Savill Building, engineered by Buro Happold, is a wooden gridshell serving as a visitor center to the Savill Garden at Windsor Great Park near London, England.

Photo courtesy of Simon Chorley

Saddle roofs, in the shape of hyperbolic paraboloids made out of plywood supported by aluminum pipes (oriented along the straight line generatrices of these surfaces), were used by Candela as units to plug the openings between the steel truss arches of the 1968 Olympic Stadium in Mexico City (Figure 12.75).

The bending disturbances at the shell boundary (see Section 12.8) correspond to local bending of the bars of the latticed roof near its supports. All other bars are only stressed by the axial forces corresponding to the membrane shell stresses (compressive or tensile) and hence use the material to great efficiency. As is the case for the thickness of thin shells, limitations in the bar sizes may be due to the requirements of buckling.

The erection of latticed domes is facilitated by the Binistar system patented by the Italian architect Dante Bini and first designed in 1985 for a triangulated dome on a hexagonal base of 131-foot diameter (40 m) in Bari, Italy. The "telescopic" steel bars of a Binistar dome consist of outer pipes and two inner rods sliding in each pipe (Figure 12.76). Short lugs, set into the inner rods in correspondence with holes in the outer pipes, can be pushed out radially by springs. The inner rods are attached to spherical connectors that allow their ends to rotate freely. To erect the dome the pipe-rod bars are mounted, over a deflated plastic balloon, into a flat triangular lattice, anchored to a base ring of reinforced concrete. As the balloon is inflated to the required dome shape, the inner rods slide inside the pipes until the lugs are shot into the pipe holes by the springs: The lugs, thus engaged into the pipes, transform the telescopic bars into rigid bars and freeze the shape of the triangulated dome (Figure 12.77). The balloon is then attached to the external latticed dome and becomes its permanent surface. As Geiger's tensile domes (see Section 6.2), Binistars may be considered membrane balloons supported by reticulated domes rather than by air pressure.

(a) Exterior view   (b) Interior view

Figure 12.75 Mexico City Sports Palace

The dome here is defined by steel trusses laid on the lines of principal curvature, the top chords of which are visible on the outside of the structure (a). The space between each truss is closed by four hypars, the surfaces of which are plywood supported by aluminum tubes. These follow the hypar straight-line generatrices and span between the larger trusses. The edges of the hypars match the angle of, and are supported by, the diagonals of the main steel trusses (b). The exterior surface is sheathed in copper.

Photo courtesy of Geometrica

Figure 12.76 Bar of Binistar dome: (a) collapsed; (b) extended

Bini's invention consists of two bars with spherical ends sliding inside a tube and separated by springs. The bars are first locked in the unextended position and allowed to spring open when in position in the dome.

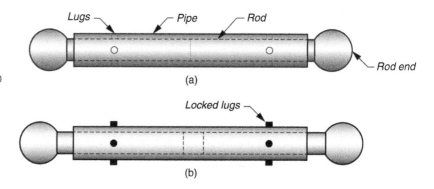

Figure 12.77 Binistar dome

The dome consists of a triangular set of bars on a hexagonal boundary initially laid on a deflated balloon. The bars are connected by unextended Bini connectors. The balloon is attached to a reinforced concrete base and is inflated into a dome shape. The connectors spring open and lock in place to form a latticed dome.

In countries with high material costs and low labor costs, thin shells of poured, prefabricated or prestressed concrete are usually economical in comparison to similar latticed steel domes. The opposite is usually true in countries with high labor costs and low material costs. The same considerations of relative materials and labor costs

explain the favor encountered by the latticed pipe domes outside the United States. While rolled-section domes can be built in sectors, using a single center support for their erection, lightweight pipe domes require more complicated scaffolds and greater amounts of labor. Since, moreover, in the United States steel pipe is, pound per pound, more expensive than rolled steel sections, the fabrication and erection costs of pipe domes do not, in general, make these structures competitive.

KEY IDEAS DEVELOPED IN THIS CHAPTER

- Membranes, shells, and domes are form-resistant structures that can carry loads because they are curved surfaces.
- If a membrane that carries loads by tension stresses is turned upside down, it becomes a shell or a dome carrying loads by compression just like an upside down cable has the shape of an arch. The materials of shells, domes, and arches must be able to withstand compressive stresses.

- As was the case for membranes, shells and domes have curvatures at every point, with maximum and minimum values in principal directions.
- Surfaces can be developable, that is, they may be flattened without cutting them, or undevelopable that have to be cut in order to be flattened.
- Curvatures may be downward (positive), dome or cylindrical shaped or upward (negative), bowl shaped at very point on the surface.
- Others may have an upward curvature in one principal direction and downward in the other.
- Curved surfaces may be generated by rotating a plane curve around an axis, by moving a line along a curve or a curve along a line.
- Shallow circular domes have arch action along both meridians and parallels. They are in compression.
- High-rise domes are compressed along meridians, but their parallels are mostly in tension. Just like for arches their base needs buttressing to prevent it from moving outward. As a consequence bending stresses are also present along with shear,
- Cylindrical shells have only one curvature. Along the curved surface, it has arch action, but a long cylinder acts similarly to a beam.
- Saddle shells have positive curvatures along one principal direction and negative in the other. Along the positive curvatures, arch action and compression prevails while the negative cable action supports the arches in tension.
- Reticulated domes are similar to space frames and are constructed of triangular or hexagonal elements.

QUESTIONS AND EXERCISES

1. Find a large orange preferably with a thick skin. Cut it in half and remove the fruit without damaging the pulp or the skin. You now have a high-rise dome. Place it on a sheet of paper and draw the outline of the base. Gently apply pressure to the top and observe the outward movement of the base. What kinds of stresses are present in the circumferential directions along the parallels? How about along the meridians?

2. Carefully cut a circular hole, about ½ inches (12 mm) in diameter, on top of the dome. Put a drinking glass with the open end on the dome and apply pressure. What kinds of stresses are present along the meridians and along the parallels at the edge of the hole?

3. Take the second half of the orange skin and cut a cap, about ½ in. tall (12 mm), to create a shallow dome. Repeat the previous two experiments.

4. Cut construction paper into a 2 foot (60 cm) long by 1 foot (30 cm) wide piece. Bend the paper into a long cylindrical barrel. Support it along the long edges with a few books. Gently apply point loads to the barrel and observe the type of deformations. Now put a small pillow on the barrel to distribute the load and put a book on the pillow. Again observe the deformations. What type of stresses can explain these deformations?

5. If you have access to "Tinker Toys" or a child's construction set, construct a Geodesic Dome similar to one shown in Figure 12.69.

FURTHER READING

Anderson, Stanford, editor, *Eladio Dieste: Innovation in Structural Art*. Princeton Architectural Press. 2004.

Garlock, Moreyra Maria, E. and Billington, David. P. *Felix Candela: Engineer, Builder, Structural Artist*. Yale University Press. 2008.

Nervi Pier Luigi, and Einaudi, Robert. *Aesthetics and Technology in Building*. Harvard University Press. 1965.

Ochsendorf, John, *Guastavino Vaulting: The Art of Structural Tile*. Princeton Architectural Press. 2010.

Sandaker, Bjorn N, Eggen, Arne P. and Cruvellier, Mark R. *The Structural Basis of Architecture*, 2nd Edition. Routledge. 2011. (Chapter 12)

Schodek, Daniel and Bechthold, Martin. *Structures*, 7th Edition. Pearson. 2014. (Chapter 12)

Tullia, Lori. *Pier Luigi Nervi: Minimum Series*. Ore Cultura Srl. 2009.

STRUCTURAL FAILURES

13.1 HISTORICAL FAILURES

The moment humanity started erecting structures, structures began to fail. In prehistoric times, corbelled domed houses of stone were built all along a band running uninterruptedly through Asia Minor, Greece, Crete, Sardinia, southern France, and England, but only a few survive intact after 2,000 to 5,000 years. The pyramids of Egypt stand lonely among the Seven Wonders of the Ancient World: the Colossus of Rhodes, an over-100-foot-high (30.5 m) bronze statue to the sun god Helios, vanished, and an earthquake destroyed the second longest lived, the lighthouse of Pharos at Alexandria (Egypt), said by some historians to be 200 feet (61m), by others 600 feet (183 m) tall.

Collapses and failures plagued some of the greatest monuments of history. The dome of Saint Sophia in Constantinople began showing signs of weakness during construction, and tradition has it that it was saved by the intervention of the emperor Justinian, who urged the architect Anthemius to complete one of its main arches that was falling, because "when it rests upon itself it will no longer need the uprights under it." Yet, after two earthquakes, the eastern arch collapsed in 557, the western in 989, and the eastern for the second time in 1346; the dome was made stable only in 1847 by placing iron chains around the base. The masterpiece of Gothic architecture, the Cathedral of Saint Pierre at Beauvais, France, had the main vaults of the choir collapse in 1284, only 12 years after it was completed, and its 502-foot tower collapsed 13 days after its erection in 1573. The great domes of Renaissance Italy, those of Santa Maria degli Angeli in Florence and of Saint Peter's in Rome, cracked, and few, if any, of the monuments of the past still extant do not show signs of weakness today.

We accept such failures because of the lack of sound structural knowledge by the builders of the times, but we are shocked and puzzled by the rash of collapses of some of our large buildings and bridges. How does our scientific age explain and countenance such disasters? If we are interested in learning why most of our buildings stand up, we also want to know why some fall down.

13.2 MAIN CAUSES OF STRUCTURAL FAILURE

In the following sections, the main causes of contemporary failures are discussed using the intuitive approach of this book, but it must be pointed out at the outset that such failures occur in a small percentage of our increasingly large, tall, and complex structures. As should be expected, our record is far superior to that of our predecessors, although the number of exceptional structures being erected grows yearly all over the world.

In the last analysis all structural failures are caused by human error, that is, are due to lack of knowledge or judgment, but for purposes of classification they may be attributed to deficiencies in design, in fabrication and erection, or in materials. One could put failures due to unexpected events, natural or human-made, in a separate category, but even these are due, most of the time, to our incomplete knowledge or lack of caution. Different causes often conspire and result in collapses. Technical investigations should try to explain them so that their repeated occurrence might be avoided, but should waste no time assigning blame, since this is the province of the law. Yet, it might interest the reader to learn the professional and legal consequences of engineering and architectural failures, which is the subject of Section 13.6.

13.3 FAULTS IN STRUCTURAL DESIGN

13.3.1 Sources

Design deficiencies may occur for a number of reasons: pure and simple mistakes in calculations, undetected errors in computer inputs, incomplete or mistaken interpretation of building codes, unfamiliarity with dynamic or "hidden loads," defective detailing, lack of redundancy, lack of coordination between the members of the construction team, and many more.

Good design can only result from experience, since design is learned through many years of continued practice. Hence, the young practitioner should not be expected to assess the relative importance of the many facets of design and should not be given responsibility for the design of essential structural components. The organization of a professional office requires clear assignment of tasks and careful supervision at all levels. Before drawings are signed, an experienced professional must carefully examine them and be satisfied that they faithfully interpret the design concept for the other members of the construction team. Unfortunately, under the time and economic pressures of our culture, such supervision and coordination are not always given the consideration they deserve and often become causes of failure.

13.3.2 Computational Errors

Contrary to what the layperson may believe, errors in calculations seldom cause deficient design: If flagrant, they are easily caught; if minor, they may be unimportant. The advent of the computer has immensely refined structural calculations (see Section 4.4) at the cost of drowning the designer in a sea of numbers: The experienced professional, aware of the common occurrence of input errors, will never accept a computer result without checking it long hand with a simplified formula or against personal experience and past records.

It must not be construed from the above that errors in calculations are never responsible for failures. In certain cases, the mistakes in design are so flagrant that their most probable cause may be shown or inferred to be an uncaught numerical mistake. According to the investigation by the National Bureau of Standards of the failure of the pedestrian bridges at the Hyatt Regency Hotel, Figure 13.1 l in Kansas City, Missouri, in 1982, in which 114 people died and 216 were injured, the collapse was due to obvious errors in

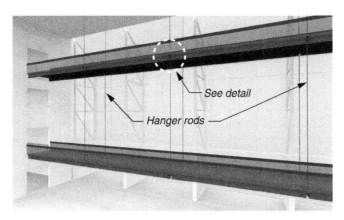

(a) Hyatt Walkways

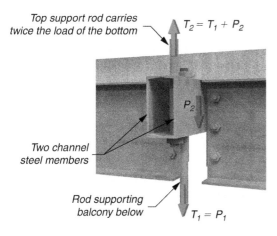

Top support rod carries twice the load of the bottom

$T_2 = T_1 + P_2$

P_2

Two channel steel members

Rod supporting balcony below

$T_1 = P_1$

P_2 = Load of upper skywalk
P_1 = Load of lower skywalk

(b) Hyatt as Constructed

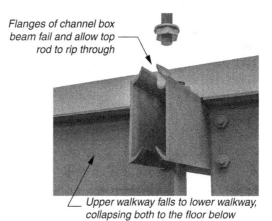

Flanges of channel box beam fail and allow top rod to rip through

Upper walkway falls to lower walkway, collapsing both to the floor below

(c) Hyatt Failure

Figure 13.1 Kansas City Hyatt Regency walkway collapse
Lack of oversight and poor communcations in the design of the suspended skywalk proved fatal. The structural engineer's drawings showed a single long hanger supporting two skywalk levels from the roof structure (a), leaving the connection design to the steel fabricator. To facilitate erection. The steel fabricator detailed (but did not calculate) the connection using two rods instead of one continuous rod, the lower rod being supported by a beam at the upper walkway level (b). This was a significant alteration, as the change caused a doubling of the load on the upper walkway support that was not detected by the structural engineers. Neither the detailer nor the engineer actually computed and sized this connection for the revised assembly. The connection was further compromised by the fact that it was constructed of a box beam made by two face-to-face channel members vs. a single box beam. On the evening of a popular dance, with both walkways full of people, the overloaded connection gave way as the channel flanges bent inward letting the rod slip through (c). In a progressive failure, both walkways then collapsed to the floor, killing 114 and injuring 216 people.

design (Figure 13.1), compounded by alterations proposed by the fabricator that were not reviewed by the engineers. The reader of the NBS report may be justified to surmise that a numerical mistake, accompanied by a lack of judgment on the part of an inexperienced designer, may have been the source of such a catastrophic error.

13.3.3 Building Codes

A thorough knowledge of building codes (see Chapter 2) requires years of patient study, careful interpretation, and frequent updating. Codes are modified every few years to take into account the accumulation of new knowledge and to clarify the meaning of their requirements. The design professional is confronted not only with a variety of codes applicable to different types of structures and materials, but also with the proliferation of codes by states, counties, and cities.

Most of the structural requirements in local codes are derived from or refer to a few widely accepted national codes. The multiple regional codes of the past are now largely consolidated into the *International Building Code* (somewhat misnamed as it is a document mainly used in the United States), but there is still sufficient local code variation to demand careful scrutiny, particularly because of their legal implications (see Section 13.6). The situation differs from that existing in other countries, where unique national codes govern at all levels. In any case, it must be remembered that the building codes make minimum recommendations and that adherence to their requirements does *not* exempt the designer from technical responsibilities.

13.3.4 Dynamic Loads: Wind and Seismic Forces

Fields in rapid evolution, like those of dynamic design for wind and earthquake loads, have requirements seldom timely updated in building codes because they demand time-consuming debates before being approved. Moreover, the practice of dealing with dynamic conditions is not common in architectural engineering because most buildings are usually subjected to static loads only (see Chapter 2). Hence, the design practitioner must not only be aware of the possibility of dynamic loads, but must keep abreast of the current literature so as to make safe and economic use of the latest information on these dangerous conditions. For example, in the design of the Toronto City Hall Building (Figure 13.2), wind forces that were wisely determined by wind tunnel tests indicated an unexpected wind load due to the channeling of the wind through the gap between its two buildings. Similarly, the heavy steel gates of one of our first missile silos were torn off their hinges and collapsed when first used, for lack of consideration on the part of the designers of the dynamic forces due to their accelerated motion.

The wind-bracing system of a building must meet conditions not only of strength but also of stiffness to minimize discomfort to its occupants and avoid damage to its curtain wall (see Section 2.4). In the Hancock Tower of Boston, Massachusetts, the excessive flexibility of the frame was

Figure 13.2   City Hall, Toronto, Ontario, Canada
The Toronto City Hall was designed in part with the aid of wind tunnel testing due to its unusual configuration. The testing revealed unexpected wind forces due to the channeling between the two buildings, and thus led to a design taking these forces into consideration.
Photo courtesy of Terri Meyer Boake

one of the main causes of the dislodgement of a high percentage of the curtain wall glass panes, and required the damping of its wind-induced motion by means of two dynamic dampers (see Section 2.6 and Figure 2.27) at a high cost and with long construction delays.

The knowledge of earthquake motions and their influence on buildings has rapidly improved during the last decades. Detailed maps of earthquake intensities in the United States (Figure 2.23) are available to the designer and must be consulted by all members of the construction team, although the technicalities of earthquake design are left to the specialist. It must be realized that data on earthquakes are gathered uninterruptedly all over the world and that as a result of these investigations earthquake requirements become continuously more demanding.

With each new seismic event, more is learned about the nature of earthquakes and how buildings respond to them. Buildings are being built with a higher degree of seismic resistance, and areas of the United States—thought until recently to be free of higher-intensity earthquakes—have proven to be subjected to them. An unexpected earthquake in Virginia in 2011, for instance, caused structural damage to the Washington Monument and National Cathedral in Washington, D.C. Even in areas of minor earthquake activity, the codes now take seismic design into account, among other considerations, in the detailing requirements of reinforced-concrete frames.

It is comforting to realize that our improved techniques of earthquake design allow high-rise buildings, like the

Torre Latino-Americana in Mexico City, to survive undamaged in earthquakes that caused the failure of low buildings of older vintage. High-rise buildings of contemporary construction that are designed for earthquake forces can in fact fare well in seismic events. The current highest building in Mexico, the 55-story Torre Mayor, is designed with hydraulic dampers in diamond-shaped frames to absorb seismic energy. It survived a large 7.6 magnitude quake in 2003 without damage. Because steel and aluminum are flexible materials that are elastic under smaller loads and deform plastically under larger ones (see chapter 3), structures with metal frames may suffer large deformations under earthquake conditions but seldom break.

Rigid structures, though, built of brick and mortar, as well as some poorly reinforced-concrete buildings and bridges, are more susceptible to earthquake damage. California's Loma Prieta 6.9 magnitude earthquake destroyed the double-decked Cypress Street Viaduct (formerly the Nimitz Freeway) in West Oakland in 1989 (Figure 13.3). A lack of sufficient confining reinforcement to prevent shear

failure at the junction of the beams and columns contributed to the collapse. The structure was designed and constructed to the codes and standards of its day some 35 years earlier, but little was truly understood about seismic action in that era. In a more recent example, during the sixth most powerful earthquake on record, the 2010 Chilean event, a 15-story apartment building toppled over and broke in two. Concrete failure between the basement and first floor has been implicated in the collapse (Figure 13.4).

13.3.5 Thermal Movements and Stress

The same caution must be used in dealing with thermal differences, the main source of locked-in stresses (see Section 2.5). Because of their dependence on a large number of parameters, the investigation of thermal stresses is advised by most codes without specific recommendations on temperature differences or methods of analysis. As a consequence, structural failures may often be attributed to neglected thermal conditions. In a concrete dome for one

Figure 13.3 Cypress Street Viaduct (Nimitz Freeway) Collapse, 1989 Loma Prieta Earthquake

The double-decked Cypress Street Viaduct failed in a progressive collapse. Like a row of dominos, the failure of one section led to overload and failure of each subsequent section. The collapse mechanism was determined to be largely the result of insufficient shear reinforcement at the lower level beam-column connection, even though it was designed and built to the codes of its day. Forty-two people lost their lives in this one failure alone in this magnitude 6.9 earthquake, centered some 60 miles (97km) south of San Francisco.

Photo courtesy of USGS

Figure 13.4 Building Collapse, 2010 Chile Earthquake

The earthquake off the coast in central Chile at magnitude 8.8 was the sixth largest ever recorded. The earthquake triggered a tsunami that devastated several coastal towns. Near its epicenter, in the city of Concepción, the forces were great enough to topple this 15-story apartment building, breaking in half as it crashed down, killing eight people.

Photo courtesy of Claudio Núñez

of the first nuclear reactors built in the United States, the stresses due to a "thermal explosion" (a sudden, large temperature rise on the interior of the reactor), originally ignored, required a modification of the design.

Most of the cracks in the brick masonry veneers of air-conditioned buildings, particularly in veneers of pre-fabricated panels, are due to the rigid connections between the exposed masonry and the interior structure that do not allow for the thermal expansions and contractions of the masonry due to daily and seasonal temperature variations of the outdoor air (Figure 13.5a). Similarly, cracking in the partitions of high-rise buildings occurs when the climate produces large differences in temperature between southern and northern facades. The thermal bending of such buildings (Figure 13.5b) is the source of high shears in the partitions, acting as webs of vertical cantilevered beams, and of cracks due to the tensile component of these shears (see Sections 5.3 and 7.1).

Unbalanced forces due to thermal differential significantly contributed to the collapse of Terminal 2E at Charles de Gaulle Airport in Paris in 2004 (Figure 13.6), just eleven months after its inauguration. The unique cylindrical structure was designed as a perforated 'squashed' elliptical shell of concrete, with an exterior glazing held away from the concrete by posts and tensioning cables. On a morning with an unusually rapid drop in temperature, shrinkage of the surrounding cables caused the posts to put pressure on the concrete. The shell was designed for these pressures; however,

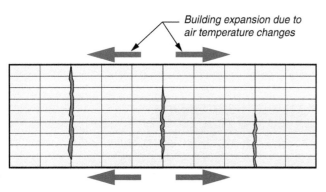

(a) Building expansion thermal cracks

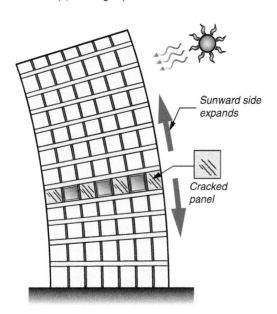

(b) Differential heating thermal cracks

Figure 13.5 Thermal cracks in building facades

Buildings are subject to nonstructural failures as well, brought on by structural actions due to thermal movements. In tall buildings, the side facing the sun will be warmed much more than the shade side, causing an expansion. Since the shade side is cooler, the expansion will be uneven, and a bending force induced between the two sides. Any materials rigidly connected will then be subject to distortion, particularly in shear along the wall face, behaving similarly to the web of a beam cantilevering out of the ground.

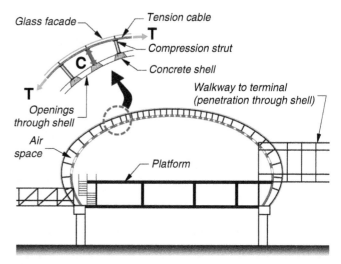

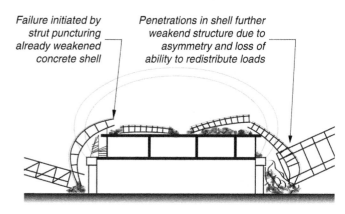

Figure 13.6 Failure of terminal building at Charles de Gaulle Airport in Paris

The unique cylindrical structure was designed as a perforated 'squashed' cylindrical shell of concrete with an exterior glazing held away from the concrete by posts and tensioning cables. The tensioning cables resisted the outward bulge of the concrete. Postfailure analysis indicated that there were faults in the construction, and it was unnoticed that the structure had been weakening. A sharp temperature drop occurred one morning that caused rapid contracture of the tension cables. This in turn caused the posts to put additional pressure on the concrete as the cables contracted. The shell was designed for these pressures; however, at one location a ramp penetrated the shell, thus creating a discontinuity. The imbalance of forces on the shell in addition to construction flaws led to its complete collapse in this area.

at one location a ramp penetrated the shell, thus creating a discontinuity. The imbalance of forces on the shell, coupled with construction deficiencies, precipitated its complete collapse in this area.

Fires are always dangerous to the occupants of a building, but they may also cause structural failures even for non-combustible materials. While steel and aluminum can withstand moderate heat, at a 1000 degrees Fahrenheit (~550 C) Aluminum can melt and steel will begin to lose its elasticity and become severely deformed (Figure 3.12). For the same reason, "rebars" in reinforced concrete must be protected with a sufficient layer of cement to provide fire resistance. Building codes (see Chapter 2) pay particular attention to this requirement.

13.3.6 Foundations

The mechanics of soils has become a reputable field of science through the investigations of the last one hundred years: It is one of the most sensitive facets of foundation design. No structural engineer will assume the responsibility of such design without a soil investigation performed by specialized laboratories that specify the bearing capacity of the soil, the maximum differential settlements to be expected, and the type of foundation most appropriate for a chosen location (see Section 4.3). Settlement differentials are of particular interest to the designer since they may cause large cracks in curtain walls and partitions and weaken the structure (Figure 13.7). A substantial number of failures in slabs on grade and curtain walls are due to foundation deficiencies related to changes in underground water levels and new construction. The National Theatre in Mexico City, founded on a soil consisting of a mixture of sand and water, subsided many feet at first when its weight squeezed the water out of the sand, but then was pushed integrally upward when a number of high-rise buildings were erected around it. In this case, the evenness of the displacements and the monolithic reinforced-concrete structure prevented damage to the building.

A dramatic foundation failure occurred in China when rain-soaked soil was excavated beside the pilings of a

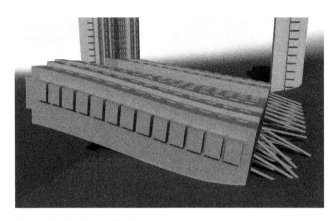

Figure 13.8 Foundation Failure

This apartment block under construction in China had the perfect storm of complicating factors. The ground was being excavated on the front side of the building, and the excavated soil was stacked up behind the building. This caused an unbalanced horizontal pressure on the building foundation, which was further compromised by soaking rains that both weakened the soil as well as added even more lateral pressure. Eventually the horizontal force was too great and the soil gave way, sliding the building forward. As it moved forward, the entire structure fell over into the excavated area in front.

multistory apartment structure (Figure 13.8). The combination of removing soil that was holding back lateral pressure from the building foundation, plus additional lateral pressure from the excavated soil piled on top of earth behind the building, led to the collapse. With the extra load on earth that was already weakened from rain saturation, lateral displacement caused the entire building to slide forward and snap the foundation pilings, then collapsing entirely into the excavated area.

13.3.7 Structural Redundancy

Some of the most spectacular failures take place in structures lacking redundancy. Statical indeterminacy is a necessary but not a sufficient condition of redundancy (see Section 4.2). For example, if a uniformly loaded fixed beam (Figure 13.9a) were to fail at one of its support sections, indeterminacy might theoretically allow it to carry the load as a cantilever (Figure 13.9b). But if it was originally designed to resist the largest bending moments in a fixed beam, which usually occur at its ends (see Section 7.3), the stresses at the cantilever support would increase by a factor of 6 and collapse the beam. If the midspan section of the same beam failed and it carried the load as two cantilevers half its length (Figure 13.9c), the stresses at the supports of these cantilevers would be 50 per cent above the maximum original stresses and dangerously near collapse, since, on an average, safety factors take into consideration a load increase of the order of 67 per cent.

Redundancy thus requires providing additional structural resistance in case of failure, particularly when the loads on a structure are supported by a large number of identical elements, as in space frames (see Section 10.9). The roof of

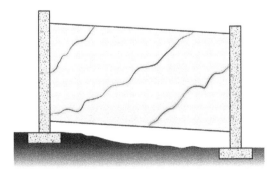

Figure 13.7 Settlement cracks in curtain wall or partition

Differential settlement along a building face will lead to shear cracks developing from tension principal stresses.

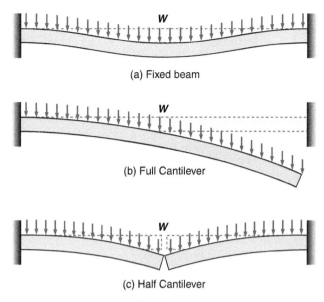

(a) Fixed beam

(b) Full Cantilever

(c) Half Cantilever

Figure 13.9   Redundancy in fixed beams

If a fixed-ended beam (13.9a) were to have a failure at one end, the structure would carry the load as a cantilevered beam (13.9b). The stresses at the remaining support would be six times higher than the original fixed-end beam. If instead the central portion of the structure failed, the remaining structure would act as two facing cantilevers (13.9c). The stresses in each of these cantilevers would be 50 percent higher than the original design stress. This has nearly used up the entire safety factor, but the structure may still have enough reserve capacity to stand, though just barely. Structural redundancy in it has allowed a redistribution of forces that would not be possible with a simply supported beam.

the Kansas City (Missouri) Arena was hung by means of 48 steel hangers from three external pipe frames (Figure 13.10). When one of the hanger connections failed during an exceptional rainstorm, the load supported by that hanger was transferred to adjoining ones that should have been able to carry the additional load. Since they were unable to do so, a chain reaction developed that collapsed a large section of the

roof, luckily without loss of life because the arena was unoccupied at the time. A similar chain reaction of failing hangers was said to cause the collapse of a suspended ceiling at the Jersey City terminal of the PATH railroad, causing two deaths. It must be noted that tensile hangers work in simple tension, distributing the load uniformly over their cross section (see Section 5.1): Hence, they do not have the reserve of strength derived from stress redistribution at a section, typical of bending action (see Section 9.3).

Ultimately, a chain is only as strong as its weakest link. The load hanging from a chain is channeled from its support along a single line as a tension member. In the case of a hanging weight, this "single-load path" (SLP) is a straight line. If a link fails, the weight falls. A chain hanging form two supports may support a distributed load along a curved line (Figures 6.6c and 6.7a). The curved chain is also an SLP structure.

The Silver Bridge connecting Point Pleasant, West Virginia, and Gallipolis, Ohio, was such a chain suspension bridge (Figure 13.11a). In this case, the links consisted of eyebars pinned to each other at their ends (Figure 13.11b). Stress corrosion combined with metal fatigue broke one of these links, causing an unstoppable progressive collapse of the bridge (Figure 13.11c).

Modern suspension bridges are supported by multistand cables, the strands sharing the load in this case as a "multiple-load path" (MLP) or "redundant" structure. If some strands fail, their share of the load is redistributed to other strands. As a consequence, the structure survives until all strands are broken.

A simple frame consisting of four bars forms an unstable parallelogram and moves under the application of a load (Figure 13.12a). With the addition of a diagonal bar, the truss is stabilized and becomes an SLP. The loads are channeled from one support along the bars to the other support (Figure 13.12b). Should any one of the bars fail, however, the truss becomes unstable or will collapse.

By adding one more diagonal, the truss will be an MLP or a redundant structure (Figure 13.12 c). In this case, the

Figure 13.10   Hanging roof of Kansas City Arena

The roof of the Kansas City Arena was suspended below exterior trusses. A lack of redundancy caused a progressive collapse of the roof structure when one of the suspension hangers failed during a heavy rainstorm. The load was transferred to adjacent hangers, which were then overloaded and subsequently failed, setting up a chain reaction of failure of the hangers that collapsed a large portion of the roof.

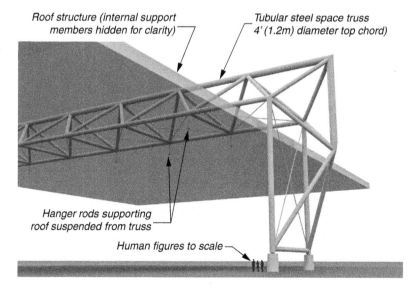

Roof structure (internal support members hidden for clarity)

Tubular steel space truss 4' (1.2m) diameter top chord)

Hanger rods supporting roof suspended from truss

Human figures to scale

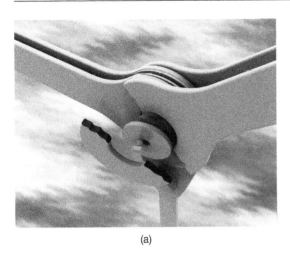

(a)   (b)

Figure 13.11 Failure of the Silver Bridge

The Silver Bridge was a chain link suspension bridge, with parallel rows of eyebars pinned together to form a suspension chain. One of the eyebars failed due to a combination of corrosion and stress. With a lack of redundancy to redistribute the load, the force in this element was then transferred to the remaining ones, precipitating a failure of the entire joint. The entire suspension structure was thus compromised and the entire bridge collapsed to the river below.

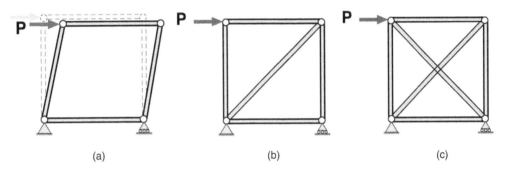

(a)   (b)   (c)

Figure 13.12 Truss Determinacy

A simple four-bar frame with pinned connections is unstable against a lateral force (13.12a). The addition of one diagonal member makes the structure stable, and it becomes a single-load path (SLP) structure (13.12b). With the addition of another diagonal, another path for the loads is provided, and it becomes a multiple-load path (MLP) structure with one redundant member (13.12c).

failure of any one bar will transfer its share of the load to the remaining bars and the truss reverts to an SLP structure.

A truss usually is supported by three reaction components, "r" (two vertical reactions at the supports to carry the load and a horizontal one to prevent sideways motion, thus $r = 3$). The number of bars, "b," and the number of joints, "j", are related by a simple formula: $b + 3 = 2j$, for a stable SLP structure (statically determinate). If $b + 3 < 2j$, this forms an unstable truss. For $b + 3 > 2j$, a stable MLP truss is created (which is, in fact, an indeterminate structure). Figure 6.30a shows a collapsed truss while Figures 6.30 b, 6.31 c, 6.31-6.35, and 6.40 b are examples of stable SLP trusses.

While most trusses are usually built as (MLP) or redundant structures, the stiffening truss of the Silver Bridge that carried the roadway was also an SLP structure. Thus, in the worst of all combinations, when the supporting SLP eyebar chain failed, the SLP roadway truss collapsed as well. A structure that had taken years to design and construct was destroyed in less than 30 seconds, killing 46 people.

13.3.8 Buckling and Redundancy

Redundancy is particularly needed whenever a chain reaction may be started by buckling. It was noted in Section 5.2 that buckling is a sudden phenomenon that occurs without warning and is usually followed by failure. The particular case of buckling involving torsion, called lateral buckling, presents the same dangerous characteristics. The published results of the structural investigations on the collapse of the space frame roof of the Hartford Civic Center in Hartford, Connecticut (spanning 300 by 270 feet (91 × 82 m) and supported on four pylons (Figure 13.13)), details the failure mechanism. The collapse of the roof during a heavy snowstorm was started by the buckling failures of an unbraced compression bar near the boundary of the frame. When the load supported by this bar was transferred to the adjoining bars, these in turn failed in buckling, precipitating a chain reaction collapse. In a matter of minutes, the 1,500 tons of steel of the entire roof dropped to the floor of the empty hockey rink—mercifully, just hours

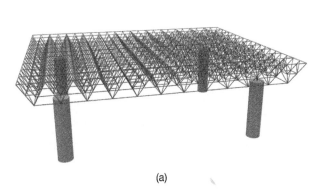

(a)

(b)

Figure 13.13 Space frame of Hartford Civic Center

The roof structure suffered a catastrophic collapse during a heavy snowstorm. The buckling failure of a single member in the nonredundant structure precipitated a chain reaction that brought down the entire roof structure.

(b) Photo: Bob Child/Ap Images

after an event had finished, thus avoiding what could have been a tragic loss of hundreds or even thousands of lives. It is not clear whether axial or torsional buckling was responsible for this chain reaction collapse.

In another example, a reticulated dome of light pipe bars (see Section 12.13), over 300 feet (91m) in diameter, collapsed in Bucharest (Romania) in the 1960s under an exceptional snowfall. The collapse was not due so much to the heavy snow load as to the weak buckling capacity of the dome, one of whose sections "snapped through," that is, caved in. The dome was rebuilt, American style, with rolled section bars, giving it greater buckling capacity.

13.3.9 Connections

As shown in Sections 9.1 and 9.2, the states of stress in the connections of structural elements are particularly complex. Hence, it will come as no surprise to the reader that they are often the source of failures. Codes and manuals give criteria for the design of connections based not only on refined mathematical stress analyses but also on series of tests on models or full-size connections. Since, because of their complexity, failures of connections occur more frequently than those of structural members, connections are designed with high coefficients of safety. This is particularly true of space frames whose connectors may have complex shapes and stress distributions (see Figure 10.48).

13.3.10 Unexpected Loads

Failures have often been attributed to unexpected natural or human-made phenomena, such as hurricanes, snowstorms, tsunamis, and fires or explosions. These do indeed bring havoc to all kinds of small and large structures, but only when the structures were not designed to resist them, when forces reach new highs (a common occurrence), or when engineers forgot the lessons of the past. The collapse of the

Tacoma Narrows Bridge in Washington State in 1940 (see Section 2.6) was identical with that of the Wheeling suspension bridge over the Ohio River in 1854 (at the time the world's longest suspension bridge), but the memory of this disaster, vividly described in the local press, had vanished. In these cases, wind-induced aerodynamic oscillations of gradually increasing amplitudes caused the collapse. In the case of the Tacoma Narrows Bridge, the great flexibility of the structure and the lack of appropriate wind bracing contributed to the failure. Descriptions of similar collapses of wooden suspension bridges had appeared in the British press early in the nineteenth century, but had also been forgotten, allowing the repetition of past disasters.

13.3.11 Ponding

In more recent collapses of large structures, the loads due to natural forces did not reach the design values recommended by the codes and must be attributed to faulty design. For example, the collapse of large flat roofs subjected to rainstorms is often due to a chain reaction phenomenon called "ponding." The flat roof caves under the weight of the rain, acquiring a concave shape that does not allow the water to reach the level of the drains. The increasing weight of rain accelerates the caving of the roof, which is transformed into a pond containing an increasing load of water that eventually may collapse the roof (Figure 13.14). Appropriate roof slope to promote runoff and appropriate distribution of drains at correct levels prevent ponding. Flat-roofed buildings are also typically designed with overflow scuppers at their roof perimeter as another safeguard against ponding.

13.3.12 Other Causes of Failure

The human-made causes of failures, such as fires and explosions, can be as damaging as those due to natural phenomena. A high-rise industrial building in New York City had two

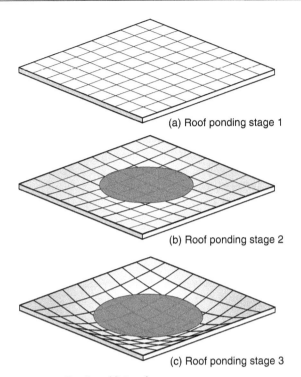

(a) Roof ponding stage 1

(b) Roof ponding stage 2

(c) Roof ponding stage 3

Figure 13.14 Ponding of flat roof
Proper slope and drainage must be provided on flat roofs to avoid the phenomenon of *ponding*. In a ponding situation, the weight of rainwater causes the roof structure to deflect under load. The deflection causes an ever-deepening, widening, and increasingly heavy pool of water to collect. With no way to remove the water, the cycle continues, and in the worst case the roof structure fails in overload. Numerous cases of ponding failures have been documented.

of its facades blown out from top to bottom by a bizarre sequence of events attributable to human error. A cylindrical, steel, compressed air tank, 10 feet (3m) high and 2 feet (0.6m) in diameter, exploded in the middle of the night, probably due to a deficiency in the tank or the malfunctioning of a pressure release valve. The tank's explosion shot up its domed top against an overhead gas pipe that sprang a leak under the impact. The gas would have been confined to the floor where the tank was located were it not for the alternate motion of the elevators in the building that sucked the gas up to the high floors and pushed it down to the low floors in a pumping action, thus spreading gas throughout the building. When in the early morning an occupant of the building put a match to his cigarette, the mixture of gas and air exploded, destroying the two facades, whose debris damaged buildings across the adjoining streets and hurt a number of their occupants.

This brief review of some of the most common causes of structural failures illustrates their great variety and suggests the responsibility of the structural engineer in the design of important structures. It will be shown in the following sections that the involvement of the other members of the construction team, and particularly of the architect, is as great as that of the engineer.

13.4 FAULTS IN COORDINATION AND SUPERVISION

The construction team of a large modern structure consists of a large number of specialists: the architect in charge and his or her team of design architects, project architect, landscape architect, and interior designers; the structural engineer and team; the mechanical engineer and team; the electrical engineer and team; the plumbing engineer and team; the soils engineer; the environmental designer; the contractor and team of specialty subcontractors; the owners' representatives; and a variety of other experts on curtain walls, roofing, costing, scheduling, and materials. Last but not least is the construction manager, who takes responsibility for the feasibility of the design and the coordination of all trades, acts as the owner's agent, and directs the execution of the job with the help of the scheduling experts, responsible for the sequence of operations and the delivery of materials. At the frequent general meetings on the progress of construction, especially on larger projects, it is not unusual to have thirty or more experts around the table, fighting their personal battles in the interest of the expeditious erection of the building.

The smooth coordination between all the facets of construction requires cooperative communication between the members of the team and implies a knowledge of the basic requirements of the other trades on the part of the representatives of each trade. The architect, uniquely positioned as team leader, must be conversant with the needs of all the trades and capable of settling the disputes arising from the mutually conflicting requirements of the consultants. In this he or she is helped by the construction manager, whose job demands practical experience and a capability for conciliation and compromise within narrow limits of acceptability. The construction documents (drawings, specifications, and contracts) are of the utmost importance in avoiding misunderstandings and mistaken interpretations that only too often lead to delays and failures. Whether a building contractor is chosen at the start of the design or by bids on contract documents, he or she must be given documents that leave little doubt as to the meaning of what is to be delivered. The time and effort spent on these documents avoids failures and costly litigation (see Section 13.6).

In spite of the interest and goodwill of the members of the team, unavoidable conflicts will arise during construction and clarifications or change orders will be issued. In this phase of construction, competent supervision becomes essential on the part of the job's superintendent, the commander of the contractor's crews, and of the architect's and/or owner's representative. It is their duty to ensure that the building is erected in strict accordance with the construction documents, since even a minor deviation from the design may be the cause of catastrophic failures. Besides any tendency on the part of the contractor to cut corners in order to maintain the schedule or to enhance profit, involuntary human error may be responsible for such mistakes. In this context, it suffices to mention the consequences of the misplacement of the reinforcement in concrete structures, the deviation from a prescribed sequence of operations in welding of steel connections, the tightening of connections that should allow thermal movements, or the improper curing of poured concrete. These are a few among the many operations that must be carefully monitored to eliminate possible failures.

Faulty rebar placement was to blame for halting construction in 2008 of what was to have been the 49-story Harmon Hotel/Condo tower in Las Vegas, Nevada. Work on the structure was stopped at the 28th floor when it was discovered that 14 floors had been constructed with reinforcing steel improperly installed in the columns. Weidlinger Engineers determined that the building (now with structurally deficient columns) would be unsafe during a major earthquake, even at a reduced height of 28 stories. After years of litigation between the owner and contractor, an agreement was reached and the building was completely torn down in 2015.

Many of the structural elements, such as steel beams and columns or prefabricated beams, columns, and slabs of concrete (often prestressed), come to the site from a fabrication shop. A check of these elements for dimensional compliance with the design is necessary to make sure that any discrepancies are within prescribed tolerances, lest the erectors be compelled to force the elements together, inducing in them dangerous locked-in stresses (see Section 2.5).

The architect and the engineers must realize that there is a wide gap between a theoretical conception and its execution in the field, and that the job of the contractor may appear less creative than theirs but is certainly as complex and demanding. A harmonious collaboration between the "legislative" and "executive" branches of the construction team is a necessity if failures are to be avoided.

13.5 FAULTS IN MATERIALS

13.5.1 Steel

The most essential properties of structural steel are strength and ductility, i.e., the capacity of plastic flow, which guarantees a reduction of stress concentrations and a reserve of strength (see Section 9.3). These properties depend on the chemical composition, the thermal treatment, and the rolling of steel shapes during manufacture, and should be checked for each batch of steel used by obtaining the corresponding data from the manufacturer: They are essential not only for beams and columns but also for nuts, bolts, and other connection components. Since the 1960s, engineers have been alerted to the dangerous phenomenon of delamination due to the welding of thick-steel sections: Unless their steel is adequately treated thermally, any parts of sections thicker than 2 or 3 inches (51 to 76 mm) have a tendency to separate into thin laminae, greatly reducing their strength. Research stemming from these disasters emphasized the importance of checking the treatment of steel used in large structures.

No operation involves greater risks in steel construction than the welding of connections: The welding materials, the type of flame, the temperature, and the speed used in this operation influence dramatically the strength of the connections. Since a number of failures of steel structures are attributable to welds, most codes require welders to be duly trained and certified.

13.5.2 Concrete

Concrete properties depend on its composition, that is, the ratios of cement, sand, stone, and water used in the mixture. Besides the type of cement and the strength of the stone, one must establish the granulometry of sand and stone, i.e., the size distribution, so as to guarantee that the voids between stones are filled by the sand grains and those between the sand grains by the cement. Both stone and sand must be carefully washed to eliminate impurities. The water–cement ratio is the most important factor in determining the concrete strength: A low ratio increases the strength but makes more difficult the pouring and vibrating of the concrete; a high ratio weakens the concrete. In all projects of any importance, the concrete mixture is designed by a concrete laboratory and its compressive strength is checked after 7 and 28 days by testing cylinders of concrete taken from daily concrete batches. The following unusual episode illustrates the need of vigilance in concrete supervision. During the construction of one of the most famous air terminals in the United States, it was noticed that concrete batches gave acceptable strength results at all times except those reaching the site in the early afternoon. When all investigations failed to discover the reason for this anomaly, the design engineer decided to follow the trucks leaving the concrete plant at lunchtime, which usually reached the site an hour later. He thus discovered that the truck drivers, before stopping for lunch, poured water into the revolving drums of the trucks so as to prevent the setting of the concrete during their leisurely meal. The concrete reaching the site had the right consistency but a reduced strength due to the higher water–cement ratio, as revealed by the tests at 7 days.

Particular care must be exercised in locating the steel reinforcing bars in the concrete, but, moreover, the steel bars must be covered with zinc or epoxy whenever highly corrosive salts may percolate through the concrete, as often happens in the slabs of garages when snow-melting salts are used on roads. The architect should be aware of the dangers due to faults in the two most commonly used structural materials, but only the metallurgist and the concrete specialist can assist in the determination of their properties and thus prevent the corresponding failures.

13.5.3 New Materials and Methods of Load Estimation

The advent of new artificial materials such as plastics, glass fiber, and carbon-reinforced composites are finding increased usage in buildings, as well as in aircraft structures. The designer is required to have a thorough knowledge of their properties and methods of construction.

As discussed, the structural properties of building materials must be carefully monitored by the methods of quality control. Nevertheless, statistical variations of these properties are unavoidable. It is therefore necessary to be informed of significant parameters such as mean values and standard deviations of properties. For instance, a strength value one standard deviation (STD) below the mean indicates that about 84 percent of materials will exceed this value. For two STD, about 98 percent will be greater.

It is also apparent that loads, particularly live loads, have even greater statistical variations than material strength. While maps such as presented in Figure 2.6 indicate highest wind velocities but do not show their frequency of occurrence. Weather reporting organizations (among them the National Oceanic and Atmospheric Administration (NOAA)) are able to provide detailed statistical information for winds at U.S. locations. Frequency distributions of wind velocities may be converted into wind pressure distributions. Here, one STD above the mean indicates that 86 percent of these pressures are below this value.

Strength and pressure (stress) frequencies may be plotted on the same figure, as shown in Figure 2.12. Infrequent high stresses may overlap infrequent low strength values. The size of the overlapping region is indicative of the probability of failure. The size of this area may be reduced by reducing stresses, using larger members or increasing strength, using better materials. The ratio of mean strength to mean stress is the mean safety factor.

Probabilistic methods such as "Stress-Strength Interference" and "Load and Resistance Factors" (LRF) designs can be used to obtain the statistical distribution of safety factors also. These technics are, however, beyond the scope of this book.

13.5.4 Fatigue Failures

Another important mechanical action that can result in structural failure is metal fatigue. It is a common experience that to break a piece of wire it has to be bent back and forth repeatedly. The material is being weakened by fatigue and eventually fails.

Structural components are subjected to repeated loading, usually from loads such as wind, snow, thermal stresses, vibrating machinery, and earthquakes. These cyclic loads are usually in the elastic range and do not result in large deformations. Nevertheless, because of their cyclic nature, they produce fatigue. Structures, however, can withstand hundreds of millions of cycles without failure but a sudden, unexpected, high load may produce fracture. In the case of highway bridges, it is often the design for fatigue resistance that becomes the controlling criterion, not simply stress.

The engine of a car, running at 60 miles per hour (97 km/h) with an engine speed of 3000 RPM, will have its pistons and piston rods going through 3000 cyclic loads per mile (1.6 km). Driving 100,000 miles (160,000 km), the rods will survive some 300 million load cycles. When the car starts burning oil, friction on cylinder walls increases the stress on the rods and they may break suddenly.

Metal fatigue cracks usually originate in highly stressed areas such as stress concentrations around rivet holes and sharp corners (Chapter 9). A dramatic aircraft failure is attributed to fatigue crack propagation along a series of rivet holes in the fuselage of a Boeing 737 Airliner owned by Aloha Airlines (Figure 13.15). A large portion of the cabin's roof flew off during flight. All passengers landed safely but a flight attendant was killed in the accident. Repeated pressurization–depressurization of the fuselage produced cyclic tensile stresses and fatigue.

Figure 13.15 Aloha Airlines Flight 234 Fuselage Failure

Fatigue cracking leading to a sudden failure along a series of rivet holes in the forward section of the plane caused the fuselage to rip open in midair. Although one flight attendant was killed, remarkably the plane was still able to land safely with all passengers.

Photo courtesy of National Transportation Safety Board

Corrosion caused by water, acid rain, and other atmospheric conditions roughen unprotected metal surfaces and also cause pockmarks that can create stress concentrations. Combined with repeated loading the component may be subjected to "stress-corrosion-fatigue." One of the eyebar links of the chain supporting the Silver Bridge (see "Structural Redundancy in Section 13.3 and Figure 13.11) failed due to stress corrosion fatigue. The chain links held to each other with flexible joints that were covered by large washers that prevented painting and allowed corrosion to take place. In addition, truck sizes and weights nearly doubled through the years since the construction of the bridge. The resulting cyclic load increase cracked the link, leading to the previously addressed progressive and complete failure.

13.6 CONSEQUENCES OF STRUCTURAL FAILURES

Structural failures result at times in bodily injury and loss of life. They are always the cause of financial losses, damaged professional standing, and even more dire consequences for one or more members of the construction team. Unless amicably settled among the parties involved, major collapses as well as minor failures lead to litigation through the courts or arbitration proceedings and, since the architect's technical consultants are often his subcontractors, the architect becomes legally and financially involved in litigation. The architect is then compelled to sue his consultants and the contractor, and the contractor, in turn, to sue the entire architectural team. The owner, who certainly has no responsibility for the failure, is usually the originator of these complex, prolonged, and costly maneuvers, in which each party is supported by a group of experts. It is not unusual for these legal proceedings to last long periods of time (sometimes years) and for the requests of damages to reach large amounts of money even when punitive damages, attributable to bad faith or fraud, are not involved. In few human endeavors is an ounce of prevention worth so many pounds of cure.

It may seem strange to read in a book on structures the recommendation that the architect should use, besides the services of competent technical consultants, those of a legal advisor. But the realities of the construction world prove that an attorney's advice on the writing of contracts, the preparation of construction documents, the approval of consultants' drawings, and, in general, on all written communications, is particularly important for the professional architect. The recommendation may even be illuminating for the layperson who reads in the daily press stories about the human, financial, and professional consequences of structural failures. It should be no wonder to him or her that forensic engineering, the art and science of expert testimony in court, has become a recognized specialty of the engineering profession. This should be no surprise when one learns, for example in the case of the Hyatt Regency Hotel in Kansas City, Missouri (see Section 13.3), that the plaintiffs' demanded damages amounting to $3 billion as a result of the collapse of the walkways and that the designing engineers of this project were found guilty of gross negligence and unprofessional conduct in the practice of engineering, and all lost their license to practice.

Is there any remedy for the occurrence of structural failures in our technological era? A measure that has this aim in mind (which has become almost universal now in the United States) is mandatory continuing education before having one's professional licenses renewed. Professional engineering societies and schools of engineering offer such courses, for example. Engineering companies and offices may offer them in-house or require their engineers to attend the courses given at universities, as well as those provided by professional societies at conferences across the United States and internationally. An ever-growing assortment of online learning tools is available as well.

The continuous and rapid evolution of the field of structures has prompted design offices to noticeably increase the use of outside consultants. Engineers are advised, particularly in important projects, to employ the use of so-called external peer reviews. These consist in requesting a reputable design firm to analyze in depth a structural design before it is submitted for execution and to point out deficiencies and suggest alternative solutions to the structural problems of the design. This valuable practice, beginning to be adopted by the best design firms in the United States, is required by law for designs of exceptional importance in France, where companies specializing in peer reviews are licensed by the state.

A number of other suggestions have been made which would require apprenticeship periods for engineering graduates and a strengthening of the state license examinations. Whatever procedure or combination of procedures may eventually be adopted, there is little doubt about the importance of updating the knowledge of the professionals in a field evolving as rapidly as that of structures, and influencing so deeply the lives of all citizens.

KEY IDEAS DEVELOPED IN THIS CHAPTER

- Structures may fail as a result of bad design, computational errors, disregard of building codes, unexpected large loads, substandard materials, and weather-related problems.
- Wind, snow, and earthquake loads require special attention due to their high variability.
- Dynamic loads and undamped vibrations frequently are the cause of structural failures in extreme wind and seismic events.
- Restrained thermal movements can cause cracks and member failure.

- Inadequate foundations may cause wall cracking, as well as the building sliding or overturning.
- Progressive collapse may occur due to a lack of redundancy and lack of strength reserve.
- Fatigue and stress corrosion may result in failure after prolonged use.
- The consequences of structural failure may be financial loss, law suits, loss of reputation, and even injury and loss of life.

QUESTIONS AND EXERCISES

1. Obtain some children's Silly Putty, normally found in the toy aisles of various stores. Try these exercises:
   a. Form it into a ball; drop it and notice how it bounces. Place the ball on the table for a few minutes and watch it deform. What is the reason for the different behaviors?
   b. Make a noodle out of the putty and pull on the ends. Watch it stretch and deform. Reshape it and start again, but pull very quickly. Is there a difference? If so, why?
   c. Reform the noodle and put in the refrigerator for half an hour. Pull on the putty and it will break easily. Why does it behave this way?
2. Obtain a wire coat hanger and cut out a straight part. Bend the wire 90 degrees and straighten it. Bend it repeatedly and count the number of times it was bent until it breaks. Repeat the process on another piece of wire, this time bending it into a U shape. Why was there a difference in the number of bends to failure?
3. Some wire coat hangers are made with a cardboard tube to put clothes on. Hang one of these on a coat rack and pull down on the middle of the tube until the structure fails. What is the purpose of the tube in this three bar truss? What kinds of stresses were present in each bar? What was the mechanism of failure?

FURTHER READING

Levy, Matthys and, Salvadori, Mario. *Earthquake Games: Earthquakes and Volcanoes Explained by 32 Games and Experiments.* Margaret K. McElderry Books. 1997

Levy, Matthys and, Salvadori, Mario. *Why Buildings Fall Down.* W. W. Norton & Co. New York, London. 1992.

Petrosky. Henry. *To Engineer is Human: The Role of Failure in Successful Design.* Vintage. 1992.

Ross, Stephen S. *Construction Disasters: Design Failures, Causes and Prevention.* McGraw-Hill (*Engineering News-Record Series*). 1984.

STRUCTURAL AESTHETICS

14.1 AESTHETICS AND STRUCTURES

There is no architecture without an aesthetic component, but is there an aesthetics of structure? And if there is, does it influence architecture deeply enough for architects to take an interest in it?

In answering these questions one may ignore the many definitions of "the beautiful" and notice instead that aesthetic tenets change with time: a piece of architecture considered a masterpiece at a given time is demoted to a second-rate achievement in another, and vice versa. The tenets of aesthetics vary but the satisfaction of aesthetic needs is one of the permanent aspirations of humanity.

Aesthetically satisfying buildings can be built that partially disregard structural laws: The example of the Parthenon was mentioned in this context in the very first section of this book. On the other hand, great structuralists have preached that one should not bother with aesthetics because if a building is correctly designed, beauty is bound to fall out from its structural correctness. Innumerable ugly buildings, correctly engineered, prove this theory to be baseless. Even if unconcerned with aesthetics, engineering geniuses like Maillart and Nervi designed wonderful structures because of their innate feeling for beauty (Figure 14.1).

In considering the influence of structure on architectural aesthetics, one must distinguish those buildings in which structure is relatively unimportant, and hence not uniquely determined, from those in which structure is essential. The appearance of a single-family house rarely depends on its structure, which may be of wood, stone, concrete, or steel, whereas that of a large suspension bridge is inherent in its funicular nature and requirements for span. Buildings between these two extremes have structures that, by influencing their appearance in varying degree, influence the aesthetic response of their users. By the same token, the constraints of architectural aesthetics influence in varying degrees the structural solutions adopted by the engineer. Totally satisfying aesthetic solutions can only be achieved through the interplay of both architecture and engineering.

Figure 14.1 The Salginatobel Bridge

This bridge over the Salgina Valley of Switzerland, designed by Swiss engineer Robert Maillart and completed in 1930, is among the most iconic of all bridges. Designated an International Historic Civil Engineering Landmark in 1991, the bridge is noted for its graceful form. The shape is antifunicular, the inverted shape of a suspended cable used as an arch in compression.

Photo courtesy of Rama Neko

14.2 SEMIOTIC MESSAGES

Since the early twentieth century the concept of the semiotic message of a completed building has become widely accepted by architects, architectural historians, and other specialists in the field of construction. Semiotics, a branch of philosophy developed over the last ninety years, considers any and all human activities and products from the particular viewpoint of nonverbal or sign communication. Communication consists of messages: Verbal communication is concerned with the transmission of meaning, and to this purpose uses the words of a given language. Verbal communication is totally dependent on the culture in which the language is understood. Nonverbal communication comprised of symbolic imagery, however, may convey a meaningful message that could also be verbally expressed. The semiotic message of a road sign may stand for the English message: "No parking" (Figure 14.2), but, in contrast with the verbal message, it is understood internationally through the now classic "P" within the diagonally slashed circle. On the other hand, a semiotic message may also be the by-product of an object or artifact whose main purpose is not to communicate, but to perform a function. Clothing, for example, is made to protect the human body, but semiotically it communicates status, as in the case of military uniforms or church garments.

Both kinds of semiotic messages are found in architecture: The push buttons in an elevator perform a specific function but also express semiotic messages related to their function of moving the elevator to specific floors. The message of a window, instead, communicates something beyond its intrinsic function of transmitting light and air; the barred windows of a jail say clearly: "This building is a jail," while the ornamented windows of a Renaissance palace state unequivocally the status of its owner (Figure 14.3).

It is obvious from these elementary examples that the semiotic message is as deeply embedded in a culture as the verbal message. A tribal native from an undeveloped land entering the lobby of a high-rise building would not understand the "meaning" of the elevator buttons; neither would he grasp the social significance of the windows of the Renaissance palace.

Finally, the meaning of a semiotic message, as much as that of a verbal one, changes with time: The pyramids communicated a religious message at the time of the Pharaohs, an incitement to glory to Napoleon and his troops, and a mixture of artistic, sociological, and structural meanings to the modern visitor.

Structure introduces in architecture two kinds of semiotic messages. When it is not visible, the structural component of the architectural message may not be apparent to the layperson, even if the building may depend essentially on it for its architectural message. On the other hand, buildings structured uniquely by the requirements of statics express a semiotic message that, while not independent of its architecture, is strictly related to structural action and acquires a meaning of its own.

14.3 ORIGINS OF THE STRUCTURAL MESSAGE

As introduced in Chapter 1, it can be argued that humans have an innate sense for the "appropriateness" of certain structural forms simply because it is something encoded in

Figure 14.2 Semiotic message of a road sign
While the text here is in English, the symbol itself conforms to international signage standards, and is thus understood around the globe, even if the text is in another language.

Figure 14.3 A Renaissance Palace
Palazzo Farnese in Rome, is now the French Embassy, but was originally built as a palace for the Cardinal Farnese family.

Photo courtesy of Stephen Adams

our genetic makeup. All structures must, by the simple reality of their physical existence, respond to the forces of the environment in which they originate. Wind and water are primary form givers, and gravity will have a differing impact at different scales of an object. These forces of nature thus shape the form of all natural things, from the most delicate spider web to the greatest tree. It is not difficult, then, to observe that at a quite unconscious level we all have an instinctive response to environmental forces in our own bodies. While wobbly, a three-year-old has already begun to master the balancing act of walking, a task that robotic scientists are still at an early stage of understanding in the early decades of the twenty-first century. As we get older, walking becomes so instinctive that on a windy day we somehow know how to lean into the wind and widen our stance in order to keep our balance. We somehow know how to not stand upright and skinny because we will topple over! Thus, one very important tool at our disposal for understanding structural forces is to become aware of how our own bodies are responding to the world around us. Notice that if you hold a heavy book out at arms length, it is much more difficult than holding it close to your body. This is a direct response to the principle of bending moment, introduced in Chapter 3. The higher the bending moment, the greater the force required to support the book, and thus ultimately the stronger we must be to hold the book for any length of time.

We see exactly the same principle at work in the branches of a tree. Observe how they become progressively thicker the closer they are to the trunk. At the greatest distance, we have the lightest features, the leaves or needles. But these progressively add weight as we look closer in to the trunk; and, furthermore, the long moment arm means that it is necessary for the branch to be thickest of all where it joins the tree. Were it not, the stresses experienced in bending would exceed its strength. Indeed, we witness the failure to accomplish this all of the time. Wet and heavy snowfalls or ice storms in northerly climates cause the failure of tree limbs every winter, as do windstorms, hurricanes, and tornados. We thus see in nature the suggestions for the "correct" shape of a cantilever.

Similarly, the trunks of trees bent by the wind confirm this behavior, while the shape of straight trunks introduces us to the accumulation of vertical loads.

And so it follows that these natural forms create within us an innate sense for the "appropriateness" of human-made structural forms. Purely structural messages originate in this intuitive understanding of structural action stemming both from our physical experience and the perception of structural forms in nature. Due to these intuitions we are puzzled by the sight of a Cretan column (Figure 14.4a) or the columns of Nervi's Papal Audience Hall (Figure 10.36), larger at the top than at the bottom, but accept as instinctively correct the tapered-up shape of a Doric column, (Figure 14.4b). Similarly, we consider inappropriate or even "ugly" a cantilever tapering toward its root (Figure 14.5a) because it contradicts a structural behavior we have seen in nature since the beginning of the human race (Figure 14.5b).

A similar reaction of shock occurs when we are confronted with a large mass in the shape of an inverted pyramid. Due to the action of gravity, mountains are shaped as right-side-up pyramids, and the Egyptian pyramids have an idealized geometrical shape identical to that of the mountains. A modern building progressively cantilevered upward (Figure 14.6a) does not "say" to the layperson how it stands up: It suggests that some "trick" was used to achieve an "unnatural" result, arousing in us a sense of uneasy surprise rather than the feeling of balance expressed by an "honest" static behavior.

Natural arches have taught us that stone spanning a gap must acquire an upward curvature because this material, strong enough in compression to support a mountain, is weak in tension. This intuitive understanding of arch action can be extended to three-dimensional structures, like domes, by referring to our experience of living in caves. Caves do not explain the required thickness of domes, but this cannot be appreciated either from their inside or from their outside, anyway.

Similarly, the vine catenaries linking tree-to-tree show us the need of downward curvatures in tensile structures, a feature that even modern engineering must use in suspension bridges.

Figure 14.4a and 14.4b
Cretan (left) and Doric (right) columns

To the architecturally untrained eye, the Cretan column may be considered to have an odd appearance with the heavier end at the top, in contrast with the Doric column being fatter at the bottom and appearing more natural.

(a) Photo: Arkady Chubykin/ Fotolia; (b) Photo: Olga Drabovich/ Shutterstock

(a)

(b)

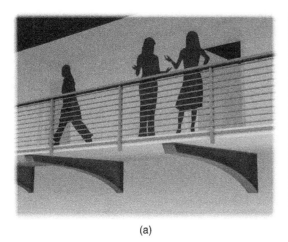
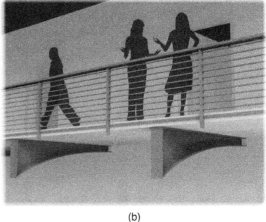

(a)   (b)

Figure 14.5a (left): "Correct" and 14.5b (right) "Incorrect" cantilevers
There is an instinctive feeling that the cantilever on the right somehow has an awkward and unnatural form.

Figure 14.6a Tempe, Arizona, City Hall (top), and 14.6b: the Great Pyramid of Cheops (bottom)
This inverted pyramid building (a) defies our common experience. While the inverted pyramid shape provides shade for the severe southwest sun, we nevertheless expect a structure to be heavier at the base in the same manner that we expect a mountain to be larger at the base. Compare this with the quintessential form and stability suggested by the ancient Egyptian Pyramid (b).

(a) Photo courtesy of Deborah Oakley; (b) Photo: Rahmo/Shutterstock

It is thus seen that, to those of us who are not structuralists, the structural message derives from atavistic experiences whose accumulation often results in aesthetic reactions. This is why the layperson considers a correctly designed cantilever (see Figure 14.5a) "elegant" and "dynamic," while a simply supported beam (Figure 14.7), correctly designed with a larger depth at midspan, is hard to understand and may be considered "ugly" by the layperson, who has seldom if ever seen such a beam in nature.

On the contrary, most natural beams are rigidly connected to their supports and have the visual profile of shallow arches, so that a haunched beam (Figure 14.8) looks "right" to the inexperienced eye. Even the expert would agree with this instinctive evaluation because he or she

Figure 14.7 A simply supported beam
Shaped for efficiency in bending may actually appear ungainly to the untrained eye.

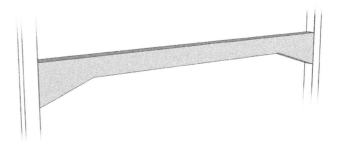

Figure 14.8 A haunched built-in beam
Also shaped for efficiency in bending, this form looks more like what we might expect to see in nature.

Figure 14.9 A highly optimized concrete beam form

Utilizing flexible fabric formwork, research professor Mark West has created beams shaped for high efficiency that approximate the moment curve of the span. This early prototype structural member exhibits an organic gracefulness of form that also minimizes material. Such moment-shaped beam can use 30% less material than conventional rectangular beams. The diagram below the photograph illustrates the bending moment of the span and the corresponding optimized beam shape.

Photo courtesy of Mark West

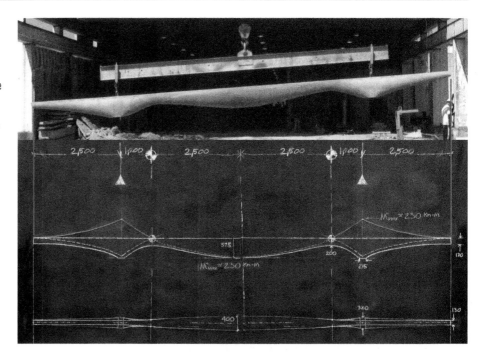

realizes that, as shown by the compressive principal stress lines in a simply supported beam (see Figure 9.9), arch action takes over when a beam of reinforced concrete is overloaded. It would seem that in this case the human eye is conditioned to see the "ultimate" shape of the beam, following the basic principle that a structure will not collapse until it really cannot help it.

Recent research in fabrication techniques has resulted in concrete beams formed in a more optimal shape that follows the flow of forces in a beam (Figure 14.9). The resulting form may seem to many to have a graceful, biomophic appearance. We instinctively gravitate toward organic forms of nature, so will structures designed with these techniques receive a universally positive reception? Only history will bear this out.

With the production of strong, inexpensive steels, frames hinged at their foot (Figure 14.10) made their appearance on the structural scene in the nineteenth century, but because their structural behavior is complex, their message is still equivocal to the layperson. While the curved shape of the arch has a strong aesthetic charge, a frame, if correctly proportioned, can be aesthetically pleasing too.

Similar reactions are elicited by those correct structural forms that have their justification in subtle physical phenomena. For example, a compressed member with a shape dictated to avoid buckling (Figure 14.11) is looked upon as a component of a machine and is hardly viewed as beautiful or ugly. In fact, for a long time, machines were considered ugly because, while their shapes are correct functionally and structurally, they could not be accepted as elements of the universe of aesthetics. Some revolutionary painters, like Picabia, did introduce machines in their oeuvre, but were careful to disassociate their images from the reality of the factory: By painting machines that did not work (Figure 14.12), Picabia made them symbols of a new order rather than real objects whose visual content was related to their role in the outside world.

The same kind of consideration applies to the three-dimensional world of structures. In nature, concave spatial elements are quite common: Seashells are not only symbols of protection, but they have a strong aesthetic content whenever ribbed in one of the ways nature strengthens them. The analogy of the large dome to the mysterious sky might fall in the same category.

14.4 SCALE AND THE STRUCTURAL MESSAGE

The semiotic message of human-made structures is not influenced by scale because the message refers to common experiences of the race that have to do with form and not size. The comparison of the dome to either the seashell or the sky is significant in this context, as is that of large tensile structures to a spiderweb, which always elicits feelings of surprise. Because of the extreme efficiency of tension fields, tensile structures, whatever their geometry and size,

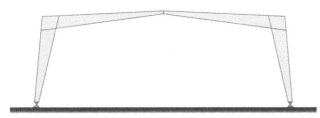

Figure 14.10 A hinged steel frame

Introduced in the 19th century, such structures have no real counterpart in nature, but are fundamental structural components in many building structures.

Figure 14.11 **Entry of the Lloyd D. George Courthouse in Las Vegas, Nevada**

This massive column is shaped to place the most material where the structure is most prone to buckling, lending a graceful curvature to the form.

Photo courtesy of Deborah Oakley

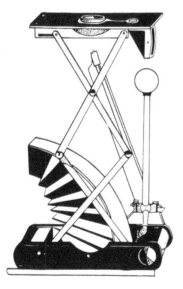

Figure 14.12 **A drawing after Francis Picabia's work** *here, this Is Stieglitz here* (1915)

The Dada artist created a number of "machinist" paintings, some of which seem to have no true function. This image depicts a broken bellows camera with an attached automobile brake from the period.

Photo courtesy of Deborah Oakley

Figure 14.13a **(Above): A spiderweb**
Figure 14.13b (Below): 1972 Munich Olympic Stadium

The lightweight spiderweb is echoed in the contemporary form-active shape of the Olympic stadium roof, conceived by noted architect Günther Behnisch and famed architect-engineer Frei Otto. Like the spider's web, the Olympic stadium epitomizes the minimal material and maximum effect of tensile structures. Prompted by the need to maximize daylighting for television coverage, the roof surface is created by an array of tiled acrylic panels, between each of which the seams consist of flexible gutters to carry away rainwater and allow for structural movement.

(a) Photo: Larry Ye/Shutterstock; (b) Photo courtesy of Terri Meyer Boake

are also light in appearance and considered intriguing and beautiful (Figure 14.13). Besides, they have shapes dictated by the loads they carry and imply to the reader of their message a correctness not obvious in compression or bending structures. If, following Gaudí's example, one uses an antifunicular (upside-down) compressive shape dictated by the corresponding tensile funicular shape (Figure 14.14), one finds that these correct shapes do not always communicate a message of beauty. On the other hand, the particularly beautiful shape of the cables of a suspension bridge becomes by inversion that of an arch shape (Figure 14.15), whose message of beauty is obvious because it is often seen in nature.

One may object that no domed structure is lighter than a cable-reinforced, pneumatic roof (see Figure 11.19) and should look elegant even to the inexperienced eye. Instead,

Figure 14.14a   (top) and Figure 14.14b (bottom)
Spanish architect Antonio Gaudi was among the first to explore the use of funicular form in compressive structures. Shown here is the model for the Güell Colony Chapel in Santa Coloma de Cervelló, near Barcelona, Spain. A series of cables supporting sacks of buckshot as weight (14.14b, bottom) was used to find the antifunicular form of the shell structure (14.14a, top) by imaging the structure upside-down. Figure 14.14a is a direct inversion of Figure 14.14b, and therefore represents a compressive structure. Gaudi made use of mirrors for this purpose, sketching his design concept directly from the inverted image of hanging model.

Photo courtesy of Terri Meyer Boake

balloon structures are so far considered unattractive because they spell heaviness: A comparison with the traditional message of the stone dome in compression does not allow us to "understand" as yet their structural action and confuses us structurally and aesthetically. Such a confusion never arises in tents because these tensile structures have their counterparts in nature, although at a reduced scale.

14.5 AESTHETICS AND STRUCTURAL "CORRECTNESS"

An understanding of structural behavior is seldom needed for aesthetic appreciation (see Section 14.1). A striking example of this unimportance is given by the general admiration for a roof shape seldom understood structurally. The hyperbolic paraboloid, one of the most efficient structural roof forms when correctly supported (see Section 12.12), is characterized by a saddle shape often encountered in nature and in art: Whether used correctly horizontally or vertically (Figure 14.16), the message of the hyperbolic paraboloid is always one of beauty. The reaction of a 12-year-old when first seeing this form indicates the associations dictated by its message: After comparing it to a horse saddle, he felt that it also looked like a bird in flight.

The roof structure of figure 14.17 clearly illustrates the tensions that can exist between structural form and aesthetic desires. From a structural standpoint, although fabricated as a truss, it is in reality a simply supported beam with maximum moment at midspan. For structural optimization, the greatest beam depth must therefore be at the middle of the span and then tapering thinner at its ends; however, in this case it has been shaped in the form of an arch with the least depth at midspan and greatest at the ends, even though technically speaking no arching behavior is occurring. We witness here the aesthetic desire for the arch shape taking precedence over "correct" structural form.

In some cases, an unconscious understanding of structural behavior may enhance aesthetic appreciation. The ribs of a slab oriented along the lines of the principal bending moments (see Figure 10.31) become a source of aesthetic satisfaction even to those who have no idea of plate theory. Similarly, the compressive lines of principal stress in a beam (see Figure 9.9) express its behavior in terms of the much more easily understood behavior of arches and cables and constitute a pleasing pattern.

One may wonder whether the puzzling message of certain structures will ever lead to their aesthetic acceptance. The shape of prestressed concrete elements governed by the tension in their invisible tendons (see Figure 3.10) may contradict human intuition. One may ask how future generations may react to masses magnetically floated in space that, for lack of reference in nature, will seem to defy some of nature's basic laws, while the airplane looks elegant like a soaring bird. We are already witness to the "improbability" of structures such as the CCTV

Figure 14.15a (top): "25th of April" Suspension Bridge crossing the Tagas River in Lisbon, Portugal, and **Figure 14.15b** (bottom): The Žďákov Bridge spanning the Vlatava River in the Czech Republic.

As with Gaudi's funicular form model above and its corresponding antifunicular compressive shell, we see in these two structures the clear analogous relationship between tensile and compressive forms. The form of the tension cable, inverted, becomes the form of the compressive arch.

(top) Photo: Sergii Figurnyi/Fotolia; (bottom) Photo: David Maska /123RF

(a)   (b)

Figure 14.16a (left) and **14.16b** (right): **Hyperbolic Paraboloid Roof Structures**

The form may be used in multiples horizontally as in the St. Aloysius Parish in Jackson, New Jersey, (14.16a) or vertically as in the Cathedral of Saint Mary of the Assumption in San Francisco, California (14.16b). The soaring interior space is one of only a handful of U.S. projects by the renowned Italian engineering genius, Pier Luigi Nervi. See also figure 3.18 for an interior view of the roof structure.

(a) Photo courtesy of Erdy McHenry Architecture and Alan Schindler; (b) Photo courtesy of Deborah Oakley

Figure 14.17   Structurally "Incorrect" Beams

Looks can be deceiving. While these roof truss structures have been created in an arching form, they are in reality simply supported beams. This can be seen by the lack of connection between the top and bottom flanges of the truss and the supports—only the webs are bolted.
As introduced in Chapter 5, the greatest forces in bending will therefore occur at midspan, in this case precisely where the structure has the least depth. This is an instance where the desire for a particular architectural appearance has overridden structural efficiency. What to the layperson is a nod to familiarity is to the structural purist an act of formal impropriety. Is one standpoint right and the other wrong? How do you react?

Photo courtesy of Deborah Oakley

Figure 14.18   The Eiffel Tower

At first decried as a mechanistic eyesore by the establishment, the Tower rapidly won the hearts of Parisians and became the very symbol of Paris, and arguably of all of France and is now clearly recognized by individuals from all corners of the world

Photo: WDG Photo/Shutterstock

Tower in Beijing (see Figure 1.7) that utterly contradict our intuitive understanding of structural behavior. The marvel of such structures is the element of surprise generated by their having no analog in nature.

It is the harmony between the visual needs of beauty and the respect of natural laws that dictated in the past and dictates today the exhibition of a building structure. Our admiration for the vaults of the Gothic cathedrals (see Figure 8.36) is the same as that for the John Hancock Company building in Chicago (see Figure 8.23).

14.6 THE MESSAGE OF STRUCTURE

It was shown in the previous sections that the semiotic message of a structure is influenced by our personal experience and the cultural experience of our society. The relative importance of these two factors is illuminated by the classic example of the Eiffel Tower in Paris (Figure 14.18). This extraordinary iron structure, designed and erected by an engineer of genius, Gustave Eiffel on the occasion of the exposition of 1889, had the utilitarian purpose of attracting visitors to the exposition and the patriotic goal of celebrating the one-hundredth anniversary of the French revolution. It was to be dismantled at the closing of the exposition. The campaign against its erection involved some of the most respected representatives of French

culture, including famous writers, poets, painters, and politicians who were incensed by the "ugliness" of the Tower. But, as is often the case with supposedly demountable structures, the Eiffel Tower was not demounted and, only 23 years after its erection, became the theme of a famous series of paintings by Delaunay, showing its acceptance from a purely aesthetic viewpoint. It did not take much longer for the tower to become not just one of the sights of that center of world culture called Paris, but its very symbol. And a few years later, the tower, all on its own structural steam, became the semiotic symbol of France. In this extraordinary case, the semiotic message stems directly and uniquely from a structure: The Eiffel Tower is a masterpiece in which nothing was conceded to decoration and nothing was used to hide its necessary sinews. Its acceptance indicates the amazing fact that a pure structure can communicate a complex symbolic message. A similar interaction between structural and aesthetic messages led to the nakedness of the towers of the George Washington

(a)   (b)

Figure 14.19a The George Washington Bridge

The proposed clad (14.19a) and, as actually constructed, unclad (14.19b) George Washington Bridge in New York City. When constructed in 1931, exposing the structural frame of a building or bridge was considered by many to be an undesirable form. Erected during the start of the Great Depression, the cost-saving measure of omitting the cladding led to a new interpretation of exposed structure as a formal expression. When completed, it was the longest suspension bridge in the world, a title that it held for only five years until San Francisco's Golden Gate Bridge opened in 1937.

Photo and artistic rendering courtesy of Max Touhey, Gothamphotog.com

Figure 14.20 The Centre Georges Pompidou

Completed in the late 1970s, the building's exposed structural and mechanical systems caused quite a stir and revolutionized the concept of what an art museum should look like. For those who know how to "read" a building technologically, however, the exposed building systems enable another layer of understanding the building. The various systems on the exterior of the building are painted different colors to distinguish their unique roles, such as intake and exhaust ducts, stairs, and elevators. To this day, the structure remains simultaneously beloved and controversial, and is one of the most visited museums in the world.

Photo: PaylessImages/123rf

Bridge, against the opposition of a large part of the New York intelligentsia and of the designing engineer himself, who wanted them clad in stone (Figure 14.19), and is another indication of the rapid change in cultural meaning of the semiotic message of structure.

The Pompidou Centre Museum (Figure 14.20) was inaugurated in Paris in 1977 a few years ago. The dismay of the art world at the erection of an art shelter, whose aesthetic message is based not only on its structure but also on its mechanical systems, must be accepted in the light of our recent past. One cannot forecast that the Pompidou Centre will become the new symbol of Paris or of modern art, but one should not be surprised if the incorporation of mechanical elements into its aesthetic message were to lead to a widening of the vocabulary of architecture and become accepted as a matter of course a few generations or even a few years hence. Art and technology are thus seen to be two facets of a culture and not two incompatible aspects of human activity.

KEY IDEAS DEVELOPED IN THIS CHAPTER

- Aesthetics of structures depend on the experience of the viewer.
- Structures that obey their natural surroundings may be more satisfying than the ones defying them.

- The shape and size of a building may indicate its purpose semiotically.
- Buildings constructed for aesthetic appeal do not always satisfy structural requirements
- Structurally correct buildings are usually pleasing to the observer.
- Structures with unusual geometric shapes indicate modernity.

QUESTIONS AND EXERCISES

1. Compare structures in nature with those in the architecturally built world. Do you find analogs in human-made structures with those in nature? An obvious example is a tree limb and a cantilevered balcony. What others can you observe?

2. Identify two structures, one that you consider to be structurally pleasing and a second that is not. What are the characteristics that differentiate them? Is the "structurally pleasing" one demonstrating visible principles of structure (e.g., a funicular hanging cable or a net-like structure)? Is the "structurally nonpleasing" one somehow in violation of structural principles or is it perhaps a matter of poor

construction? Try to become aware of what exactly it is that makes a structure aesthetically pleasing to you.

FURTHER READING

Anderson, Stanford, editor, *Eladio Dieste: Innovation in Structural Art*, 2004. Princeton Architectural Press.

Billington, David, P. *The Art of Structural Design: A Swiss Legacy*. Princeton University Art Museum. 2003.

Billington, David P. *The Tower and the Bridge: The New Art of Structural Engineering*. Princeton University Press. 1985.

Holgate. Alan, The *Art in Structural Design: An Introduction and Sourcebook*. Oxford University Press. 1986.

Holgate. Alan, *The Art of Structural Engineering: The Work of Jorg Schlaich and His Team*. Axel Menges. 1997.

Rowland Mainstone, *Developments in Structural Form*, 2nd Edition. Architectural Press. 1999.

John Ochsendorf, *Guastavino Vaulting: The Art of Structural Tile*. Princeton Architectural Press. 2010.

CONCLUSION: UNDERSTANDING OF STRUCTURAL PRINCIPLES

15.1 INTUITION AND KNOWLEDGE

The preceding chapters have attempted a qualitative presentation of structural principles on the basis of general experience with forces, materials, and deformations. A casual reading of the foregoing pages may illuminate the structural actions most commonly encountered in architecture; a careful reading may clarify, in addition, more sophisticated types of structural behavior seldom considered by the layperson. In any case, the purely intuitive approach used to introduce these principles cannot be expected to lead to quantitative knowledge in a field as complex as structures. For this, an analytical, mathematical presentation is needed, of the kind required for an understanding of any branch of physics.

Intuition is an essentially synthetic process that brings about the sudden, direct understanding of ideas more or less consciously considered over a period of time. It becomes a satisfactory road to knowledge on two conditions: It should be based on a large amount of prior experience, and it should be carefully verified. Pure—that is, *unchecked*—intuition is misleading most of the time.

Intuition may be greatly refined by experience. One of the best tools for refining structural intuition is the use of models demonstrating the diversified actions considered in this book. Since all structural actions involve displacements, and displacements are the visual result of these actions, models are ideally suited for the intuitive presentation of structural concepts. This is why, at times, the reader has been invited to build elementary models that demonstrate the structural behavior of simple elements more convincingly than any drawing ever will. The suggested exercises at the end of each chapter are a springboard for such inquiry.

On the other hand, it cannot be overemphasized that intuition without experience is a dangerous tool, since it leads to unchecked assumptions. The reader should be wary of what he or she "seems to feel should happen" in a given physical situation, and, in particular, of the suggestions from the purely geometrical aspects of a structure. For example, at first it is hard to believe that the straight sides of a stiffener-supported cylindrical barrel move inwards under load, because the curved section of the cylinder suggests arch action, and arches are "known" to push outwards, and yet such is the case.

15.2 QUALITATIVE AND QUANTITATIVE KNOWLEDGE

Qualitative knowledge should often be a prerequisite to quantitative analysis, since interest in a field is seldom aroused without some prior understanding. It is hoped that the reader interested in structures may obtain from the preceding chapters that minimum understanding of structural behavior required to arouse his or her interest, and be led by it to a serious study of the subject.

Structures are best presented in the language proper to the quantitative analysis of measurable phenomena: mathematics; not the complex mathematics required for an understanding of the more advanced aspects of science, but the simple mathematics of algebra, trigonometry, and, sometimes, elementary calculus. In this context, one cannot overemphasize the importance of numerical computations, and the fact that the calculator and the computer have made them fast, reliable, and painless for all. No thorough knowledge of structures may be acquired without the use of these mathematical tools. Mathematics does not explain physical behavior; it just describes it. But mathematical descriptions are so efficient that a short formula may clearly and simply express ideas that in verbal form would require pages of complex statements.

The availability of structural knowledge, made possible by the use of mathematics, has produced impressive results. Structures, which in the past could have been conceived and built only by architectural geniuses, are designed, at present, by modest engineers in the routine of their office work. This democratization of structural knowledge, while putting advanced structures within the reach of the average architect, introduces the danger of architectural misuse by the practitioner who lacks a solid structural foundation.

There is little doubt in the minds of both engineers and architects that modern structural concepts are used properly only when the architect has a thorough understanding of structures. This does not imply that all architects should become mathematicians; it simply suggests that those practitioners who wish to express themselves through structural forms should first learn to use the tools of quantitative analysis. They will be amazed to find, later on, that their cultivated intuition will often reach "correct" structural solutions without too many mathematical manipulations.

15.3 THE FUTURE OF ARCHITECTURAL STRUCTURES

The twenty-first century has ushered in the digital age in all segments of society. We are similarly entering a new age of architectural structures. Complex structural analysis and design at one time only a dream is now enabling designers to generate and verify structures thought impossible as recently as the latter twentieth century—less than the span of time since the last publication of this text. New high-tech materials such as carbon fiber are joining the ranks of the traditional materials. These developments have arisen simultaneously with computer-controlled manufacturing that has allowed the fabrication of structures with remarkably small tolerances—a prerequisite for complex geometry since even the slightest deviation from a theoretical position in space creates structural elements that cannot be fit together (Figure 15.1). It is very conceivable that by the middle of the twenty-first century, robotic assembly and even self-assembling systems will begin to be the norm as well.

New freedom brings new responsibility and new challenges, though. Forms that are highly irregular defy any intuitive understanding of their proportioning and require the closest collaboration between architect and engineer from the very start. In some regards it can be argued that nearly *any* structural form can be constructed today, but we must guard against mere arbitrary whim.

In the past, aside from structural and construction limits, economy was most often the main restraint on design. New challenges such as environmental sustainability and carbon neutral design, however, confront us with new limits. Recognizing the significance of climate change and the undesirable influence fossil fuels have on it, architecture will need to contribute its share in alleviating their effects. Energy efficient structures with improved insulation, heat reflecting/absorbing coverings, solar paneled walls and roofs, building locations/ orientations and new materials are just a few examples of the many environmentally friendly solutions. A future edition of this book may consider such problems in detail.

Architecture at its highest form can excite, enliven, and inspire, and pure structural pragmatism is antithetical to this aim. Nevertheless, just because one *can* design nearly any form does not mean we *should*, especially if it comes at the cost of structural and material inefficiency (and thus excess).

Design practice is rapidly evolving to address these challenges while at the same time making complex forms possible in the first place. The building industry is beginning to model the assembly of structures in the manner led by the aerospace industry starting in the 1980s. It has taken this long for the architectural/engineering and construction industry (AEC) to catch up simply because of the complex nature of the building industry, with the many different entities involved (architects, engineers, fabricators, contractors—typically all from different companies). New approaches of Integrated Project Delivery (IPD), facilitated by the

(a)

(b)

Figure 15.1   The Museo Soumaya

The Museo Soumaya in Mexico City is an example of daring architectural design and structural integration in the digital age. The irregular form and tight construction tolerances would have been impossible until recent developments in computer aided design, analysis, fabrication and construction management techniques. The building under construction (a) reveals the layers of structure: The primary structure consists of the widely spaced, white tubular members following the curvature of the form (each one is unique), surrounding a central structural core. Atop the frame is a double-layered space frame that defines the surface, onto which the final sheathing and façade are applied. The finished building (b) is covered with hexagonal mirrored aluminum tiles that form a rain screen. The hexagonal shape allows the tiles to follow the building contour, and the spaces between tiles allow rainwater to pass through and drain behind the actual façade surface.

Photo courtesy of Geometrica

development of Building Information Modeling (BIM), are emerging as key components of this process. Though years away from the realization of its full potential, the goal of BIM is to digitally represent all aspects of a project in all disciplines, with all associated material and technical data tagged with each part, permitting designers to check and verify all components long before fabrication even begins, and to ensure that during construction errors and associated changes are minimized. The digital architectural model is becoming linked with the digital structural analysis model and environmental performance simulations in a symbiotic loop, enabling the exploration of a greater range of design alternatives. It portends to be the biggest revolution in building design and construction since the development of the modern materials of steel and concrete that literally built the contemporary world.

Nevertheless, however complex and challenging the future may be, and however sophisticated our computational tools become, at the core of *all* structural understanding are elemental principles that are fundamentally invariant. In a sense, they are akin to the nucleotides that comprise DNA, which in turn is the basis for all life as we know it. Whether simple or complex, *all* structures must be stable and strong. *All* structures obey the basic principles of linear and rotational static equilibrium. A true and deep understanding of these elemental structural principles comes only with time and continued practice, and combines intuitive knowledge plus at least elemental mathematical application. Such foundational understanding is the root of the development of any architecture in which the structure is integrally a part of the design as opposed to an afterthought.

So, to conclude this introductory overview of structures in this book, it is even more true today than when the closing paragraph in the first edition was written some 50 years ago. The field of structures is evolving rapidly under the pressure of the growing needs of society. The structuralist and the architect must strive, by all means at their disposal, toward mutual understanding and fruitful collaboration. May the technician and the designer work together to the greater glory of architecture and in the greater service to humankind.

KEY IDEAS DEVELOPED IN THIS CHAPTER

- A casual reading of this book can bring one to a basic conceptual understanding of structural principles. A more careful reading can clarify sophisticated principles seldom considered by the layperson.
- Intuition alone cannot lead to quantitative knowledge. A complete understanding of structures requires an analytical approach, plus physical experimentation with models to check one's intuitive assumptions.
- The mathematics required for a basic quantitative understanding are elementary algebra and trigonometry, and sometimes basic calculus.
- Mathematics is a dense symbolic language. A simple formula may encapsulate a concept that takes many pages to verbally describe.
- Architectural practitioners who wish to express themselves through structural form should learn the tools of quantitative analysis in addition to the intuitive presentation in this text.
- New challenges of environmental sustainability and carbon neutral design (among others) confront us and call for efficient structural systems like never before.
- New high-tech materials, computer-aided design, structural analysis, and building simulation, Computer Numerically Controlled (CNC) fabrication, Building Information Modeling (BIM), and Integrated Project Delivery (IPD) are making possible previously impossible structural forms.
- New structural forms are presenting new opportunities and challenges. Those that are highly irregular defy a purely intuitive understanding of their behavior, and require the closest collaboration between architect and engineer from the very start.

STRUCTURES BIBLIOGRAPHY

STRUCTURAL CONCEPTS

Ching, Francis D. K., Onouye,Barry S., and Douglas, Zuberbuhler. *Building Structures Illustrated: Patterns, Systems, and Design*, 2nd Edition. Wiley. 2013.

Cowan, Henry J., and Wilson, Forrest. *Structural Systems*. Van Nostrand Reinhold Company. 1981.

Engel, Heino. *Structure Systems*, 3rd Edition. Hatje Cantz. 2007.

Gordon, James E. *Structures: Or Why Things Don't Fall Down*. Da Capo Press. 2003.

Hanaor, Ariel, *Principles of Structure*. Blackwell Science. 1998.

Heller, Robert, A., and Pap, Arpad A. *Mechanics of Structures and Materials*. Video Tape Series. Virginia Tech. Intellectual Properties. 1980.

Levy, Matthys and, Salvadori, Mario. *Earthquake Games: Earthquakes and Volcanoes Explained by 32 Games and Experiments*. Margaret K. McElderry Books. 1997.

Levy, Matthys and, Salvadori, Mario. *Why Buildings Fall Down*. W. W. Norton & Company. 1992.

Lin, T. Y., and Stotesbury, S. D. *Structural Concepts and Systems for Architects and Engineers*, 2nd Edition. Van Nostrand Reinhold. 1988.

MacDonald, Angus J. *Structure & Architecture*, 2nd Edition. Architectural Press. 2001.

Millais Malcom. *Building Structures: From Concept to Design*, 2nd Edition. Spon Press. 2005.

Moore, Fuller. *Understanding Structures*. McGraw-Hill. 1998.

Moussavi, Farshid. *The Function of Form*. ACTAR, Harvard Graduate School of Design. 2009.

Robbin, Tony. *Engineering a New Architecture*. Yale University Press. 1996.

Salvadori, Mario. *The Art of Construction: Projects and Principles for Beginning Engineers and Architects*. Chicago Review Press. 1990.

Salvadori, Mario. *Building: From Caves to Skyscrapers*. Antheum. 1979.

Salvadori, Mario. *Why Buildings Stand Up: The Strength of Architecture*. W. W. Norton & Company. 2002.

Torroja, Eduardo. *Philosophy of Structures*. University of California Press. 1962.

Wentworth Thompson, D'Arcy. *On Growth and Form: The Complete Revised Edition*. Dover Publications. 1992.

BOOKS ABOUT STRUCTURAL ENGINEERING, THE ART OF STRUCTURES, AND THE HISTORY OF STRUCTURES

Boake, Terri Meyer. *Understanding Steel Design*. Birkhäuser. 2011.

Boake, Terri Meyer. *Diagrid Structures*. Birkhäuser. 2014.

Boake, Terri Meyer. *Architecturally Exposed Structural Steel*. Birkhäuser. 2015.

Anderson, Stanford, editor. *Eladio Dieste: Innovation in Structural Art*. Princeton Architectural Press. 2004.

Berger, Horst. *Light Structures—Structures of Light: The Art and Engineering of Tensile Architecture*. Birkhäuser Basel. 1996.

Billington, David P. *The Art of Structural Design: A Swiss Legacy*. Princeton University Art Museum. 2003.

Billington, David P. *The Tower and the Bridge: The New Art of Structural Engineering*. Princeton University Press. 1985.

Charleson Andrew. *Structure as Architecture: A Sourcebook for Architects and Structural Engineers*, 2nd Edition. Routledge. 2014.

Charleson Andrew. *Seismic Design for Architects: Outwitting the Quake*. Elsevier Press. 2008.

Faber, Colin. *Candela: The Shell Builder*. The Architectural Press. 1963.

Garlock, Moreyra Maria, E. and Billington, David. P. *Felix Candela: Engineer, Builder, Structural Artist*. Yale University Press. 2008.

Holgate, Alan. *The Art in Structural Design*. Oxford University Press. 1986.

Holgate, Alan. *The Art of Structural Engineering: The Work of Jorg Schlaich and His Team*. Axel Menges. 1997.

Mainstone, Roland. *Developments in Structural Form*, 2nd Edition. Routledge. 2001.

Nervi, Pier Luigi, and Einaudi, Robert. *Aesthetics and Technology in Building*. Harvard University Press. 1965.

Nervi, Pier Luigi. *Structures*. McGraw-Hill. 1956.

Nordenson, Guy, editor. *Seven Structural Engineers: The Felix Candela Lectures*. The Museum of Modern Art, New York. 2008.

Ochsendorf, John. *Guastavino Vaulting: The Art of Structural Tile*. Princeton Architectural Press. 2010.

Otto, Frei, and Rasch, Bodo. *Finding Form: Towards and Architecture of the Minimal*, 3rd Edition. Axel Menges. 1996.

Petrosky. Henry. *To Engineer is Human: The Role of Failure in Successful Design*. Vintage. 1992.

Rice, Peter. *An Engineer Imagines.* Ellipsis London Press Ltd. 1998.

Ross, Stephen S. *Construction Disasters: Design Failures, Causes and Prevention.* McGraw-Hill (*Engineering News-Record* Series). 1984.

Saint, Andrew. *Architect and Engineer: A Study in Sibling Rivalry.* Yale University Press. 2008.

Tullia, Lori. *Pier Luigi Nervi: Minimum Series.* ORE Motta Cultura srl. 2009.

MORE TECHNICAL BOOKS ON STRUCTURES

Adriaenssens, Sigrid, Block Phillippe, Veenendaal, Diederik, and Williams, Chris, editors. *Shell Structures for Architecture: Form Finding and Optimization.* Routledge. 2014.

AISC *Steel Construction Manual,* 14th Edition. American Institute of Steel Construction, 2011.

American Institute of Civil Engineers—Structural Engineering Institute, *Minimum Design Loads for Buildings and Other Structures (ASCE/SEI 7-10),* ASCE, 2010

Allen, Edward and Zalewski, Waclaw. *Form and Forces: Designing Efficient, Expressive Structures.* John Wiley & Sons, Inc. 2009.

Ambrose, James and Tripeny, Patrick. *Building Structures,* 3rd Edition. Wiley. 2011.

Becker, Hollee Hitchcock. *Structural Competency for Architects.* Routledge. 2014.

Hanaor, Ariel. *Principles of Structures.* Blackwell Science. 1998.

International Code Council, *2012 International Building Code.* ICC, 2011.

Kaufman, Harry. *A Structures Primer.* Pearson. 2010.

Ochshorn, Jonathan. *Structural Elements for Architects and Builders.* Butterworth-Heinemann. 2009.

Onouye Barry S., and Kane, Kevin. *Statics and Strength of Materials for Architecture and Building Construction,* 4th Edition. Prentice Hall. 2011.

Place, Wayne. *Architectural Structures.* Wiley. 2007.

Sandaker, Bjorn N., Eggen, Arne P., and Cruvellier, Mark R. *The Structural Basis of Architecture,* 2nd Edition. Routledge. 2011.

Schaeffer, Ronald. *Elementary Structures for Architects and Builders,* 5th Edition. Prentice Hall. 2006.

Schodek, Daniel and Bechthold, Martin. *Structures,* 7th Edition, Pearson. 2014.

Schueller, Wolfgang. *The Design of Building Structures.* Prentice-Hall. 1995.

Underwood, Rod and Chiuini, Michele. *Structural Design: A Practical Guide for Architects.* Wiley. 2007.

INDEX